Ecological Studies

Analysis and Synthesis

Edited by

W.D. Billings, Durham (USA) F. Golley, Athens (USA)
O.L. Lange, Würzburg (FRG) J.S. Olson, Oak Ridge (USA)

Volume 35

John A. C. Fortescue

Environmental Geochemistry

A Holistic Approach

With 131 Figures

Springer-Verlag   New York   Heidelberg   Berlin

John A. C. Fortescue
University of Missouri
Environmental Trace Substances Research Center
Route 3
Columbia, Missouri   65211

Financial support in the preparation of this text by the Canadian Forestry Service is gratefully acknowledged.

Library of Congress Cataloging in Publication Data

Fortescue, John A   C
  Environmental geochemistry.

  (Ecological studies ; v. 35)
  Bibliography: p.
  Includes index.
  1. Geochemistry—Environmental aspects.
  2. Environmental chemistry. I. Title. II. Series.
  QE515.F69   551'.0154   79-13595

The use of general descriptive names, trade names, trademarks, etc. in this publication, even if the former are not especially identified, is not to be taken as a sign that such names, as understood by the Trade Marks and Merchandise Marks Act, may accordingly be used freely by anyone.

All rights reserved.

No part of this book may be translated or reproduced in any form without written permission from Springer-Verlag.

© 1980 by Springer-Verlag New York Inc.

Printed in the United States of America

9 8 7 6 5 4 3 2 1

ISBN 0-387-90454-9 Springer-Verlag   New York
ISBN 3-540-90454-9 Springer-Verlag   Berlin   Heidelberg

In gratitude to F.A.N. and all at Q.
and
to M.F. for her help

Foreword

It is the policy of the federal Canadian Forestry Service to sponsor research initiatives from the private sector that are judged to be pertinent to its mandate and offer particular promise towards the optimal management of Canadian forest resources. This book is based on such an initiative. It represents the philosophy of the author himself and is in no way constrained by the views of the sponsoring agency.

Over the past two decades Dr J. A. C. Fortescue has become well known at a number of research centers throughout the world. He has pioneered the approach to environmental understanding that is comprehensively developed in this text. The limitations of traditional compartmentalized approaches are deprecated and the case is made for a holistic rethinking of basic concepts and principles. *Landscape Geochemistry* is the disciplinary outcome that gives expression to this rethinking. It may be viewed as the minimum scale of conceptual approach necessary in the environmental sciences to solve present-day problems and to exploit future opportunities.

In forest-resource management, for example, production and protection practices become increasingly more complex and intensive, and in a setting where external environmental stresses become increasingly more severe and diversified. Under such conditions more has to be known about the basic functioning of ecosystems and of the dimensions of their resilience and behavior patterns. Historically, of course, forestry has by no means progressed in isolation from related sciences, but the approach being promulgated by the author places the ramifications of this interdependence on a more visible and rational basis, from which advances should be facilitated for all environmental sciences.

Ottawa, Canada
October, 1979

P. J. RENNIE

Preface

This book is written for three groups of readers. The first, and perhaps smallest, group consists of exploration geochemists, professional geochemists, and graduates and undergraduates who have geochemistry as the focal point of their chosen profession. The second group of readers includes undergraduates and graduates of environmental science, environmental geology, geography, ecology, soil science, forestry, epidemiology, and other disciplines who require an introductory text in environmental geochemistry. Teachers in these disciplines can use this book as a guide around which they may structure a semester course in their speciality. The third group of readers is made up of both scientists and non-scientists who have some general or specific interest in the geochemistry of the environment as it relates to a problem of pollution (or potential pollution), land use decision making, or environmental health. This third group of readers should commence with Chapter 1, examine the synopsis at the end of the book, and then read Chapter 20 before attempting the remainder of the text.

This book revolves around holistic thinking in environmental geochemistry. It is a general, introductory statement of the current status of a discipline for environmental geochemistry, the scope of which should appeal particularly to young scientists who wish to make contributions to a discipline that is in the formative stage; new opportunities for specialization are evident at almost every turn. This book is not written for the destructive critic. He must wait until the house is built before he pulls down the scaffolding!

As described by its founders F. W. Clarke, V. I. Vernadski, V. M. Goldschmidt and their co-workers and followers, landscape geochemistry is a scientific discipline firmly rooted in general geochemistry. The discipline is concerned with the geochemistry of the environment and, in particular, with the role played by chemical entities in the synthesis or decomposition of environmental materials of all kinds. Landscape geochemistry has very close relationships to modern systems ecology, environmental geochemistry, and the science of the total environment. It differs from today's environmental geochemistry, however, in that it is based on a series of seven interrelated principles and concepts that together provide a unique approach to the study of the environment. Some of these concepts, such as element abundance, are common to all geochemistry; some, such as the study of biogeochemical cycling, are shared with ecology; and some, such as geochemical classification of landscapes, are unique to landscape geochemistry. The concept of the Earth's daylight surface is considered important in landscape geochemistry because it is at or near this surface that the circulation of chemicals in the environment reaches a maximum intensity owing to the presence of the living biosphere. The landscape geochemistry approach is holistic because it involves the study of all elements in the lithosphere, hydrosphere, pedosphere, biosphere, and atmosphere, as well as all chemical interactions.

In this book I share with the reader an approach to environmental geochemistry that has gradually evolved during my career and is based on three sources of thinking: (1) the ideas of B. B. Polynov, A. I. Perel'man, and M. A. Glazovskaya (and other Russian workers) who pioneered the discipline of landscape geochemistry; (2) the ideas and examples of environmental geochemistry that have been described by scientists outside Russia; and (3) my 25 years of experience as a student, research worker, and teacher.

I first saw the need for a holistic approach to the geochemistry of the environment when I was an undergraduate at the University of British Columbia; I was studying under Professor H. V. Warren who worked with research in geochemical prospecting (Warren et al., 1955). During graduate work at the University of Oxford, under Dr. L. Leyton of the Department of Forestry, I applied a holistic, conceptual model to the study of the circulation of chemical elements in glacial terrain in Southern Norway. This activity convinced me that it is desirable in environmental geochemistry to treat rocks, soils, and vegetation as a single system. Post doctorate research at the Geochemical Prospecting Research Centre, Imperial College of Science and Technology, London, England (under Professor J. S. Webb) involved the production of regional, multielement, geochemical maps of an area in the Namwala Concession of Zambia. This experience showed how geochemical relationships between rocks and stream sediments occur in an area that has been subjected to several cycles of tropical weathering and indicated how local landscape features may modify distribution patterns for particular elements in stream sediments.

During the summer of 1955 I gained practical experience in geochemical prospecting based on soils around mineral deposits in Southern Norway when working for the Statens Råstofflaboratorium under Director Alsak Kvalheim in Trondheim. In 1959 and 1960, I had practical experience with regional geochemical prospecting in Northern Norway based on stream sediment surveys carried out by Bjorn Bølviken and his co-workers. This demonstrated the general effectiveness of soil and stream sediment survey techniques and showed the extent to which the depth of the glacial cover can govern the extent of geochemical anomalies for copper (and other elements) in Norway.

Other practical geochemical prospecting experience was obtained in the summers of 1956 and 1957 for Kennco Explorations (Canada) Ltd. The 1956 New Brunswick experience, under the direction of Dr. H. E. Hawkes, was concerned with the study of relationships between airborne electromagnetic anomalies and geochemical anomalies in stream sediments in areas underlain by graphite shale or massive sulfide deposits. During this time I was involved with the discovery of the geochemical anomalies which led to the finding of the Murray deposit. The 1956 summer experience convinced me of the importance of attention to details of element distribution patterns in the interpretation of stream sediment surveys regardless of the levels of concentration of elements involved. In the summer of 1957 I worked with Dr. J. J. Brummer on a regional stream sediment geochemical mapping project in the Guichon Batholith area of Southern British Columbia. This experience showed that, even in semi-arid areas, geochemical anomalies in stream sediments could be discovered in the vicinity of porphyry copper deposits—some of which were deeply buried.

I spent eight years working for the Canadian Government. From 1962 until 1967 I was in the Geochemistry Section at the Geological Survey of Canada, under R. W. Boyle, and afterwards at the Petawawa Forest Experiment Station of the Canadian Forestry Service with International Biological Program. During this time at the Geological Survey of Canada, I described a general holistic approach to the study of mineral exploration data and a guide to prospecting methods research called "Exploration Architecture," based on a lecture delivered in Norway in the fall of 1959 to the Trondheim Geological Club (Fortescue, 1965).

My main activity at the Geological Survey was to introduce a holistic approach to plant prospecting methods research (Fortescue, 1970). Based on the principles of landscape geochemistry this involved: (1) the development of a sequential system for the selection of drilled but undisturbed mineral deposits where research was desirable; (2) the use of a generalized conceptual model for collection of like set of samples of environmental materials around such deposits; (3) the development of a movable geochemical laboratory for the processing of botanical, soil, and plant samples in the field; (4) a computer based system for geochemical data processing; (5) standard methods for plotting morphological and chemical data; and (6) standard methods for the analysis and synthesis of findings from different areas. This experience demonstrated the practicality of using the holistic landscape geochemistry for the development of standard methods in exploration geochemistry research and, in particular, the advantages of using a conceptual model for the selection of sites for study.

It was once evident that the same general approach could be utilized in forestry and other areas of environmental geochemistry. While at Petawawa, I was introduced by Dr. C. G. Marten to the then new study of "systems ecology," explained in the first volume of *Ecological Studies* (Fortescue and Marten 1970). This technique showed how the circulation of chemical elements in forest ecosystems could be studied from a truly holistic viewpoint, as well as the potential of using computer models to describe the circulation of elements in ecosystems in the real world and their potential for simulation forest growth during the future evolution of a given forest stand. In 1970 systems ecology was about to provide a new paradigm for environmental ecology.

In 1970 I accepted a post in Geological Sciences at Brock University, St. Catherines, Ontario. At the time the Department was largely concerned with Quaternary and environmental geology and it was thought that landscape geochemistry would enhance the departmental program. During the next seven years, students and co-workers and I made fundamental and applied studies of environmental geochemistry in the Niagara Peninsula—based on principles of landscape geochemistry. The research included: (1) the use of local and regional stream sediment sampling for mapping effects of man's activity on the geochemistry of the environment; (2) the use of stream sediments to explain a natural source of heavy metals in an agricultural area; (3) studies of the trace element content of sewage sludge; (4) study of lead pollution of highways and orchards; and (5) preliminary studies of relationships between trace element distribution and disease incidence in man. These activities convinced me that environmental geochemistry as a whole was hampered by the lack of a formal and logical discipline which would be used for making comparative studies in the general field of environmental geochemistry. It appeared that the development of environmental geochemistry required formal statements of basic concepts and principles as a background for sophisticated studies in environmental chemistry—which resulted from the "environmental revolution."

As an attempt to fulfill this need, I wrote this book. It developed from a series of papers on landscape geochemistry (Fortescue, 1974a,b) and work with Dr. P. M. B. Bradshaw (of Barringer Research, Toronto) on a compilation of case histories in exploration geochemistry (Bradshaw, 1975). A generous contract awarded by the Canadian Forestry Service in January 1976 allowed me to commence a research text on landscape geochemistry in September 1976. After April 1977, a Visiting Research Professorship at the Environmental Trace Substances Research Centre, University of Columbia, Missouri, allowed a further six months for research and an enlarged research text was delivered to the Canadian Forestry Service in mid-February 1978. This book, an abridgement of the research text, was prepared for publication during the spring and summer of 1978 and is published by written permission of the Canadian Forestry Service.

Acknowledgements. I take this opportunity to thank all the scientists who worked with me over the years on different aspects of what in this book is called landscape geochemistry. In particular, I would like to thank Mr. F. Armitage (late of the Canadian Forestry Service) and Dr. P. J. Rennie of the same organization who encouraged me during the preparation of the original research text and Dr. D. E. Reichle and Dr. J. S. Olson of the Environmental Sciences Division of the Oak Ridge National Laboratory who provided informed criticism of the project as it developed. I am most grateful for formal permission to publish information from the research text (Environment Canada Contract No. 5505: KL 020-5-0047) that was used during the preparation of this book. I would also like to express appreciation to T. Clevenger, Associate Director, Environmental Trace Substances Research Center, University of Missouri, Columbia, for allowing me to prepare at the Center during 1978 for the publication of this book.

Thanks are also due to Dr. R. W. Boyle of the Geological Survey of Canada who arranged to have books by Perel'man (1966, 1972) translated from Russian into English. These were, as we shall see, a major source of information used during the preparation of this book.

Escondido, California
September, 1979

JOHN A. C. FORTESCUE

Contents

Part I. Introduction   1

1. General Overview   3

    Landscape Geochemistry   3
    The Purpose of this Book   3
    The Scope of this Book   5

2. Outline of Historical Development of Geochemistry   7

    The Development of Modern Geochemistry   7
    The Problem of Unconnected Data Bases   9
    Some General Concepts of Geochemistry   9
    Landscape Geochemistry   14
    Summary and Conclusions   17
    Discussion Topics   17

3. A Philosophy for Environmental Geochemistry   19

    Introduction   19
    The Hierarchy of Space   19
    The Hierarchy of Time   20
    The Hierarchy of Chemical Complexity   20
    The Hierarchy of Scientific Effort   22
    The Principle of Successive Approximations   25
    Summary and Conclusions   27
    Discussion Topics   27

4. Other Approaches to Environmental Geochemistry   29

    Environmental Science and Geochemistry   29
    Environmental Science and Landscape Geochemistry   30

Examples of other Approaches to the Study of Environmental Geochemistry	31
An Ecological Approach to Environmental Geochemistry	31
The Land Classification Approach to Environmental Geochemistry	34
Summary and Conclusions	39
Discussion Topics	40

Part II. The Basics of the Discipline of Landscape Geochemistry   41

5. Definitions of Concepts and Principles   43

Introduction	43
An Overview of Landscape Geochemistry	48
Summary and Conclusions	49
Discussion Topics	49

6. Element Abundance   51

Introduction	51
The Abundance of Elements in the Environment	52
The Absolute Abundance of Elements in the Lithosphere	52
The Abundance of Elements in the Hydrosphere	57
The Abundance of Elements in the Atmosphere	60
The Abundance of Chemical Elements in the Biosphere	65
General Summary Statement on the Absolute Abundance of Elements in the Four Geospheres	71
The Relative Abundance of Elements in the Four Geospheres	72
The Partial Abundance of Elements	74
The Concept of Element Abundance and Holism	78
Summary and Conclusions	78
Discussion Topics	78

7. Element Migration in Landscapes   79

Introduction	79
A Study of the Absolute Mobility of Elements in a Landscape	79
The Migration Rate Equation	81
A Classification Scheme for Migrant Elements	85
Element Migration and Holism	91
Summary and Conclusions	93
Discussion Topics	93

8. Geochemical Flows in Landscapes   95

Introduction	95
The Flow of Substances through Landscapes	95
Geochemical Flow Patterns within Landscapes	96
Elementary Landscape Cells	98
General Principles for the Flow of Elements through Landscapes	101
Landscape Geochemical Flows and Holism	106

	Summary and Conclusions	107
	Discussion Topics	108

9. Geochemistry Gradients — 109

	Introduction	109
	Geochemical Gradients	111
	Ecological Gradients	115
	Summary and Conclusions	120
	Discussion Topics	121

10. Geochemical Barriers — 123

	Introduction	123
	Kinds of Geochemical Barriers	123
	Effects of Geochemical Barriers	124
	Typomorphic Elements	127
	Epigenitic Processes	128
	Geochemical Barriers and Holism	129
	Summary and Conclusions	132
	Discussion Topics	132

11. Historical Geochemistry — 133

	Introduction	133
	Polynov's Concept of the Weathering Process	133
	The Role of Groundwaters in Landscape Development	140
	The Geological Substrate of Landscapes	141
	Details of Landscape Development during Pedological and Ecological Time	142
	Historical Geochemistry and Holism	152
	Summary and Conclusions	152
	Discussion Topics	153

12. Geochemical Landscape Classification — 155

	Introduction	155
	A Geochemical Classification of Landscapes	155
	Basic Principles of the Geochemical Classification of Landscapes	156
	General Characteristics of the Four Landscape Groups	156
	Hierarchical Geochemical Classification for Landscapes	157
	General Comments on the Perel'man Geochemical Classification of Landscapes	159
	Glazovskaya's Geochemical Classification of Landscapes	161
	Glazovskaya's Conceptual Models for Landscape Features	166
	Summary and Conclusions	171
	Discussion Topics	171

13. Chemical Complexity and Landscape Geochemistry — 173

	Introduction	173
	Example I. The Isotope Level of Complexity	173

Example II. The Element Level of Complexity — 175
Example III. The Partial Element Level of Complexity — 176
Example IV. The Persistent Chemical Level — 180
Example V. The Nutrient Level — 182
Example VI. The Tissue Level — 183
Example VII. The Organism Level — 186
The Relationships between Landscape Geochemistry and the Hierarchy of Chemical Complexity — 188
Discussion Topics — 190

14. Scientific Effort and Landscape Geochemistry — 191

Introduction — 191
The Descriptive/Empirical Level of Scientific Effort — 191
The Statistical Level of Scientific Effort — 197
The Systems Analysis Level of Scientific Effort — 201
The Systems Simulation Level of Scientific Effort — 207
Summary and Conclusions — 209
Discussion Topics — 212

15. Space and Landscape Geochemistry — 213

Introduction — 213
Units — 213
Examples of Approaches to the Study of Environmental Geochemistry in Space — 213
Summary and Conclusions — 226
Discussion Topics — 226

16. Time and Landscape Geochemistry — 229

Introduction — 229
Retrospective Monitoring in Geological Time — 232
Retrospective Monitoring in Pedological Time — 234
The Uniformity of Environmental Conditions during Ecological Time — 237
Summary and Conclusions — 242
Discussion Topics — 246

17. Landscape Geochemistry as a Totality — 247

Introduction — 247
A Sequential Description of Landscape Geochemistry — 247
The Terminology of Landscape Geochemistry — 250
Summary and Conclusions — 252
Discussion Topics — 252

Contents xvii

Part III. Applications of Landscape Geochemistry — 253

18. Practical Applications of Landscape Geochemistry — 255

Introduction — 255
Practical Applications of Landscape Geochemistry — 263
Summary and Conclusions — 289

19. A New Paradigm for Environmental Geochemistry — 291

Introduction — 291
Paradigms in General Geochemistry (1900–1978) — 291
A New Paradigm for Environmental Geochemistry — 294
Five Types of Landscape Geochemistry — 298
Summary and Conclusions — 304

Part IV. Summary and Conclusions — 307

20. Overview — 309

Introduction — 309
Environmental Science, Environmental Geochemistry, and Landscape Geochemistry — 309
Two General Concepts of Environmental Science — 309
Environmental Geochemistry and Landscape Geochemistry — 310
The Seven Basics of Landscape Geochemistry — 311
Five Types of Landscape Geochemistry — 313
General Conclusions Regarding the Future Development of Landscape Geochemistry — 314

Synopsis — 317

Part I. Introduction — 317
Part II. The Basics of the Discipline of Landscape Geochemistry — 318
Part III. Applications of Landscape Geochemistry — 319
Part IV. Summary and Conclusions — 320

References — 321

Author Index — 333

Subject Index — 337

Part I. Introduction

"Landscape Geochemistry as a distinct scientific discipline originated in this country in the 1940s. It arose then and here not by accident but from certain general 20th century trends in science, and its roots lie in certain lines of Russian scientific thought that arose around the turn of the century."

    A. I. Perel'man *Landscape Geochemistry,* Vysshaya Shkola, Moscow
                    1966, (G. S. C. Translation 676, Ottawa) p. 1.

1. General Overview

"In the simplest terms, geochemistry may be defined as the science concerned with chemistry of the earth as a whole and of its component parts."

Brian Mason, *The Principles of Geochemistry,* 3rd ed. (New York: John Wiley & Sons, 1966), p. 1.

Landscape Geochemistry

Geochemistry can also be defined as the study of the role chemical elements play in the synthesis and decomposition of natural materials of all kinds. Landscape geochemistry deals with the study of the environment that occurs at or near the daylight surface of the Earth. The environment results from the interaction of the lithosphere with the hydrosphere, atmosphere, and biosphere during geological, pedological and ecological time together with the cumulative effects of man's activities during the past few hundred years.

In this book, landscape geochemistry is described as a conceptual and holistic approach to the study of the geochemistry of the environment. It is holistic for three reasons:

1) it involves the circulation of all chemical elements in the environment, not just certain groups of elements such as "nutrients" or "toxic elements.";

2) it involves the circulation of elements in all kinds of environments (including, for example, adjace "terrestial," "bog" and "aquatic ecosystems") found within a particular area of country and;

3) it may include reference to all levels of detail at which scientific investigations are carried out in relation to an area of country including local, regional, and global studies.

The Purpose of this Book

This book is designed to introduce readers to the general thinking that lies behind studies of environmental geochemistry. In the short term, this book should lead to an overhaul of the concepts, principles, and terminology used by environmental geochemists. In the long term, the approach should lead to an "Occam's Razor" for the appraisal of studies of environmental geochemistry and to an increase in scientific rigor of new studies designed with the principles of landscape geochemistry in mind.

Today much thinking in environmental geochemistry is rooted in concepts and principles that were described before (1) the advent of the computer, which allowed for the mathematical analysis and synthesis of environmental data on a scale previously thought to be impossible, and (2) the advent of modern instrumental methods of analysis (which have resulted in the simultaneous collection of accurate isotope, element, and biochemical data). The far-reaching implications of these two advances will eventually lead to a rethinking of fundamental concepts and principles.

This book should stimulate the beginnings of this rethinking and lead to the consideration of environmental geochemistry as a totality within which standard data and information are collected from data bases that are systematically related in both space and time.

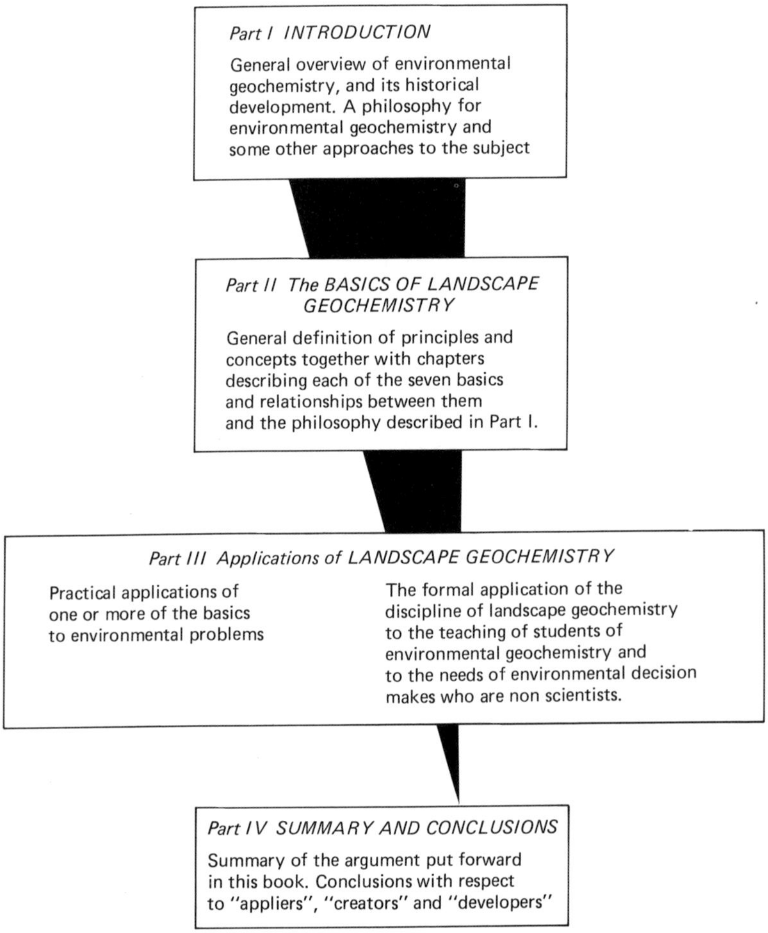

Part I INTRODUCTION

General overview of environmental geochemistry, and its historical development. A philosophy for environmental geochemistry and some other approaches to the subject

Part II The BASICS OF LANDSCAPE GEOCHEMISTRY

General definition of principles and concepts together with chapters describing each of the seven basics and relationships between them and the philosophy described in Part I.

Part III Applications of LANDSCAPE GEOCHEMISTRY

Practical applications of one or more of the basics to environmental problems

The formal application of the discipline of landscape geochemistry to the teaching of students of environmental geochemistry and to the needs of environmental decision makes who are non scientists.

Part IV SUMMARY AND CONCLUSIONS

Summary of the argument put forward in this book. Conclusions with respect to "appliers", "creators" and "developers"

Figure 1.1. General organization of this book.

The Scope of this Book

This is not a handbook of environmental geochemistry concerned with "state of the art" summaries of important fundamental or applied aspects of the subject. Such summaries would each require a separate book because of the very large data base that exists for environmental geochemistry. Instead, this book, organized in four parts (Figure 1.1), is an attempt to analyze and synthesize the data of environmental geochemistry using relatively few concepts and principles. When applied by a geochemist experienced in a particular aspect of the subject to the "state of the art" of his special field of activity, these basic concepts and principles should lead to a new paradigm for environmental geochemistry. This book is written for progressive thinkers in environmental geochemistry—the "appliers, developers, or creators" of science (Bunge, 1962). Throughout this book, the inclusion of quotations taken from the writings of others is designed to inject a leaven of wisdom into a book designed primarily as an intellectual exercise leading to a new view of environmental geochemistry.

2. Outline of Historical Development of Geochemistry

"Modern geochemistry studies the distribution and amounts of chemical elements in minerals, ores, rocks, soils, waters and the atmosphere and the circulation of elements in nature, on the basis of their atoms and ions."

V. M. Goldschmidt, *Geochemistry,* Alex. Muir, ed. (Oxford, England: Clarendon Press, 1954), p. 1.

The Development of Modern Geochemistry

Prior to 1905 geologists, chemists, and other scientists had established estimates for the chemical composition of many rocks, minerals, and waters. A major and lasting contribution to the then infant science of geochemistry was made by the American geochemist F. W. Clarke (1847–1934) who collected, classified, and synthesized this information in his classic book *The Data of Geochemistry,* originally published in 1908. So well did he complete his task that the book went through five editions in his lifetime, and the last was published sixteen years after the first (Clarke, 1924). This source of geochemical information has been used as a starting point for training generations of geochemists and is still quoted as a reference.

A second pioneer of geochemistry was a Russian scientist, V. I. Vernadski (1863–1945). As a young man Vernadski studied under such thinkers as D. I. Mendeleev (who first described the periodic table) and V. V. Dokuchayev (the founder of modern soil science) at the University of St. Petersburg. In the beginning Vernadski studied mineralogy and geology; later on, as a geochemist, he made substantial contributions to our knowledge of the lithosphere, hydrosphere, atmosphere, and biosphere. He was both lucid and prolific as a writer of books on geochemistry. Outside Russia, he is best known for his book *The Biosphere* (Vernadski, 1927). He also wrote other important books including one on general geochemistry (Vernadski, 1924) and one on natural waters (Vernadski, 1934). His book on general geochemistry included chapters on such modern topics as the geochemistry of carbon, radioactive elements, and the geochemical cycles of manganese, bromine, and iodine (Vinogradov, 1963). It was Vernadski who defined the terms lithogeochemistry, atmogeochemistry, hydrogeochemistry, and biogeochemistry (Rankama and Sahama, 1950). He founded a laboratory for biogeochemical research in Moscow in 1927; his book on natural waters laid foundations for the systemic study of chemical hydrology and oceanography (Vinogradov, 1963). Vernadski's contributions to geochemistry were extensive and, largely as a result of his activities and those of some of his pupils [notably A. E. Fursman (1883–1945) and A. P. Vinogradov (1895–1975)], during the past 50 years geochemistry has become a major natural science in Russia and elsewhere.

The third pioneer of geochemistry was a Norwegian, V. M. Goldschmidt (1888–1947). Like Vernadski, Goldschmidt started out as a geologist. His first major contribution was his doctoral thesis on the application of the principles of physical chemistry to the study of contact metamorphism in the Oslo region of Norway completed in 1911. [Mason (1966) called this study "a basic contribution to geochemistry."] Most of Goldschmidt's professional career was in Norway; during 1928–1935, he was in Germany. His researches, and those of his numerous co-workers whom he inspired embraced all aspects of geochemistry. In each instance, monumental contributions were made. A succinct summary of his activities and those of his co-workers, prior to 1937, is given in the published account of a lecture he delivered in London (Goldschmidt, 1937). Some idea of the scope of his activities may be gained from the list of topics covered in the lecture:

1) elements in meteorites
2) a geochemical classification of the elements
3) rules for the incorporation of elements in minerals based on atomic radii
4) a discussion of the geochemistry of the Rare Earths
5) the role of ionic potential in sedimentary geochemistry
6) the geochemistry of ocean waters
7) the distribution of chemical elements in coal ashes
8) the enrichment of elements in the humus layer of deciduous forests.

When he died in 1947, as a result of privations during World War II, Goldschmidt was writing a book on geochemistry. It was eventually published (Goldschmidt, 1954), though only partly completed, and edited by Alex Muir. Unfortunately the chapters Goldschmidt had planned to write concerning the synthesis and analysis of the behavior of elements in natural environments were never written; the book, however, does contain chapters on each of the elements in the periodic table. The book is characterized by seminal thoughts gleened from Goldschmidt's deep understanding of his subject. For example, the section on "mercury" ends with these prophetic words: "Mercury has been found to be concentrated in a number of marine animals, for example fishes." (Goldschmidt, 1954, p. 279).

It is evident from these outlines that the three principal originators of geochemistry—Clarke, Vernadski, and Goldschmidt—were holists with respect to the study of the chemistry of the environment. They considered the role all elements play in the synthesis and decomposition of natural materials of all kinds. This holistic approach is also evidenced by the three objectives for geochemistry that were stated by Goldschmidt over 40 years ago as follows (Rankama and Sahama, 1950):

1) to establish abundance relationships of elements and nuclides in the earth,

2) to account for the distribution of elements in the geochemical spheres of Earth,

3) to detect laws governing the abundance relationships and distribution of the elements.

The Problem of Unconnected Data Bases

Unfortunately, the enthusiastic holism of the pioneer thinkers in geochemistry was not always imitated by later generations of geochemists, especially those outside Russia. During the 30-year period since Goldschmidt's death, geochemistry in general and environmental geochemistry in particular has tended to evolve in a fragmented manner without reference to general geochemical concepts in individual investigations. This has been caused partly by the difficulty of making numerous chemical analyses for elements of interest within environments chosen for study and partly because of the lack of general concepts to interpret and compare the environmental data with after it had been collected. Consequently, the problem of unconnected data bases has become important, particularly in relation to the geochemistry of landscapes. This problem was described clearly by the Russian landscape geochemist Perel'man (1966 p. 161):

> The way to elucidate the geochemical features of landscape and to establish the basic laws of supergene element migration is not simple accumulation of facts on the chemical compositions of plants, soils, rocks, and other parts of the landscape but it is correlation of data obtained in purposeful and combined study of particular landscapes.

Some General Concepts of Geochemistry

Unlike ecology, which has the useful concept of the ecosystem as a central focus (which may be related to local, regional, or global ecological studies), geochemistry lacks a central concept. Probably the closest approach to a central concept is the geochemical cycle (Figure 2.1) as described by Mason (1958); it is based on the more fanciful version in the first edition of his book six years earlier (Mason, 1952). The geochemical cycle is designed to summarize the circulation patterns for elements in nature during geological, pedological, and ecological time. It is a conceptual model with two parts, one geological, and the other environmental. The geological part (major geochemical cycle) occurs deep within the earth. It commences when sedimentary rocks are buried and undergo diagenesis and includes deep-seated processes of metamorphism and metasomatism that eventually result in the formation of crystalline rocks. The environmental part (minor geochemical cycle) commences when crystalline rocks are exposed to weathering processes at or near the daylight surface of the Earth; it continues with a transport stage during which the products of weathering migrate in the environment; and it ends with the formation of sedimentary rocks which, after deep burial, again enter the major geochemical cycle. Mason (1958) pointed out

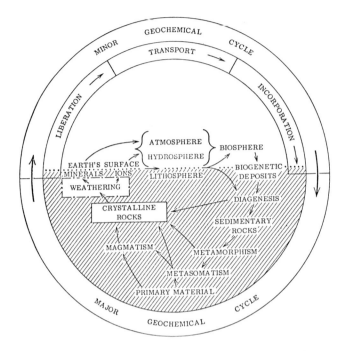

Figure 2.1. Diagram of the geochemical cycle (Fortescue, 1967b after Mason, 1958).

that the concept of the geochemical cycle has limitations. In 1966 he noted that "Like any ideal cycle, the geochemical cycle may not be realized in practice; at some stage it may be infinitely halted or shortcircuited or have its direction reversed" (p. 287). But, in spite of these limitations, the general principle of the geochemical cycle is useful because it focuses attention on the overall ability of chemical elements to circulate during geological time.

Unfortunately, the concept of the geochemical cycle cannot be applied in detail in relation to the behavior of most elements in landscapes. This is because each element plays a unique role in environmental geochemistry, and generalizations concerning the behavior of elements with similar chemical (or mineralogical) properties can only be carried so far. For example, chlorine and bromine are two elements that have rather similar chemical properties, and both are active migrants in the environment. However, because chlorine is nearly a hundred times more abundant than bromine in the Earth's crust, it may accumulate and modify environmental conditions in landscapes in a way that bromine never does because bromine is always relatively rare.

In general, different isotopes of the same element behave almost identically in the environment, although there is significant fractionation under favorable conditions. For example, fractionation of strontium isotopes occurs during the major geochemical cycle (Wedepohl, 1971). However, the advent of man-made radioactive nucleides in landscapes, as a result of atmospheric nuclear explosions, has provided environmental geochemistry with a unique opportunity to study the details of the behavior of particular elements in the environment at the global, regional, and local scale of intensity.

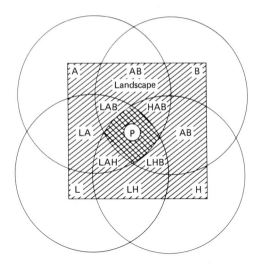

Figure 2.2. Diagram showing the possible interactions between the lithosphere (L), the hydrosphere (H), the atmosphere (A), and the biosphere (B). When they interact they form the "pedosphere" (P) which forms a part of the "landscape" (cross hatched) (modified after Mattson, 1938).

A second general conceptual model in geochemistry is that of the "pedosphere" as described by Mattson (1938) (Figure 2.2). This concept is important because it relates the idea of the four geospheres (as described by Vernadski) to the events that occur in landscapes during geological and pedological time.

Mattson (1938) noted that when the four geospheres interact at or near the daylight surface, a fifth geosphere—the pedosphere—results. The precise nature of the pedosphere varies with the climate, geological conditions, and time at which the pedosphere has been forming. The pedosphere may be envisaged as occurring, in addition to terrestial areas, in marsh and bog areas and in the bottoms of lakes. Examples of the different kinds of environments that result from the interaction of the four geospheres during pedological time are listed on Table 2.1. One advantage of this concept is that it facilitates the description of relation-

Table 2.1. Examples of different kinds of landscapes generated by interactions of two or more geospheres during geological and pedological time as indicated on Figure 2.2 (partly after Stålfelt, 1972).

Code on Figure 2.2	Example of Landscape
LA	Barren desert where the dispersed system consists of mineral particles and air. There is no water or life.
AB	Uppermost part of the soil consisting of plant cover.
HB	A pond with countless organisms.
LH	Waterlogged sand or clay under sterile conditions.
LAH	Extreme conditions in some saline soils.
LAB	Deposits of guano by migratory birds.
HAB	Organic soils and forest litter.
LHB	Waterlogged soils and lake bottoms.

ships between global geochemistry of the geospheres on the one hand and regional and local geochemistry on the other. Mattson (1938) discussed the base used for the expression of environmental geochemistry data. He noted that for certain purposes the weight percent (i.e. percent or parts per million) is adequate, but for other environmental study, the use of a percent by volume may also be of considerable importance. Thus the time may come when landscape geochemistry may be described routinely with reference to either base.

It is likely that as the study of the geochemistry of landscapes becomes more systematic and detailed, the estimates for the abundance of all elements in each of the five geospheres will become more important standards of reference. For example, general estimates will focus attention on relationships between the local, regional, and global aspects of the rate of migration of elements in landscapes during geological, pedological, ecological, and technological time (see Chapter 3).

A third general concept, which was described originally for mineral exploration by Fortescue (1967b), is the prism of landscape. A landscape prism is envisaged as a volume unit with vertical sides which is centered on the pedosphere and extends downwards to the unweathered bedrock and upwards through the biosphere (or hydrosphere in lakes and rivers) to the atmosphere. Like the infinitesimal in calculus, the prism-of-landscape concept is abstract and general which, for purposes of convenience, is usually shown to have a small horizontal extent. In order to simplify the drawing of such prisms (which are designed to show morphological relationships between plant and soil cover types and geological and geochemical features at a particular point in a landscape) the diagrams are plotted on isometric paper. They may describe a single solid prism (Figure 2.4) or an exploded view, as illustrated in Figure 2.5.

In this book the landscape prism concept is used in a number of different ways. For example, a landscape prism drawn accurately to represent a particular point in a particular landscape is called a "tactical landscape prism" (Fortescue

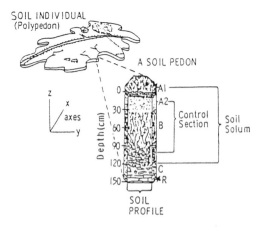

Figure 2.3. A soil pedon is a natural unit in the landscape characterized by position, size, profile, and other features (Buol et al., 1973)

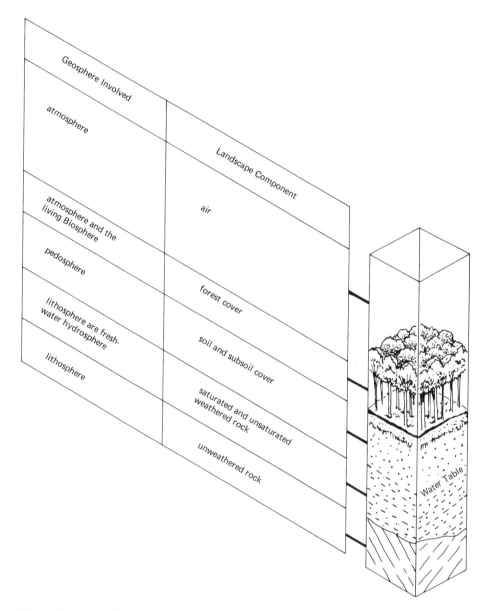

Figure 2.4. A landscape prism (grand strategic) drawn to show relationships between the geospheres and landscape components in a typical forest ecosystem.

and Bradshaw, 1973). If the prism represents information common to a region it is called a "strategic landscape prism" and may be used to describe general features common to all or most landscapes in the region. "Grand strategic landscape" prisms are more general and are used to express general features of landscapes that may be located anywhere on the Earth's surface. Polynov's elementary landscape types (Figure 2.5) are examples of grand strategic landscape prisms.

It should be noted that the use of volume units for the description of landscapes is not confined to landscape geochemistry. For example, in pedology and ecology the concept of the "tessera," was described by Jenny (1958):

> The entire landscape can be visualized as being composed of such small landscape elements. This picture is comparable to the elaborate mosaic designs on the walls of Byzantine churches which are made up of little cubes or dice or prisms called tesseras. We shall use the same name, tessera, for a small landscape element. (p. 5)

This concept is very similar to the concept of the landscape prism although viewed from an ecological standpoint. Similarly pedologists have described volume units for the description of the morphology of soils in terrestial and bog landscapes. In this case the smallest unit recognized is a "pedon" (Figure 2.3) described by the U.S. Soil Survey Staff (1960) as follows:

> A pedon is the smallest volume that can be called "a soil" . . . Its lower limit is the vague and somewhat arbitrary limit between soil and "not soil." The lateral dimensions are large enough to permit study of the nature of any horizons present, for a horizon may be variable in thickness or even discontinuous. Its area ranges from 1 to 10 square meters, depending on the variability in the horizons. Where horizons are intermittent or cyclic and reoccur at linear intervals of 2 to 7 meters (roughly 7 to 25 feet), the pedon includes one-half of the cycle. Thus each pedon includes the range of horizon variability that occurs within these small areas. Where the cycle is less than 2 meters or where all horizons are continuous and of uniform thickness, the pedon has an area of 1 square meter. Again, under these limits, each pedon includes the range of horizon variability associated with that small area. The shape of the pedon is roughly hexagonal. One lateral dimension should not differ appreciably from any other (p. 73).

An important aspect of the concept of the pedon is that it enlarges the idea of a soil profile to include a lateral as well as a vertical extent and thereby puts a natural limit on the size of the volume considered. A soil body containing more than one pedon is called a "polypedon" which was defined by Johnson (1963) as

> one or more contiguous pedons, all falling within a defined range of a single soil series. It is a real, physical soil body, limited by "not-soil" or by pedons of unlike character in respect to criteria used to define soil series. Its minimum size is the same as the minimum size of one pedon, one square metre; and it has no prescribed maximum area. Its boundaries with other polypedons are determined more or less exactly by definition. (p. 213) (Figure 2.3)

For reasons previously mentioned, the concept of a landscape prism is more general and flexible than either that of the tessera or the pedon (and polypedon). Consequently, this unit is used in this book.

Landscape Geochemistry

Parallel with the development of general geochemistry in Russia by Vernadski and his co-workers, a second more specialized discipline—landscape geochem-

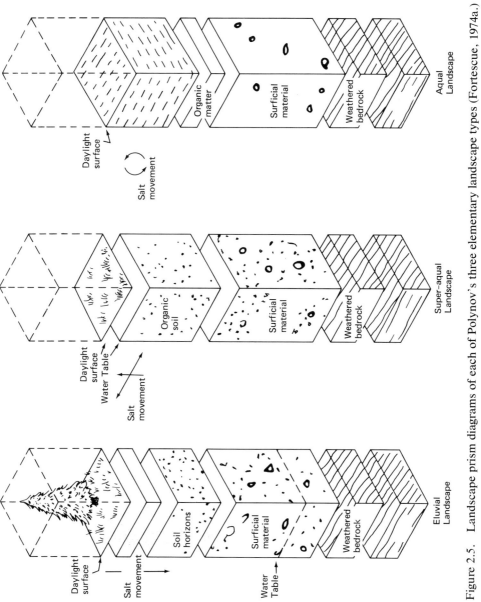

Figure 2.5. Landscape prism diagrams of each of Polynov's three elementary landscape types (Fortescue, 1974a.)

istry—was evolved largely as the result of the ideas of B. B. Polynov (1867–1952). Parfenova (1963) summarized the fundamentals of Polynov's thinking:

> His earlier works on soil science were mostly of a physiographic nature. He tried to uncover the relationships among vegetation, soils, underlying rocks, water, geomorphology and geology of a locality by using complex investigations according to Dokuchayev's program. These studies led to new ideas concerning the landscape as a dynamic system which changes constantly as the result of the struggle between the opposite directions of natural processes concerning the relict and newly arising features of the landscape, which makes it possible to judge the direction in which it develops, and the palaeogeography of landscapes and the necessity for an historical approach to their study. The concept of elementary landscapes is found for the first time in these studies. (p. 111)

Two ideas lie at the heart of Polynov's approach to the study of landscapes: (1) the concept of "elementary landscape types," and (2) the notion of the relative mobility of chemical elements in the environment. According to Polynov (1951), there are three elementary landscape types that may often occur together in the same area of country. These are eluvial landscapes* in which the water table is always, or nearly always, below the daylight surface; super-aqual landscapes in which the daylight surface and the water table are coincident; and aqual landscapes in which the solid matter is below a layer of water, as in lakes and rivers (Figure 2.5).

Thus, Polynov's starting point for the geochemical classification of landscapes relates directly to interactions between the lithosphere, the hydrosphere, and the daylight surface of these geospheres in contact with the atmosphere. In his opinion, these relationships, together with the effects of the living biosphere, govern the migration of elements in landscapes.

Table 2.2. The Relative Mobility of Nine Elements and Ions in Landscapes (Polynov, 1937).

	Average Composition of Igneous Rocks	Average Composition of the Mineral Residue of River Waters	Relative Mobility of Elements and Compounds
SiO_2	59·09	12·80	0·20
Al_2O_3	15·35	0·90	0·02
Fe_2O_3	7·29	0·40	0·04
Ca	3·60	14·70	3·00
Mg	2·11	4·90	1·30
Na	2·97	9·50	2·40
K	2·57	4·40	1·25
Cl'	0·05	6·75	100·00
SO_4''	0·15	11·60	57·00
CO_3''	—	36·50	—

---

* Perel'man (1966, 1972) uses the term "autonomic" to describe what are here called "eluvial" landscapes. The older term is used throughout this book to avoid confusion.

Table 2.3. A Migration Series for Elements in Landscapes (Polynov, 1951).

Migratory Class of Elements	Order of Migration	Index of Size of Migration
Energetically removed	Cl(Br,I)S	$2n \times 10$
Readily removed	Ca, Na, Mg, K	n
Mobile	SiO_2(silicate), P, Mn	$n.10^{-1}$
Inert (slightly mobile)	Fe, Al, Ti	$n.10^{-2}$
Practically immobile	SiO_2	$n.10^{\infty}$

Polynov (1937) described ideas on the movement of elements in landscapes in his classic book *The Cycle of Weathering*. He described estimates of the relative mobility of nine elements and ions in landscapes based on comparisons of the average composition of igneous rocks and the chemical composition of the mineral residue from river waters (Table 2.2). On this scale, his calculations showed that the anions of chlorine and sulphur were the most mobile followed by an intermediate group including calcium, magnesium, sodium, and potassium and a relatively immobile group which included silicon, aluminum, and iron. Later on, Polynov (1951) provided a more comprehensive classification of migrant elements in waters as indicated on Table 2.3.

The discipline of landscape geochemistry has gradually evolved from the basic concepts and ideas of Polynov as first described over 25 years ago (Polynov, 1937).

Summary and Conclusions

Holism is not new in general geochemistry but the description of general concepts in environmental geochemistry may be difficult. How problems of this type were solved by the Russian landscape geochemists is described in Part II where six general concepts for landscape geochemistry are described in addition to abundance (which is common to all geochemistry).

Discussion Topics

1) Can the scope of a scientific discipline be judged on the basis of the activity of a small number of its seminal thinkers?

2) Should general concepts be discussed at the beginning of the advanced level of study of a scientific discipline?

3) Which of the two ideas and concepts due to Polynov described in this chapter is more important in environmental geochemistry?

3. A Philosophy for Environmental Geochemistry

"A philosophy does not spell out the detailed action one should take in specific instances; rather it deals with underlying principles, concepts, and general methods which are relevant to whole classes of problems . . . a philosophy forms an intellectual superstructure or overall strategy which molds and guides the development of discipline, discipline provides an intermediate intellectual structure or strategy which molds and guides way to attact categories of problems. The practitioner, when dealing with an immediate and particular problem, must develop from his knowledge of discipline a specific attack or tactic which resolves the problem."

Morris Asimow, *Introduction to Design* (Englewood Cliffs, New Jersey: Prentice-Hall, Inc., 1962), p. 3.

Introduction

According to the *Concise Oxford Dictionary,* philosophy "deals with ultimate reality, or with the most general causes and principles of things." The landscape geochemistry approach to the study of environmental geochemistry has two component parts: one, which is common to all studies of the geochemistry of the environment, is called its philosophical base; the other, which deals with the concepts and principles of the subject, is called its discipline. This chapter outlines the scope of the philosophy on which the discipline (to be described in detail in Part II) is based.

A philosophy for environmental geochemistry is conveniently described with reference to four kinds of graded organizations called hierarchies. They are (1) the hierarchy of space, (2) the hierarchy of time, (3) the hierarchy of chemical complexity, and (4) the hierarchy of scientific effort. Any study of the behavior of a chemical in the environment includes reference to two or more of these hierarchies. Such a study may be of value either as an "isloated incident" or as a basis for comparative purposes. In general, studies of the geochemistry of the environment carried out without reference to the concepts and principles of landscape geochemistry belong to the former category and those designed in relation to landscape geochemistry belong to the latter category. In either case the philosophical basis is the same.

The Hierarchy of Space

Many studies of the behavior of chemicals in the landscape relate to a particular area (or volume) of country. Consequently, the first hierarchy has to do with the measurement of the migration of chemicals in space. Fortunately, the choice of units is not a problem since the metric system is used almost universally by environmental scientists.

There are three levels of intensity at which environmental geochemistry is studied: (1) local (usually less than 100 km^2 in area); (2) regional (usually more than 100 but less than 1,000,000 km^2 in area); and (3) global (which involves information obtained from a continental area or from the whole surface area of the globe). Geochemical surveys carried out during the search for mineral deposits often occur at the local or regional levels of intensity. Studies of the fate of radioactive debris resulting from nuclear explosions in the atmosphere involve global phenomena which affects all landscapes.

The Hierarchy of Time

In any landscape, chemical substances move about at different rates for different reasons. The rate of movement is expressed as the time taken for a particular substance to move from A to B (e.g., in cm/second, mph, or km/1,000,000 years). It is often desirable to distinguish between different processes that result in the movement of elements from one part of a landscape to another. Four time scales are commonly used in this book: (1) geological time (related to events and processes that usually occur slowly during thousands or millions of years); (2) pedological time (taken for the development of a soil at a particular place and may involve one or more cycles of weathering associated with major changes in climate); (3) ecological time (which indicates the development of a landscape under a given plant cover type); and (4) technological time (which indicates man's effect on the environment either by physical disturbance or more complex activities). In general, at a given site, geological processes are slower than pedological processes, and ecological processes are slower than technological ones. Exceptions do occur; for example, a landslip is a geological event that occurs at an instant in time and results in a period of slow pedological and ecological adjustment before the landscape returns to normal. But in spite of exceptions it is often convenient to refer to the different time scales with respect to particular landscapes and processes that occur within them.

The Hierarchy of Chemical Complexity

A number of units describe the ways in which chemical entities move in landscapes. Some are relatively simple and are used by geochemists and other environmental scientists in contrast to others that fulfill the needs of particular disciplines only. The most common unit is the absolute abundance of the element on a weight percent basis (i.e., as a percent or part per million). Such estimates for rocks and soils are usually made on a "dry weight basis." Similarly, the abundance of elements in biological materials is often made on an "oven dry weight" basis (usually after drying at between 80°C or 100°C). An important aspect of the chemical complexity of natural materials involves the study of the behavior of particular isotopes. Such isotopes may be either radioactive (e.g., ^{137}Cs, ^{90}Sr, or ^{131}I) or nonradioactive (e.g., ^{32}S and ^{34}S). Simple ions such as Na^+ or Cl^- are also determined in waters (or atmospheric samples) as are complex ions such as SO_4^{--} or HCO_3^-. Persistent chemical substances, such as pesticide residues (e.g., DDT, PCB's), may also be studied as they migrate through landscapes. One element may migrate in several forms in the same landscape at the same time at different rates. Each of these forms is then referred to as a "chemical species" and the change from one species to another is called "speciation." For example, Nye and Peterson (1975), using high voltage electrophoretic techniques on aqueous extracts from seleniferous soils of Eire, showed that selenite was the predominate ionic species present in contrast to similar analysis of aqueous extracts of seleniferous soils of South Dakota where selenate was found to be the most common ion. Studies of speciation of elements and organic chemicals are becoming increasingly important in both fundamental and applied landscape geochemistry.

Pedologists, geologists, and geochemists are often concerned with less than the total amount of an element, or chemical substance present in a landscape component or components. Thus the water extraction of selenium from soils carried out by Nye and Peterson (1975) would provide estimates of the partial abundance level of selenium in the soils. Partial abundance data is only important for comparative purposes when the conditions of sample collection and preparation are specified together with the details of the extractant used and the technique by which the extraction was carried out. For example, in environmental geochemistry, maps may be drawn describing the distribution and amount of nickel extracted from oven dried 80 mesh stream sediment material by a cold solution of 1:10 (by volume) hydrochloric acid after shaking for a 12 h period (Fortescue and Terasmae, 1975). Partial abundance data is very important at the local and regional level of geochemical investigations but generally less important on a global scale owing to difficulties in making comparisons. It should be noted that partial abundance data may or may not involve speciation of the chemical substances and ionic forms present in an extract. When, during geochemical surveys, speciation has been carried out on partial extracts from particular materials the interpretation of the distribution patterns obtained may vary from species to species. This principle provides answers to certain problems (for example, in exploration geochemistry) that cannot be obtained in any other way. Another example is the form in which mercury is present in the environment. It is well known that methyl mercury is toxic to life, whereas the same level of concentration of metallic mercury (or mercury sulphide) may be tolerated by living organisms without ill effects.

Let us now consider the behavior of a particular element as an example illustrating the principle levels in the hierarchy of complexity. In Table 3.1 six levels of complexity are listed relative to the behavior of the element sulphur in the environment. Because each element behaves differently in the environment, data for other elements would be similar but not identical to that shown for sulphur.

Table 3.1. Six Levels of Chemical Complexity of the Element Sulphur in the Environment.

Level	Example
Isotope	^{35}S as a radioactive tracer, ^{32}S and ^{34}S fractionation by organisms (Bowen, 1966).
Element Abundance	Abundance of S in Earth's crust 340ppm (Ronov and Yaroshevsky, 1972).
Simple Ion	S^{--} in natural reducing environments.
Complex Ion	SO_4^{--} in natural environments.
Partial Abundance	9ppm S was extracted by CA $(H_2PO_4)_2$ soln. in HOAc from soil in which Alfalfa was growing (Reisenauer et al. 1973).
Organic Matter	Abundance of sulphur in land plants estimated to be 3,400 ppm (Bowen, 1966).
Persistent Chemical	SO_2 pollution of the atmosphere.

The Hierarchy of Scientific Effort*

The hierarchy of scientific effort has two aspects; the first deals with the chemical data from the environment (i.e., its precision and accuracy) and the second involves the type of experimental approach used.

Let us consider these separately. Unfortunately it is not always possible to determine the level of accuracy of methods of chemical analysis used in relation to particular projects. This is because not all published papers list the quality control data derived from their analytical program. However, a terminology for the effectiveness of the scientific effort involved in chemical analysis may be easily described. In this book, chemical data are said to be quantitative data (QD) if the accuracy of the chemical technique used to obtain it was governed by data from International Reference Standard Materials (done at the same time as the unknowns) and the coefficient of variation of the data is less than 1%. Under similar conditions, semi-quantitative (SQD) data are obtained if the coefficient of variation is less than 10%. Data that have not been obtained in relation to parallel analysis of International Standard Substances and are likely to have a coefficient of variation less than 100% are called the test data (TD); data that are not accompanied by information related to standard materials or the coefficient of variation are called appraisal data (AD). Most of the chemical data referred to in the examples quoted in this book are either test data or appraisal data, but in the future, landscape geochemists may insist on either quantitative, or semi-quantitative data for the description of the chemical behavior of elements and other chemical entities in landscapes.

With respect to field investigations, four levels of scientific effort may be recognized although many investigations described in the literature may include reference to more than one of these levels. *The descriptive level* (which is often

---

*A universal constraint to the scientific effort applied for the solution of a problem is availability of funds. This aspect of scientific effort is not considered in this book even though it often dictates the kind of scientific effort used to solve a specific problem.

empirical) includes all studies of the chemistry of the environment not controlled by a mathematical design plan. All the information in Clarke's classic book *The Data of Chemistry* (Clarke, 1924) is of the descriptive-level type because it was obtained before the general application of statistical methods for the design of experiments. Since then many scientists who collect chemical and descriptive information from the environment have adopted statistical methods that result in investigations at the *statistical level*. A third level of scientific effort that is becoming popular in environmental geochemistry involves the methods of *systems analysis*. The use of systems modeling is now common in ecology and in certain applied aspects of environmental geochemistry. The fourth and most complex level of scientific effort is *systems model simulation*. This involves the application of mathematical models, derived from the application of systems analysis, of particular environments to other problems involving changes of

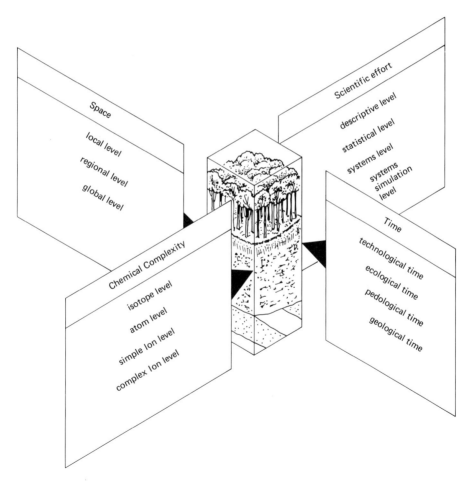

Figure 3.1. A generalized landscape prism representing the environment in relation to the four hierarchies of environmental geochemistry.

Table 3.2. A List of Terms Used to Describe the Hierarchies of Space, Time, Chemical Complexity, and Scientific Effort as Components of a Philosophy for Environmental Geochemistry.

Hierarchy	Term Used	Remarks
SPACE	Local Level	less than 100 sq km in area
	Regional Level	less than 1,000,000 sq km in area
	Global Level	on a continental or global scale
TIME	Technological Time	usually less than 200 years
	Ecological Time	usually greater than 500 years
	Pedological Time	usually greater than 5,000 years
	Geological Time	usually greater than 10,000 years
CHEMICAL COMPLEXITY	Isotope Level	both radioactive and nonradioactive isotopes
	Element Level	usually based on wt percent or volume percent
	Simple Ion Level	important in air and waters
	Complex Ion Level	important in air and waters
	Organic Matter Level	total abundance usual on oven dry basis
	Persistant Chemical Level	important with respect to pesticides
	Partial Abundance Level	conditions of sample preparation and extraction technique most important
SCIENTIFIC EFFORT		
Chemical Data	Quantitative Data (QD)	known to be within 1% of amount present in sample
	Semi-Quantitative Data (SQD)	known to be within 10% of amount present in sample
	Test Data (TD)	known to be within 100% of amount present
	Appraisal Data (AD)	collected without reference to standard substances
Field Effort	Descriptive Level	information collected without statistical plan
	Statistical Level	information collected with a statistical plan
	Systems Model Level	information collected with reference to a systems model
	Simulation Model Level	information collected from a simulation model

environments in response to physical, chemical, or biological changes during technological, ecological, pedological, or even geological time. Consequently, systems model simulation is the ultimate goal of much scientific effort in both fundamental and applied landscape geochemistry.

The terminology used to describe the philosophy of environmental geochemistry in this book is summarized in Table 3.2 and illustrated in Figure 3.1. Each of the four hierarchies is described in relation to the discipline of landscape geochemistry in a separate chapter in Part II.

The Principle of Successive Approximations

A general consideration of the philosophy of environmental geochemistry (and the hierarchies included in it) in relation to global geochemistry focuses attention on the manner in which geochemical data and information are collected. In order to bridge the gap between global geochemistry and local geochemistry, it is necessary to refer to the general principle that connects them—the principle of successive approximations is used to focus attention on relationships of this type.

A good example of the use of the principle of successive approximations (in relation to space and scientific effort)—occurs in prospecting for mineral deposits in virgin country. The problem is to locate very small areas of ground where economically minable mineral deposits occur, in areas of country that may be several million times as large as the deposits. This problem is solved by making surveys (geological, geophysical, geochemical, etc.) of the whole area by techniques that indicate "areas of interest" within which mineral deposits are most likely to occur. In this way interest may be concentrated on, say, 10% of the total area. The "areas of interest" are then surveyed by more intensive techniques designed to locate and describe "landscape anomalies" that are due to the presence of mineralization. After evaluation the most promising anomalies are subjected to a third stage of intensity of scientific effort—the detail level—in order to discover if the anomalies indicate economically minable deposits. In this example, three successive approximations are carried out sequentially using different techniques but leading to a common goal. It should be noted that this approach to prospecting was developed to locate mineral deposits as rapidly, reliably, and cheaply as possible. When rightly planned, successive approximations are a most effective way of solving certain complex environmental problems.

The principle of successive approximations may also be used for the solution of other problems in environmental geochemistry. For example, the development of a new technique for environmental study may be seen to pass through three distinct stages or approximations. The first of these is concerned with the *feasibility* of an idea and is usually established with relatively small resources in a landscape chosen for its suitability to such an experiment. The second approximation often requires substantial resources of time, manpower, and money and during it, the feasible idea passes through a *development* stage. At the end of the development stage the third approximation is reached when the technique resulting from the original idea has been proven reliable in all types of situations where it is applicable. The technique is then said to be in the *established* stage. For example, during World War II it was discovered that a magnetometer mounted in an aircraft could be used to detect the presence of submarines. Later it was found that airborne magnetics was also a most valuable tool for geological surveying including, occasionally, the direct detection of mineral deposits. This activity represents the development stage of the technique. Today airborne magnetic surveys are a standard tool used in geological investigations of many kinds and, as a consequence, may be said to have reached the established stage of development.

Recently environmental scientists have formalized the principle of successive approximations in a "phased approach" for performing environmental pollution source assessment within industrial plants. Hamersma et al. (1976) described this approach as follows:

> The phased approach required three separate levels of sampling and analytical effort. The first level utilizes quantitative sampling and analysis procedures accurate within a factor of two or three and: 1) aprovides preliminary environment assessment data, 2) identifies problem areas, and 3) formulates the data needed for the prioritization of energy and industrial processes, streams within a process, components within a stream and classes of materials for further consideration in the overall assessment. The second level, Level 2, sampling and an analysis effort, after having been focussed by level 1, is designed to provide additional information . . . Level 3 utilizes Level 2 or better sampling and analysis methodology in order to monitor the specific problems identified in Level 2 so that the critical components in a stream can be determined exactly as a function of time and process variation for control device development. (p. 1–2)

Dorsey et al. (1977) provided a flow diagram for the systematic analysis of Level 1 (Figure 3.2).

At Level 3, special techniques would often be needed and would require development prior to routine operation. This example shows the formal use of successive approximations in the examination of complex environmental materials.

Field and laboratory techniques, as well as data processing methods and conclusion drawing criteria, should be examined to discover if they are well-established and pertinent to the problem discussed. In the past, considerable confusion has occurred in the minds of many environmental scientists because data and information obtained at one level of a series of successive approximations has been assumed to relate to a more rigorous level. Also, there is a tendency in some quarters to dismiss appraisal level data as misleading when, under favorable circumstances, it can lead to important and valid conclusions at the level at which it was intended. For example, when one considers the environmental effects of pollution from a point source, appraisal data based on rapid surveys of partial abundance of elements in stream sediments (obtained in a few days) can be used to establish the extent of geochemical anomalies due to the pollution. More intensive and careful environmental monitoring in the vicinity of such anomalies is almost always required to relate them directly to the incidence of diseases in plants, animals, or man.

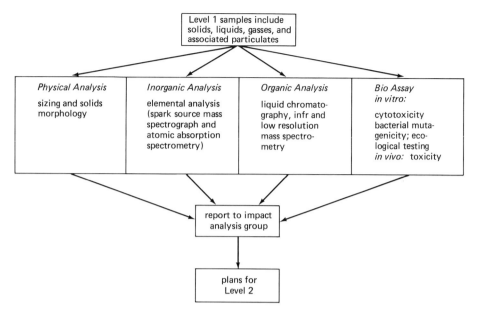

Figure 3.2. Flow chart for Level 1 scheme (Dorsey et al., 1977)

Summary and Conclusions

In order to describe common attributes of environmental and landscape geochemistry, it was decided to provide a philosophy that is general and common to both. The philosophy described in this chapter provides a firm and comprehensive foundation on which specifics of all aspects of our subject can be described, synthesized, and analyzed. More specific relationships between the seven concepts of the discipline of landscape geochemistry and each of the four hierarchies described in this chapter appear in Chapters 13 (Chemical Complexity), 14 (Scientific Effort), 15 (Space), and 16 (Time).

Another basic ingredient of the philosophy of environmental geochemistry is the principal of successive approximations. This complements the idea of the four hierarchies because it provides a connection between them. Consequently, consideration of the implications of the principle of successive approximations is essential when environmental geochemical investigations are in the planning stage. In general, the principle of successive approximations complements the notion of the hierarchies, and together they form a logical foundation for the discussion of environmental geochemistry.

Discussion Topics

1) Should each scientific discipline be linked to a particular philosophy?
2) Are hierarchies as described in this chapter helpful in environmental sciences?
3) Should scientific discipline use well-known terminology or generate its own jargon?

4. Other Approaches to Environmental Geochemistry

"We have seen how unreliable an observer's report of a complex situation often is. Indeed, it is very difficult to observe and describe accurately even simple phenomena."

W. I. B. Beverage, "The Art of Scientific Investigation" (New York: Vintage Books, 1957), p. 136.

Environmental Science and Geochemistry

It is significant that prior to the late 1970s, introductory texts in geochemistry omitted any reference to "environmental geochemistry" even though during the past twenty years the study of the geochemistry of the environment has become of major concern to the scientific community as a whole. For example, the chapter entitled "Objectives and History of Geochemistry" in the introductory text by Wedepohl (1971) includes no reference to environmental geochemistry or to exploration geochemistry. Other scientists have also remarked on the lack of formal interest by geochemists in environmental geochemistry. For example, in a preface to the published proceedings of a symposium held at the 162nd meeting of the American Chemical Society in Washington, D.C., on September 15, 1971, Kothny (1973) stated:

> This volume represents perhaps the first effort to fit geochemistry into environmental science. In studying environmental geochemistry we gain insight into the origin, transition and concentration of a particular element. Also we may assess the real impact of man-made alterations in natural environments. In fact, the natural cycle of a few elements is altered in smaller environments, but given enough time, they are incorporated back into the whole terrestial cycle. As the result of this alteration, anomalous metallic concentrations are discovered near cities, freeways, smelters, power plants and industrial plants. (p. vii)

It was at a meeting organized by chemists in 1971 that geochemists were reminded of the holistic approach to the geochemistry of the environment and man's activity that had been an integral part of the thinking of the founders of geochemistry (e.g., Clarke, Vernadski, and Goldschmidt—as described in Chapter 2).

Definitions of environmental science are difficult to devise because of its broad scope and diverse components. A good definition was provided by Oliver and Manners (1972):

> The emergence of environmental science as a field of study in its own right during the last decade is a reflection of the inadequacy of existing disciplines to deal with the complexity of the total environment system. Essentially, environmental science attempts to identify, define and analyze those physical and biotic processes that actively influence, or are influenced by, man's actions. . . . Thus, environmental science seeks to combine the approach of such disciplines as biology, geology, pedology, and climatology, and attempts to avoid the artificial division of the environment into quite distinct and separate compartments. (p. 337)

It is evident that the philosophy for environmental geochemistry described in Chapter 3 fits easily into this definition of environmental science.

Environmental Science and Landscape Geochemistry

"Holism" has often been used in relation to landscape geochemistry and to environmental science in general: it is basic to both. The term was defined in 1926 by J. C. Smuts (Smuts, 1961):

> We are all familiar in the domain of life with what is here called wholes. Every organism, every plant or animal, is a whole, with a certain internal organization and a measure of self direction, and an individual specific character of its own. This is true of the lowest micro-organism no less than of the most highly developed and complex human personality. What is not generally recognized is that the conception of wholes covers a much wider field than that of life, that its beginnings are traceable already in the inorganic order of nature, and that beyond the ordinary domain of biology it applies in a sense to human association—like the State, and to the creations of the human spirit in all its greatest and most significant activities. Not only are plants and animals wholes, but in a certain limited sense the natural collections of matter in the universe are wholes; atoms, molecules and chemical compounds are limited wholes; while in another closely related sense human characters, works of art and the great ideals of the higher life are or partake of the character of wholes. (p. 98)

Thus, the philosophy of landscape geochemistry as described in Chapter 3 is holistic in its approach to the study of environmental geochemistry because it includes, in addition to chemical complexity, the hierarchies of space, time, and scientific effort. A consideration of the implications of the holistic philosophy for environmental geochemistry leads to the inevitable conclusion that in order for relationships to be readily established between the chemistry of landscape components on a global scale, some kind of standardization of the approach is helpful. Landscape geochemistry, as described in this book, focuses attention on this very difficult problem and provides a series of concepts which, if developed further, can lead to a logical, rational, and effective discipline for environmental geochemistry. Perel'man's principle of unconnected data bases (described in Chapter 2) is another way of drawing attention to the need for a formal discipline of landscape geochemistry.

Examples of other Approaches to the Study of Environmental Geochemistry

Landscape geochemistry is not the only scientific discipline in environmental science that includes reference to geochemistry and chemistry of the environment. In order to illustrate how geochemistry is involved in other disciplines, two examples are included here. One involves a brief reference to the role of geochemistry in ecology and the other to the role of geochemistry in land use planning. These two disciplines were selected because they are both holistic and are concerned with the study of the environment at all the different levels of intensity included in the hierarchies described previously.

An Ecological Approach to Environmental Geochemistry

Until about 20 years ago, ecology often focused attention on either a plant or animal and the habitat in which it lived. More recently, owing largely to broadened interest in the problem of environmental degradation, the definition of ecology has been broadened considerably to include many aspects of science. For example, Southwick (1976) has defined ecology as follows:

> Ecology is a vast and encyclopaedic subject. It can be confusing and discouraging at the outset because it seems so diffuse. After all, it must take into account the life habits of over two million different kinds of animals and plants, and consider all manner of influence and interaction among them. Thus, ecology must include not only the life sciences, but chemistry, physics, geology, geography, meterology, climatology, hydrology, paleontology, archaeology, anthropology, and sociology as well. In fact, so extensive is the scope of ecology that it seems to have no limits at all, and ecologists could claim domain over all the natural and social sciences. . . . Fortunately, the innumerable facts of ecology can be distilled into some relatively basic and simple principles. The logical starting point is the ecosystem. An *ecosystem* is any spatial or organizational unit which includes living organisms and non-living substances interacting to produce an exchange of materials between the living and the non-living parts. (p. 161)

This definition is of interest for two reasons: (1) the writer includes so many scientific disciplines within it and (2) he was able to focus his attention on the concept of the "ecosystem," which is a most convenient way of drawing attention to the study of ecology at the local, regional, or global scale of investigations. It may very well be that some of the popularity of the term "ecology" in the public mind today stems from the concept of the ecosystem which can be easily understood by scientists and laymen alike.

The term ecosystem was originally defined by Tansley (1935):

> The fundamental concept appropriate to the biome considered together with all the effective inorganic factors of its environment is the ecosystem which is a particular category among physical systems that make up the universe. In an ecosystem the

organisms and the inorganic factors alike are components which are in relatively stable dynamic equilibrium. Succession and development are instances of the universal processes tending towards the creation of such equilibriated systems. (p. 306)

At the time this definition was written, ecology was at a descriptive stage of development, broadly similar to landscape geochemistry today. More recently ecology has become both quantitative and, to some extent, predictive and includes all four levels of scientific effort (see Table 3.2) with major stress placed on systems analysis and systems simulation modeling. Burgess (1975) summarized recent progress in ecology with respect to the Eastern Deciduous Forest biome project of the U.S. International Biological Program:

Ecosystems models consisting of systems of non-linear difference or differential equations are being developed by a number of research groups. These models describe the ecosystem process by process incorporating mathematical analogs for all mechanisms considered important. Because of our lack of understanding, the approach has been most useful for indicating specific deficiencies in our knowledge and suggesting new directions for research.

The success of such models in structuring research in the various biome studies suggests some optimal approaches. During model development lack of understanding of specific processes if often encountered. In such cases alternative models can be proposed, each incorporating experimentation until clear differences in behavior are found. (p. 25)

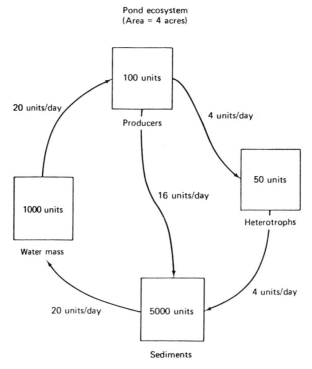

Figure 4.1. Biogeochemical pools and flux rates for a hypothetical nutrient cycle in a lake ecosystem (Collier et al., 1973, p. 430).

Table 4.1. Absolute Flux Rates, Turnover Rates, and Turnover Times for the Hypothetical Biogeochemical Cycle Shown on Figure 4.2 (Collier et al., 1973).

Transfer Route From To	Absolute Flux Rate (units/ acre/day)	Turnover Rate		Turnover Time (days)	
		Departure Pool	Recipient Pool	Departure Pool	Recipient Pool
Water mass—Producers	5	0.02	0.20	50.	4.
Producers—Sediments	4	0.16	0.0032	6.25	312.5
Producers—Heterotrophs	1	0.04	0.08	25.	12.5
Heterotrophs—Sediments	1	0.08	0.0008	12.5	1250.
Sediments—Water mass	5	0.004	0.02	250.	50.

Another concept closely related to that of the ecosystem that is frequently used by ecologists to describe the circulation of elements—usually nutrients—in ecosystems is the "biogeochemical cycle." This term, made popular in ecology by Hutchinson (1944, 1950), is derived from the writings of the Russian geochemist Vernadski (1927) and relates to the cyclic path followed by nutrient elements during the growth of living matter in an ecosystem. All living organisms absorb certain elements from the environment in which they live in order to complete their life cycle. After the organisms die, the same elements are usually recycled by other organisms. The ecosystem models referred to by Burgess et al. (1975) are concerned with the details of the circulation of chemical substances and energy through particular ecosystems situated in the eastern deciduous forests of North America.

A clear idea of the concept of a biogeochemical cycle may be obtained from a hypothetical example described by Collier et al. (1973). In this case the flow of a single element in a lake ecosystem is described by a series of arrows and boxes in a conceptual model (Figure 4.1). In this case the arrows indicate flux rates and the boxes indicate pools for the nutrient in the lake viewed as a system. A biogeochemical pool is envisaged as the quantity of a particular nutrient in some biotic or abiotic ecosystem component, and a flux rate is the quantity of the nutrient passing from one pool to another per unit time, per unit area, or per unit volume of the system. The turnover rate is the flux rate into or out of the pool divided by the amount of a nutrient in the pool. Turnover rates for the hypothetical system (Figure 4.1) are listed in Table 4.1. In this example the highest turnover rate is from the water mass to the producers (photosynthetic organisms), and this component is therefore likely to be subject to the greatest short-term fluctuations. The turnover time is the quantity of the nutrient in the pool divided by the flux rate (Table 4.1). In this case the shortest turnover time is from the input to the producer pool (4 days). The movement from the sediment to the water to the producers to the heterotrophs and back to the sediment is a true cycle, or nearly so, in ecosystems that are undisturbed by man's activities. When the cycle is upset, for example by the addition of nutrient chemicals to the lake water in sewage effluent, the nature of the biogeochemical cycle may be altered drastically. Within a particular ecosystem, each of the nutrient elements follows a different biogeochemical cycle; but for purposes of convenience the cycling of

Comments on the Ecological Approach to Environmental Geochemistry

The hierarchies of space, time, chemical complexity, and scientific effort are all involved in ecological studies. Two of the concepts central to modern ecology (i.e., the ecosystem and biogeochemical cycling) are important for the geochemistry of landscapes. The living biosphere is the central theme of ecology and the behavior of chemical elements in the environment is of secondary importance. The opposite is the case in landscape geochemistry in which the behavior of the elements and other chemical entities in the landscape is of central importance and the role of the biosphere, either as living or dead organic matter, is but one of the contributing factors to the circulation of chemical elements. In ecology the main focus of attention is the circulation of elements by ecosystems today (or during ecological time) in contrast to landscape geochemistry which also considers the pedological and geological processes of the landscape that result in the circulation of elements. Another difference between ecology and landscape geochemistry is in the methodological approach adopted on the local scale of intensity. Most ecological studies are carried out in specific types of ecosystems whereas in landscape geochemistry areas of country (often involving several types of ecosystems and the transitions between them) are considered as a whole.

The Land Classification Approach to Environmental Geochemistry

So far we have considered the ecological approach to the study of the geochemistry of the environment. We have also considered relationships between ecology and landscape geochemistry and noted that there are important concepts in ecology that are common to landscape geochemistry as well. In this section we briefly consider a branch of science—land classification—that involves the use of numerous concepts for purposes of classification, some of which are based on the chemical properties of landscapes.

During the past fifty years foresters, soil scientists, geographers, and other scientists have proposed many schemes for the holistic description of areas of country. Such schemes, which are usually described at the regional level of intensity, are aimed at the classification of lands in relation to a specific purpose. One of the most comprehensive and best known of these systems is the one described by Sukachev and Dylis (1964) which was based on the ideas of Sukachev: this approach is chosen as an example of a land classification system even though it is not formalized in great detail with respect to a single region. The approach is based on the concept of the "biogeocoenose" which was defined as follows:

A biogeocoenose is a combination on a specific area of the Earth's surface of a homogeneous natural phenomena (atmosphere, mineral strata, vegetable, animal, and microbic life, soil and water conditions) possessing its own specific type of interaction of these components and a definite type of interchange of their matter and energy among themselves and with other natural phenomena, and representing an internally-contradictory dialectical unity, being in constant movement and development. (p. 26)

This definition, although similar to the definition of the ecosystem given by Tansley (1935), is broader in scope and may be used as the conceptual basis for land classification systems in large areas of the country. For example, Krajina (1972) described a forest ecological classification for British Columbia based on the concept of the biogeocoenoses. The interactions between the various components of a biogeocoenose are indicated in Figure 4.2.

The purpose of lands classification in forestry was described succinctly by Burger (1976) as follows:

Forest management deals with productivity systems, that is, ecosystems or biogeocoenoses, or total site from which it attempts to extract efficiently and continuously

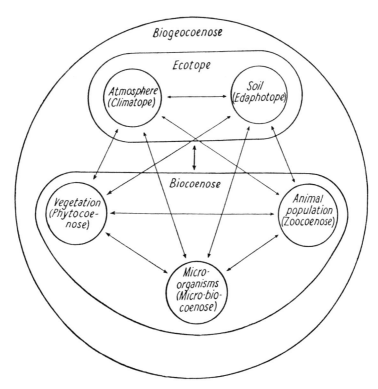

Figure 4.2. Diagram of the interactions of components of a biogeocoenose (Sukachev and Dylis, 1964, p. 27).

a product, preferably of highest possible yield. The role of site classification in forest management is to provide a framework for organizing our knowledge of ecosystems and for transferring experience and knowledge from one locality for the application in the management of another. (p. 1)

Other systems of land classification have been devised to relate to either single or multiple types of land use. For example, in Canada, systems for peatlands (Zoltoi et al., 1975; Stanek, 1977) and arctic landscapes, (Lavkulich and Rutter, 1975) have been described. A comprehensive system for the province of Ontario was described by Hills and his co-workers several years ago and summarized recently by Burger (1972). It is to be expected that as man's use of the environment becomes more intensive, land classifications systems will increase in importance and effectiveness so that "tradeoffs" between different kinds of land use in particular areas of country will be based on scientific rather than political principles.

Geochemistry may play two roles in land classification systems. First, it may be used as a diagnostic tool for the mapping of land types; second, geochemical information of importance may be collected as a result of land classification activities. A simple example of the former was included by Storie (1964) in a land classification scheme in California (Figure 4.3). In this case the geochemical nature of the rock and soil combined was used to define "soil depth classes"

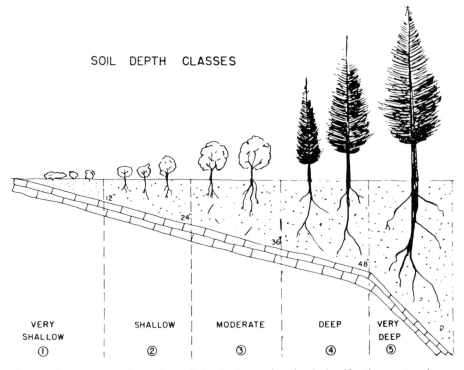

Figure 4.3. Diagram illustrating soil depth classes in a land classification system in use in California (Storie, 1964).

Table 4.2. Annual Cycling of Four Nutrient Elements in Forest Stands (kg/ha) (Bakuzis, 1971).

Process	Species	Nitrogen	Phosphorus	Potassium	Calcium
Uptake	Pine	45	5	7	29
	Beech	50	13	15	96
	Oak	87.7	6.5	79	95.1
Return	Pine	35	4	5	19
	Beech	40	10	10	82
	Oak	55.6	3.1	58.6	82.8
Storage	Pine	10	1	2	10
	Beech	10	3	5	14
	Oak	27	3	13.9	0.8
Removed by thinning	Oak	5.1	—	—	10.5

which were used in mapping units in the classification system. An example of geochemical information collected with respect to different plant cover types in the same area of country was described by Bakuzis (1971) in a discussion of forest management. He stressed the importance of variation in (1) nutrient uptake, (2) nutrient storage, (3) nutrient return, and (4) nutrient losses, in relation to different kinds of woodlands, and provided data for nitrogen, phosphorus, potassium, and calcium in pine, beech, and oak stands to support his contention (Table 4.2). Data of this type obtained from a single locality (or region) may readily be incorporated to global level estimates for the uptake of nutrients—such as that described by Rodin and Bazilevich (1967).

Although land classification systems appear attractive as a basis for the description of the behavior of elements in landscapes and might reasonably be incorporated in the landscape geochemistry approach without modification, this is not the case (except on the local level of intensity in some areas). This limitation is due largely to the different data bases used by each system of land classification. For example, Jourdant et al. (1975) described the holistic principles on which one land classification scheme was based:

> The philosophy of the classification system rests on the recognition that, although all environmental factors (landform, climate, organisms, soil and groundwater within a framework of time) influence each other, the landforms (including structure and composition as well as of terrestial and water bodies) influence the other factors to a far greater degree than they are influenced by them. Thus landform classification and mapping are the framework within which climatic, vegetational, pedologic and hydrologic data are described, characterized and classified. (p. 487)

This system, Biophysical Land Classification, is for all of Canada, but, as we have seen, particular Canadian provinces also have other systems: for example, the Hills system in Ontario (Burger, 1972) and the system described by Krajina (1972) for British Columbia. Each system employs a different terminology and uses different criteria for the detailed description of particular localities. Consequently, the landscape geochemist interested in local, regional, and global aspects of the circulation of elements finds these systems important at the local level but of limited value as a basis for broad-scale comparisons. The landscape geochemical approach provides a system of land classification that may be readily applied anywhere in the world and may, under favorable circumstances, be related directly to systems of land classification already in the use in particular regions.

Comments on the Land Classification Approach to Environmental Geochemistry

As in the case of the ecological approach, the hierarchies of space, time, chemical complexity, and scientific effort are all involved in the land classification approach to environmental geochemistry. As expected, the main preoccupation of scientists involved in land classification is the delineation of spatial relationships between different component parts of landscapes. To this end consideration of ecological, pedological, geological, and technological time taken in the formation of landscapes are considered individually or collectively as possible data bases for the description of land types or units. Similarly, certain aspects of the chemical complexity of landscapes may also be used as criteria for the delineation of boundaries between land units. Descriptive and statistical methods are commonly used for the collection of information relating to the classification of lands. Currently, the systems model approach and the systems simulation approach are of less interest in relation to land classification. In their place are the sophisticated techniques of remote sensing that combine a number of alternative methods of imagery with information obtained at the different levels in the hierarchy of space [Everett and Simonett (1976), (Table 4.3)].

Geochemistry is only one aspect of the total environment that is considered a basis for the classification of land. It is indeed unfortunate that land classification mapping, unlike topography, has not yet been standardized for use on a global scale; however, this may occur in the future as a result of the contribution made by global remote sensing. Until then, particular systems of land classification will be an aid to landscape geochemistry and not an integral part of our subject.

Table 4.3. Summary of the Impact of Satellite Information and Data on Science in General (Everett and Simonett, 1976).

Stimulating the development of new hypotheses and theories in the natural and social sciences.

Providing opportunities for partitioning other aggregated natural and social data (e.g. census data).

Improving multistage sample designs in research and applications development, the selection of sites for more detailed scientific study, the calibration of other components of an integrated resource management system, and the evaluation of the interaction of static and dynamic components of the environment.

Providing a basis for extrapolation, interpolation, and refinement of scattered scientific observations, and for the making of unique observations feasible only with satellite systems.

Stimulating the development of new science models which may be raised with ERS data to higher efficiencies than are now feasible.

Opening, through storage of a unique, internally consistent longitudinal data set, unparalleled opportunities for developing quantitative hindcasting and forecasting models.

Summary and Conclusions

A brief outline of the ecological approach and the land use approach to the study of the geochemistry of the environment has been given. It was noted that the ecological approach to environmental geochemistry shares concepts of the ecosystem and biogeochemical cycling with landscape geochemistry, and very close relationships occur between the two disciplines. Landscape geochemistry differs from ecology in that it considers the circulation of all elements in the landscape, not just nutrients or those that are toxic to life. Also, it is centered on the circulation of elements in landscapes as a result of all processes, not only those that relate directly to the living biosphere. Although the land classification approach to environmental geochemistry might be seriously considered as a starting point for the systematic and general study of the circulation of chemical elements in the environment, a close look at systems of land classification reveals that the migration of elements often plays a small part in such systems and that land units are frequently delineated by concepts unrelated to geochemistry. Another drawback with land use systems currently in use, in relation to landscape geochemistry, is that although they may be useful on a local or regional level, they are generally not comparable at the global level of study. A possible solution to this problem is the direct and detailed application of data obtained from satellites to problems of land classification and evaluation. While the same philosophy underlies the collection of environmental geochemical data from both these approaches, there is currently no scientific discipline based on concepts and principles that relates specifically to the circulation of chemical elements in the total environment.

Several pitfalls lie in the way of the establishment of such a scientific discipline. Some of them were described vividly by Stone (1975) in relation to forest soil study:

> Science tends to be one of particulars, at least in its real life applications. Such applications are strongly conditioned by local economic as well as environmental variables, and these are nowhere twice the same. Thus, it is easy to fall into various kinds of provincialisms, narrowing horizons to only one locality, one forest, one kind of forestry, or one small aspect of science. Or to retreat from complexity and become a disciple of some single approach or universal method. These are dead end streets. We need a much more comprehensive vision of our science and of what we wish to accomplish with it. (p. 9)

In order to avoid these pitfalls, it is desirable that landscape geochemistry fulfill the following criteria:

1) it must be applicable to all landscapes including those in their natural state and those that have been modified to a greater or less extent by man's activities during technological time.

2) its concepts and principles must utilize wherever possible units of measurement that exist already and provide data and information that can readily be used for comparative purposes.

3) its concepts and principles must be flexible enough to incorporate existing information on the geochemistry of the environment with a minimum of effort.

4) it must allow for the study of substances at various levels in the hierarchy of chemical complexity that migrate as entities within landscapes and for the description of circulatory patterns of these entities at the local, regional, or global levels of intensity. It must also allow for the study of migration patterns with respect to technological, ecological, pedological, and geological time by techniques that include descriptive, statistical modeling, systems analysis modeling, and simulation modeling levels of scientific effort.

If these criteria are satisifed, the newly proposed discipline will take its place beside ecology as one of the important branches of environmental science.

Discussion Topics

1) Is the principle of successive approximations applicable to general environmental science?

2) What other approach to environmental goechemistry is as important as the two described in this chapter?

3) Does the formal description of criteria desirable for a discipline within the environmental science contribute to holism?

Part II. The Basics of the Discipline of Landscape Geochemistry

"The science of landscape was created by representatives of various sciences, some of which are formally remote from geography (geochemistry, agronomy, etc.). But regardless of the differences in the development of their scientific careers and irrespective of the disciplines with which they were initially connected, these scientists eventually came to the conclusion, albeit in different ways, that a need existed for studying the Earth's surface as a single entity, and also for studying the relationships between certain natural phenomena."

        A. I. Perel'man, *Landscape Geochemistry,* Vysshaya Shkola, Moscow 1966, (G. S. C. Translation 676, Ottawa) p. 15.

5. Definitions of Concepts and Principles

"The purpose of scientific thought is successful prediction, which is a prerequisite of understanding. Predictions are made by constructing and extrapolating conceptual models. The construction of a model is a creative act that admits no standard procedure, but the validation of the model follows a regular process called the *scientific method*. Models are never certain and are always subject to revision, but each new model includes the successful parts of older models. Thus, scientific knowledge is cumulative."

Marshall Walker, *The Nature of Scientific Thought* (Englewood Cliffs, N.J.: Prentice-Hall, Inc., 1963,) p. 13.

Introduction

Two of Polynov's contributions to landscape geochemistry were outlined briefly at the end of Chapter 2. This chapter is concerned with definitions of the seven concepts of the discipline of landscape geochemistry. In Part II, each concept is described in a separate chapter (Chapters 6–12.). At the end of Part II, relationships between the basics of the discipline and the four hierarchies that constitute the philosophy are treated in a series of four chapters (Chapters 13, 14, 15, and 16).

The basics of landscape geochemistry were all described by the Russian landscape geochemists Polynov (1937, 1951), Perel'man (1966, 1967, 1972), Glazovskaya (1962, 1963), Kozlovskiy (1972), and their co-workers. The Russian thinking in landscape geochemistry has been illustrated where possible by examples taken from the western literature. In some cases, additions have been made by myself (e.g., "partial abundance" and "speciation" of chemical entities in the concept of "Element Abundance").

Concept I. Element Abundance

Landscape geochemistry is concerned with the absolute, relative and partial abundance of chemical elements and compounds as they circulate in landscapes. The units used to describe the circulation of elements in the lithosphere, or biosphere, are usually based on weight percent, while in water, or air, either weight percent or volume percent are commonly reported. The reference standard currently used by landscape geochemists to express the relative abundance of elements in natural materials is the Clarke which is the estimate of the weight percent of each element in the earth's crust. Eventually, similar estimates may be used as standards of reference for the relative abundance of elements in the hydrosphere, atmosphere, and the biosphere as well.

Two aspects of the geochemistry of an element that largely determine its behavior in the Earth's crust are its chemistry and its abundance in nature, or more specifically, in the Earth's crust. Outside Russia, geochemists have tended to stress the total or absolute abundance of elements in natural materials as a basis for discussion of element mobility. Within Russia, things are a little different. First, A. E. Fersman defined the Clarke as the unit of measurement for the abundance of an element in the Earth's crust (Mason, 1966) and later Vernadski proposed a second unit—the Clarke of Concentration (KK)—to express the relative abundance of elements. The Clarke of Concentration is the abundance of an element expressed as weight percent divided by its Clarke (Mason, 1966).

The partial abundance of an element is the amount of that element extracted from a sample of a natural material that has been collected and processed by a standard technique by an extracting solution of a uniform composition for a stated time at a particular temperature. Partial abundance data are of considerable interest in landscape geochemistry—particularly in relation to soil and sediment material derived from streams or lakes. Many different kinds of partial procedures and extractants may be used in relation to the same element. For example, Dahnke and Vasey (1973) listed seventeen different extractants that had been used by one or more of fifty-nine writers to determine the amount of nitrogen "available" to plants in soils. Initial levels of NO_3 nitrogen was one of the most successful tests of this purpose. The concepts of total, relative, and partial abundance of elements are of fundamental importance in landscape geochemistry.

Concept II. Element Migration

> The abundance of an element in a particular natural material (or landscape component) is a measure of the level of concentration of that element in the material at the time of collection. Polynov and the other Russian landscape geochemists have also been interested in the absolute and relative rate of movement of elements, ions and chemicals in landscapes. This has led to the definition of a number of descriptive parameters for the movement rate of elements through landscapes. These include the "migration series for elements" (described in Chapter 2). Parameters for element migration are of importance at the local, regional, and global levels of environmental investigations.

The problems of element migration into and out of landscapes have been of interest to geochemists ever since the study of the chemistry of the environment began over 100 years ago; many scientists have made a study of the migration rates of elements from particular landscapes. For example, Likens and Bormann (1975) provided quantitative estimates for the absolute migration rate of elements and ions from the Hubbard Brook catchment area. The innovation introduced by the Russian workers in relation to element migration is the definition of parameters to describe the relative mobility of elements in the living and nonliving parts of the environment and the establishment of a classification system for migrant elements. This will lead to the establishment of general behavior patterns for particular groups of elements and, eventually, to facilitate the establishment of systems and simulation models in landscape geochemistry.

Definitions of Concepts and Principles

Concept III. Geochemical Flows

A Russian geochemist, Kozlovskiy (1972), formalized concepts and principles to describe in general terms the flow patterns that occur for chemicals and substances in landscapes. He recognized three principal flow types. The *main migration cycle* has a predominantly vertical movement and corresponds to the more narrow concept of biogeochemical cycling. The *landscape geochemical flow* refers to the continuous flow of matter through landscapes (for example, the atmosphere in a terrestial landscape, or water in an aqueous landscape) and may involve the physical or chemical modification of the landscape resulting in the removal of matter or in the addition of matter to the landscape. When matter is removed from or added to a particular landscape, an *extra landscape flow* results. Such a flow is positive if it involves the addition of material to the landscape and negative if it involves the removal of material from the landscape.

The concepts of geochemical flows and the principles associated with them as described by Kozlovskiy (1972) provide a starting point for discussions of the behavior of elements during their circulation through landscapes. Such circulation patterns are governed partly by the absolute abundance of the elements in landscapes, by the partial abundance in particular landscape components, and by the factors that govern mobility. It is convenient to describe the geochemical flows in relation to the landscape prism concept as shown on Figure 5.1.

Concept IV. Geochemical Gradients

An important aspect of soil science and field ecology is the study of gradients that occur in soil cover types and plant cover types as a result of gradual or abrupt changes in landscape conditions. Similar gradients that may or may not be related to plant or soil cover types relate to the variation in the distribution and amount of chemical elements in landscapes and are called geochemical gradients. Geochemical gradients may occur at the local, regional, or global scale of intensity and usually relate to the flow of water, or air. Global gradients involve variations in the input of solar energy to landscapes. Although many gradients are natural, others relate directly to man's activities and are frequently studied in relation to the pollution of the environment.

The importance of the concept of geochemical gradients is that it allows for the systematic suggestion of the cause and effect of gradients on landscapes, regardless of the type of landscape involved. This facilitates comparable interest in relation to problems of space, time, and chemical complexity in the environment. The formalization of the gradient concept in environmental geochemistry seems long overdue. A large number of studies of what are here called geochemical gradients are described in the literature as "isolated incidents."

Concept V. Geochemical Barriers

In some landscapes where there is a change in local conditions, certain elements may accumulate and others be preferentially removed due to a combination of mechanical, physical, chemical, or biological processes working together or individually. Such changes in local conditions are called geochemical barriers. Such barriers are of considerable fundamental and practical importance in landscape geochemistry. In

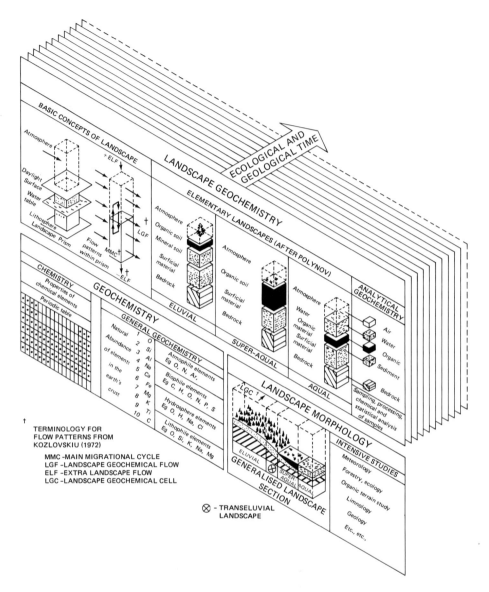

Figure 5.1 Chart showing some basics of landscape geochemistry (Fortescue, 1974b).

extreme cases geochemical barriers may result in the formation of economic mineral deposits although most geochemical barriers have less dramatic geochemical effects in the environment.

A gold placer deposit in a stream is an example of the effect of a mechanical geochemical barrier. A more subtle type of geochemical barrier occurs at the edge of a bog or swamp where certain elements accumulate and others increase in migration rate due to changes in Eh/pH condition of groundwaters. Like the concept of geochemical gradients, the concept of geochemical barriers has often been described by examples that appear in the literature of environmental geochemistry. A task of landscape geochemistry is to encourage the systematic study of geochemical gradients and geochemical barriers by conceptual and mathematical models.

Concept VI. The Historical Geochemistry of Landscapes

It should be noted that the geochemical classification of landscapes is based on present day features of the landscape. These are often a mixture of components that evolved during geological, pedological, ecological, and technological time. Because of the dynamic nature of these processes, which result in the current abundance patterns for elements in landscapes (including geochemical barriers and gradients), a full understanding of the behavior of elements in a particular landscape may often be governed largely by its geochemical history. Consequently the concept of the geochemical history of landscapes is of fundamental importance in landscape geochemistry. For example, the presence of relict structures derived from previous weathering cycles may govern the circulation of groundwaters in landscapes today. A clear understanding of the migration patterns for elements in landscapes today is often impossible if the study of the historical development of the landscape is ignored.

Landscape geochemistry, because of its holistic outlook, may utilize observations from all parts of an area of country (i.e., land, bog, and lake) in order to construct the sequence of geochemical events during its evolution. Such studies, when combined with geochemical landscape classification, are very important to impact statements and the disposal of wastes in landscapes. The study of the fate of particular nuclides (generated by atomic explosions in the atmosphere) is of considerable interest in historical landscape geochemistry particularly for events during technological time.

Concept VII. Geochemical Landscapes

If one considers the elements in the lithosphere being acted on by the hydrosphere, atmosphere, and biosphere during geological time in relation to the concepts of (1) the pedosphere, (2) element migration, (3) geochemical flows, (4) geochemical gradients and (5) geochemical barriers, it is evident this leads to the geochemical description of landscapes at the local, regional, or global levels of intensity. This has led Polynov (1951) and the Russian landscape geochemists to describe the concept of a geochemical classification for landscapes.

Following the lead given by Polynov with his definition of elementary landscape types, Glazovskaya (1963) and Perel'man (1966) have described schemes for the geochemical classification of landscapes at the local, regional, or global levels. Although the terminology devised to describe landscapes at the local or regional level is somewhat complicated, its importance lies in that it can be applied to any landscape in either its natural state or after disturbance by man's activities. Consequently, the geochemical classification of landscapes is of interest in both fundamental and applied landscape geochemistry and in environmental geochemistry. Briefly, relationships between the level of the water table (or hydrosphere) and the daylight surface on the one hand and the nature of the geological substrate on the other determine the general conditions of a particular landscape on the local scale. On a regional or global scale,

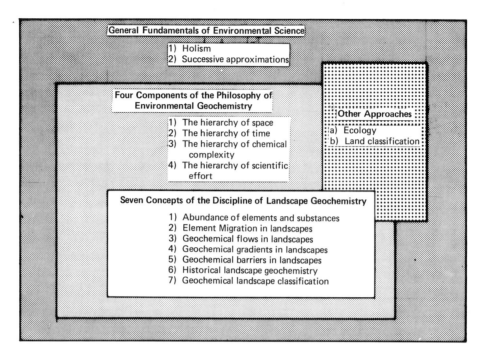

Figure 5.2. Relationships between general fundamentals of environmental science, the philosophy of environmental geochemistry, and the discipline of landscape geochemistry.

climate is also an important factor in the classification scheme. In general, the classification is based on the eluvial (terrestial) landscapes of an area to which the swamps and bogs (superaqual landscapes) and the rivers and lakes (aqual landscapes) are considered subordinate.

Although still at an early stage of development, the geochemical classification of landscapes is of considerable fundamental (and practical) importance in landscape geochemistry and environmental geochemistry. Clearly, as the emphasis shifts in the study of landscape geochemistry from predominantly descriptive and statistic levels of scientific effort to systems and simulation modeling levels, the importance of geochemical classification of landscape within which the systems models can be fitted becomes most important. The Russian conceptual approach to the geochemical classification of landscapes is described in Chapter 12 and an example of the practical application of the approach is included in the chapter on exploration geochemistry (Chapter 18).

An Overview of Landscape Geochemistry

So far (1) the philosophy for environmental geochemistry has been described; (2) some general principles for environmental science, which relate directly to environmental geochemistry, have been discussed; (3) two approaches to the geochemistry of the environment, other than landscape geochemistry, have been outlined; and (4) the seven concepts of landscape geochemistry have been defined. Relationships between these different entities have been brought together in Figure 5.2. This diagram can be used in several ways. Two examples are: (1) the items in it can be considered one by one during the design of projects in environmental geochemistry or landscape geochemistry; or (2) the diagram can be used as a starting point for listing other fundamentals of environmental science that are relevant to both environmental geochemistry and landscape geochemistry.

Summary and Conclusions

In this chapter, definitions and general information have been provided for each of the seven concepts of the discipline of landscape geochemistry, as it is described in this book. As they are defined, other basic concepts will be added. Figure 5.2 shows relationships between the discipline and philosophy of landscape geochemistry and general environmental science; relationships of this type are of particular importance in terms of Type E Landscape Geochemistry as described in Chapter 19.

The discipline of landscape geochemistry has the potential of fulfilling the constraints of the four criteria listed at the end of Chapter 4.

Discussion Topics

1) Do diagrams such as Figures 5.1 and 5.2 contribute to the understanding of the fundamentals of a scientific discipline?

2) After reading this chapter, rank the seven concepts in order of decreasing importance in environmental science.

3) Are all the seven concepts of equal importance in exploration geochemistry? If not, describe which is of least interest in the short term and which in the long term.

6. Element Abundance

"With the data now before us we are in a position to compute the relative abundance of the chemical elements in all known terrestrial matter."

F. W. Clarke, *The Data of Geochemistry,* 5th ed. U.S. Geological Survey Bulletin No. 770, (Washington, D.C., 1924) p. 35.

Introduction

There is general acceptance of estimates for the abundance of elements in the lithosphere (Earth's crust) although there is less agreement on the abundance level of elements and other chemical entities in the other geospheres.

In order to place the concept of absolute element abundance in perspective, it is convenient to consider the relative masses of the four principal geospheres. Rankama and Sahama (1950) estimated that, if the mass of the biosphere was considered as 1 unit, then the atmosphere would be 300 units, the hydrosphere (mostly the oceans) would be 69,000 units, and the upper lithosphere would be in the order of 10^6 units. Hence from the viewpoint of general geochemistry, the lithosphere is the most important geosphere because it is concerned largely with processes that occur during the major geochemical cycle. But in landscape geochemistry, where the primary concern is with the minor geochemical cycle, the migration rate of elements in the environment, as evidenced by partial abundance data on mobile fractions, is frequently of greater importance than the absolute abundance of elements. Absolute abundance estimates for each of the four principle geospheres also have another important role in landscape geochemistry: that of datum for the calculation of relative abundance ratios which, in turn, are used to locate geochemical anomalies in the environment.

Hence in general geochemistry the role of geology is of paramount importance, but in landscape geochemistry this role is shared by geomorphology, soil science, biology, ecology, limnology, forestry, and other disciplines. If the current rapid progress toward the routine collection of quantitative, multielement, geochemical data on environmental samples of all kinds continues, internationally accepted estimates for the absolute abundance of all elements in each of the geospheres will soon be available. This will lead to the routine computerized screening of all absolute abundance data for geochemical anomalies. When located, these will lead to the identification of problem areas in either fundamental or applied landscape geochemistry. Anomalies now studied by exploration geochemists would be special cases of more general problem areas.

The Abundance of Elements in the Environment

Before discussing the geospheres individually, it is important to stress the complexity of study of the geochemistry of the environment and the problem of selecting natural materials for chemical analysis (Green, 1972);

> The abundance of elements in a sample do not change, only the data do. Concentration of an element in a geosphere may vary in space and time because of differentiation, viscosity, solutional, radioactive, shock, gravity, aeolian, pH/Eh, pressure, thermal, mechanical or other effects. In terms of man's stay on Earth, only lead, carbon, and certain radioactive isotopes of H, C, Sr and the transuranium elements have experienced change in the atmosphere, hydrosphere, surface soils and human artifacts.
>
> The major geospheres in which elements are concentrated are more or less agreed upon but tens of categories within these spheres can be created by the geochemist depending upon the relationship he wishes to quantify. The geochemist should define the limits of his category; in some cases the data he collects may more precisely define these limits. But the category must make geochemical sense. For example, in shales, the elemental abundance data for red versus black facies would be desirable from the standpoint of prospecting for sedimentary ore deposits. Likewise of significance in petroleum exploration would be compilations of abundances of "chelatable" metals in the following categories; iron oxides of shallow water oxidizing environments to the iron carbonates of the less oxidizing to the iron sulphides of the deep basinal reducing environments. (p. 268)

Green (1972) also provided a table of abundance categories for elements that includes criteria based on geological and environmental parameters (Table 6.1).

Several conclusions can be drawn from Green's approach to environmental abundance geochemistry as indicated in Table 6.1. It focuses attention first on the difficulty of finding a base for the expression of abundance data in the lithosphere, second, on the question of abundance estimation in the lithosphere being quite different at the detailed level from what might be expected in the other geospheres; and third, on the lack of a clearly defined, natural hierarchical system for the description of lithogeochemical abundance data beyond the use of rock types, rocks, and minerals. At the present stage of development of the concept of absolute abundance data it is desirable to consider element abundances in specific materials directly in relation to the global abundance level.

The Absolute Abundance of Elements in the Lithosphere

Table 6.3 lists estimates for the abundance of 40 elements in order of their abundance of the lithosphere together with estimates for 10 other elements that are of interest in landscape geochemistry. The abundance of an element in the lithosphere on a weight percent basis is called its Clarke. It is interesting that the Clarke for the most abundant element, oxygen, is 456,000 ppm, compared with 7 ppm for the 40th element samarium. Also some unfamiliar elements, such as zirconium, lanthanum, and thorium, are relatively abundant in the Earth's crust compared with others that are more familiar, such as tin, molybdenum, mercury, and gold, which are rare.

Table 6.1. Some Geochemical Abundance Categories (Green, 1972).

Category Based on	Examples
1. Geosphere	Lithosphere, hydrosphere, atmosphere
2. Subgeosphere	Metamorphic, igneous, sedimentary rocks
3. Major unit within subgeosphere	Basalt
4. Minor unit within subgeosphere	Tholeiitic basalt, alkalic basalt
5. Geochemical "break"	
(a) Major element	Low and high calcium "granites"
	Low and high iron concentrations in chondrites
(b) Volatile element	Low to high S, C, and H_2O
(c) Minor or trace element	Low to high rare earth basalts
6. Time interval	Early Paleozoic low Ca/Mg ratios in shales
7. Major province	Basinal, alpine, volcanic, orogenic
	Oceanic: pelagic, tectonically active margins; plateau, abyssal
	Continental: shield, shelf, geosynclinal margins; deltaic
	Lunar mare, lunar higland
	Metallogenic
8. Position	
(a) Level or depth	Aqueous or solid bodies, atmosphere, core samples, soil horizons
(b) Structural	Fold axes and limb zones
(c) Boundary	Dike or batholith margin and interior, sample exterior and interior, zoned crystals, banded rocks
9. Facies	Mineralogical, in subgeospheres
	Regressive and transgressive
	Oxidizing to reducing
	Quiet water to turbulent water
	High Eh to low Eh
	Abiotic and biotic
	High to low Al in clinopyroxenes
10. Heat and pressure history including metamorphic grade index	Lignite to graphite
11. Other environmental conditions	Fractured to massive materials
	Altered to fresh materials
	Acid to basic fluids
	Mineralized to barren rocks
	Low to high gravity
12. Texture	Vesicular to nonvesicular
	Welded to nonwelded
	Coarse to fine grained
	Columnar to nonclumnar
13. Size fraction	Coarse to fine mesh
14. Constituent	Phenocryst and matrix
	Mineral and bulk sample
	Leached and unleached sample
15. Miscellaneous	Radiation responsive to radiation unresponsive
	Biologically processed or produced to nonbiologically processed or produced
	Shocked to nonshocked materials

Let us consider the absolute abundance of elements in the principal components of the Lithosphere. Ronov and Yaroshevsky (1972) described details of the abundance of elements in the three principal components of the Earth's crust which are (1) the stratified sedimentary shell, (2) the granitic shell (which occurs below continents and thins out at ocean boundaries) and (3) and basaltic shell which has a different structure under oceans and continents. A schematic subdivision of the Earth's crust on the basis of volume is indicated in Figure 6.1, and numerical values for the principal elements in each of the components together with the mass of each component are listed in Table 6.2.

Ronov and Yaroshevsky (1972) estimated that about 75% of all sedimentary rocks on the continents are found in geosynclinal areas and 25% on platforms. The thickness of the rocks in geosynclinal basins was some 10 km compared with 1.8 km on the platforms. Clay and shale were the most widespread sedimentary rocks on the continents (42%) together with sandy or arenaceous rocks (20%), volcanic rocks (19%), and carbonate rocks (18%). All other rock types, mostly evaporites, constitute approximately 1% of the total.

Table 6.2. Volumes, Masses, and Average Chemical Composition of the Earth's Crust, i.e. Lithosphere (Ronov and Yaroshevsky, 1972).

Types of Crust	Shell	Volumes (10^6km^3)	Average Thickness (km)	Mass (10^{24}g)	SiO_2	TiO_2	Al_2O_3	Fe_2O_3
Continental	Sedimentary	500	3.4	1.29	49.90	0.65	12.97	2.99
					0.645	0.008	0.167	0.039
	"Granitic"	3000	20.1	8.20	63.94	0.57	15.18	2.00
					5.243	0.047	1.245	0.164
	"Basaltic"	3000	20.1	8.70	58.23	0.90	15.49	2.86
					5.066	0.078	1.348	0.249
	Total Continental	6500	43.6	18.19	60.22	0.73	15.18	2.48
					10.954	0.133	2.760	0.452
Subcontinental	Sedimentary	210	3.2	0.52	49.90	0.65	12.97	2.99
					0.258	0.003	0.067	0.016
	"Granitic"	590	9.1	1.61	63.94	0.57	15.18	2.00
					1.029	0.009	0.244	0.032
	"Basaltic"	740	11.4	2.14	58.23	0.90	15.49	2.86
					1.247	0.019	0.332	0.061
	Total Subcontinental	1540	23.7	4.27	59.35	0.73	15.07	2.55
					2.534	0.032	0.643	0.109
Oceanic	Sedimentary (layer I)	120	0.4	0.19	40.73	0.61	11.45	4.60
					0.077	0.001	0.022	0.009
	Volcanic-sedimentary (layer II)	530	1.8	1.45	45.62	1.02	14.24	3.37
					0.661	0.015	0.206	0.049
	"Basaltic"	1520	5.1	4.40	49.72	1.37	16.57	2.35
					2.188	0.060	0.729	0.103
	Total Oceanic	2170	7.3	6.04	48.44	1.26	15.85	2.67
					2.926	0.076	0.957	0.161
	Total Crust	10,210	20.0	28.50	57.60	0.84	15.30	2.53
					16.414	0.240	4.360	0.722

[a] For each shell the upper line is weight percent, the lower line is mass (10^{24}g).

With respect to the "granitic shell" acidic granitoides and metamorphic rocks make up some 85% with basic and ultrabasic rocks being the remainder of the shell's volume (Ronov and Yaroshevsky, 1972). The "basaltic" shell consists of two parts. The continental crust is composed of strongly metamorphosed rocks (of both acid and basic composition) together with a significant proportion of magmatic rocks, and the oceanic crust is characterized by the occurrence of ultrabasic rocks. These latter are considered as outcrops of the mantle and are typically seen in zones of deep faulting, such as in the mid-ocean rift valleys (Ronov and Yaroshevsky, 1972).

Broad divisions of the lithosphere are often of global, regional, or local significance in relation to landscape geochemistry. For example, certain rock types weather more rapidly than others in a given climate, and some contain trace elements at higher concentrations than others. Thus crustal basalts are characterized by a relatively low concentration of potassium, rubidium, strontium, barium, phosphorus, uranium, thorium, and zirconium compared with oceanic basalts that are high in silicon, titanium, uranium, thorium, and zirconium

Table 6.2. (cont.)

				Components, wt % and Mass (10^{24}g)[a]							
FeO	MnO	MgO	CaO	Na$_2$O	K$_2$O	P$_2$O$_5$	Corg	CO$_2$	S^n	Cl	H$_2$O +
2.80	0.11	3.06	11.70	1.70	2.04	0.16	0.48	8.20	0.18	0.21	2.90
0.036	0.001	0.039	0.151	0.022	0.026	0.002	0.006	0.106	0.002	0.003	0.037
2.86	0.10	2.21	3.98	3.06	3.29	0.20	0.17	0.84	0.04	0.05	1.53
0.234	0.008	0.181	0.326	0.251	0.270	0.016	0.014	0.069	0.003	0.004	0.125
4.78	0.19	3.85	6.05	3.10	2.58	0.30	0.11	0.51	0.03	0.03	1.00
0.416	0.016	0.335	0.526	0.270	0.224	0.026	0.009	0.044	0.003	0.003	0.087
3.77	0.14	3.05	5.51	2.99	2.86	0.24	0.16	1.20	0.05	0.06	1.37
0.686	0.025	0.555	1.003	0.543	0.520	0.044	0.029	0.219	0.008	0.010	0.249
2.80	0.11	3.06	11.70	1.70	2.04	0.16	0.48	8.20	0.18	0.21	2.90
0.015	0.001	0.016	0.061	0.009	0.010	0.001	0.002	0.043	0.001	0.001	0.015
2.86	0.10	2.21	3.98	3.06	3.29	0.20	0.17	0.84	0.04	0.05	1.53
0.046	0.002	0.036	0.064	0.049	0.053	0.003	0.003	0.013	0.001	0.001	0.025
4.78	0.19	3.85	6.05	3.10	2.58	0.30	0.11	0.51	0.03	0.03	1.00
0.102	0.004	0.082	0.130	0.066	0.055	0.006	0.002	0.011	0.001	0.001	0.021
3.82	0.16	3.14	5.97	2.90	2.79	0.23	0.16	1.57	0.07	0.07	1.43
0.163	0.007	0.134	0.254	0.124	0.119	0.010	0.007	0.067	0.003	0.003	0.061
0.97	0.47	2.94	16.29	1.13	2.01	0.15	0.26	13.27	—	—	5.17
0.002	0.001	0.006	0.031	0.002	0.004	0.0003	0.0005	0.025	—	—	0.010
4.23	0.31	5.43	13.67	2.04	1.03	0.14	0.12	6.04	—	—	2.74
0.061	0.0045	0.079	0.198	0.030	0.015	0.002	0.002	0.088	—	—	0.040
6.85	0.17	7.52	11.48	2.80	0.21	0.13	0.01	—	0.03	0.03	0.66
0.306	0.0075	0.331	0.505	0.123	0.009	0.006	0.0004	—	0.001	0.001	0.029
6.11	0.21	6.89	12.15	2.57	0.46	0.13	0.04	1.87	0.02	0.02	1.31
0.369	0.013	0.416	0.734	0.155	0.028	0.008	0.003	0.113	0.001	0.001	0.079
4.27	0.16	3.88	6.99	2.88	2.34	0.22	0.14	1.40	0.04	0.05	1.37
1.218	0.045	1.105	1.992	0.822	0.667	0.062	0.040	0.399	0.012	0.014	0.389

Table 6.3. Estimates for the Abundance of Elements in the Four Principal Geospheres.

Element	Abundance in the Upper Lithosphere[a] (Earth's Crust)	Abundance in the Hydrosphere[b] Fresh	Abundance in the Hydrosphere[b] Marine	Abundance in the Atmosphere[c]		Abundance in the Biosphere[d]
1) O	456,000	889,000	857,000	755,100		780,000
2) Si	273,000	6.5	3	—	(4.0)	21,000
3) Al	83,600	0.24	0.01	—	(3.0)	510
4) Fe	62,200	0.67	0.01	—	(3.0)	1,100
5) Ca	46,600	15.00	400	—	(2.0)	51,000
6) Mg	27,640	4.1	1,350	—	(1.0)	4,100
7) Na	22,700	6.3	10,500	—	(1.1)	2,100
8) K	18,400	2.3	380	—	—	31,000
9) Ti	6,320	0.009	0.001	—	(0.01)	81
10) H	1,520	111,000	108,000	0.035	(300.)	105,000
11) P	1,120	0.005	0.07	—	—	7,100
12) Mn	1,060	0.012	0.002	—	(0.01)	110
13) F	544	0.090	1.30	—	(0.01)	51
14) Ba	390	0.054	0.03	—	—	310
15) Sr	384	0.080	8.10	—	—	210
16) S	340	3.7	885.0	—	(3-50)	5,100
17) C	180	11.0	28.0	460	(as CO_2 164,000)	180,000
18) Zr	162	.003	.00002	—	—	—
19) V	136	.001	.002	—	(0.001)	—
20) Cl	126	7.8	19,000	—	(1.2)	2,100
21) Cr	122	.0002	.00005	—	(0.002)	—
22) Ni	99	.01	.005	—	(0.002)	5
23) Rb	78	.002	.12	—	—	51
24) Zn	76	.01	.01	—	(0.07)	51
25)	68	.01	.003	—	(0.02)	21
26) Ce	66	—	.0004	—	—	—
27) Nd	40	—	—	—	—	—
28) La	35	—	.00001	—	—	—
29) Y	31	—	.0003	—	—	1.1
30) Co	29	.0009	.0003	—	(0.0007)	2.1
31) Sc	25	—	.000004	—	—	—
32) Nb	20	—	.00001	—	—	—
33) N	19	.23	.50	755,100	(9.73×10^{10})	31,000
34) Ga	19	.001	.00003	—	—	—
35) Li	18	.0011	.18	—	—	1.1
36) Pb	13	.005	.00003	—	(0.2)	5.1
37) Pr	9.1	—	—	—	—	—
38) B	9.0	.013	4.6	—	—	110
39) Th	8.1	.00002	.00005	—	—	—
40) Sm	7.0	—	—	—	—	—
Gd	6.1	—	—	—	—	—
Hf	2.8	—	.000008	—	—	—
Br	2.5	0.2	65	—	—	15
U	2.3	0.001	.003	—	—	—
Sn	2.1	0.0004	.003	—	(0.01)	5.1
Be	2.0	< 0.001	0.0000006	—	(0.0001)	—
As	1.8	0.0004	.003	—	(0.01)	3.1
Ho	1.3	—	—	—	—	1.1
Hg	.086	0.00008	.00003	—	—	—
Au	.004	< 0.00006	.000011	—	—	—

(a) Data on the lithosphere (Ronov and Yaroshevsky, 1972) (wt. percent basis).
(b) Data on the hydrosphere (Bowen, 1966) (wt. percent basis).
(c) Data on the atmosphere (Mason 1966) (wt. percent basis: - volume percent basis in brackets, Bowen, 1966).
(d) Data on the biosphere (Perel'man, 1966) after Vinogradov (wt. percent oven dry material).

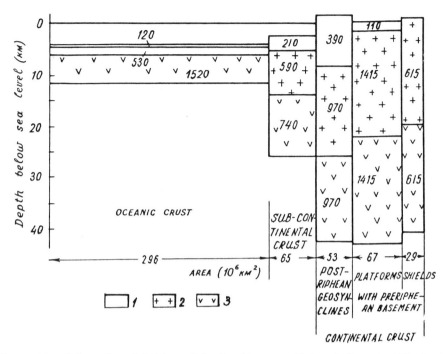

Figure 6.1. Schematic subdivision of the Earth's crust. Values indicate in $10^6 km^3$ (1) sedimentary layer, (2) "granitic layer" and (3) "basaltic layer" (Ronov and Yaroshevsky, 1972).

(Ronov and Yaroshevsky, 1972). Such information is of growing importance in landscape geochemistry because the accumulation of particular elements in landscape developed under past or present climatic conditions may seriously influence the nutrition and health of plants, animals, and man. One limitation of Table 6.2 from the viewpoint of landscape geochemistry is that the data are listed on the basis of "weight percent oxide" rather than "weight percent element." Although this procedure is currently acceptable in petrology it has limitations relative to landscape geochemistry.

Because of the preponderance of mineral matter of the lithosphere in terrestrial landscapes, estimates for the absolute abundance of elements in the Earth's crust as a whole provide a convenient starting point for calculation of the relative abundance of elements in other natural materials. From the global viewpoint, estimates for the absolute abundance of elements in the lithosphere are also used as a first approximation for the abundance of elements on landscapes in which the nature of the underlying geological formations is known.

The Abundance of Elements in the Hydrosphere

Goldschmidt estimated that for every square centimeter of the Earth's surface there are 273 liters of water including 268.45 liters of sea water, 0.1 liter of fresh water, 4.5 liters of water in continental ice, and 0.003 liters in water vapor (Rankama and Sahama, 1950, p. 265). Consequently, when one considers the hydrosphere as a geosphere it is usually with respect to the oceans rather than in relation to fresh water. Some idea of the chemical composition of the oceans is obtained from the estimate listed on Table 6.3 where sodium and chloride ions are the major components of sea water.

In landscape geochemistry the chemical composition of the oceans, which comprise 97% of the total water on the globe (Hordon, 1972), is of marginal significance except that they are a source for "cyclic salts" that are carried from sea to land in coastal regions. Of the 3% of natural waters related to landscapes, Hordon estimated that 75% occurs in polar ice and glaciers, 14% in groundwaters between 762 and 3810m above sea level, 11% in groundwater below 762m, 0.3% in lakes, 0.06% as soil moisture, 0.035% in the atmosphere, and 0.03% is in streams. For this reason it is difficult to provide a single estimate for the distribution of elements in fresh waters as a whole. Some general estimates of this type have been made such as those listed in Tables 6.4 and 6.5 but they are of little use as a basis for global abundance comparisons.

Another difficulty with fresh waters is that they change in chemical composition with season and with the discharge rate. For example, Gibbs (1972) described variations in the flow and content of potassium, magnesium, sodium, calcium, silicon, sulphate, chloride, and bicarbonate and total salinity in the Amazon River in South America. He concluded that, as a whole, the Amazon River basin could be modeled "as a constant rate of supply diluted through the year by a varying amount of water" (Figure 6.2).

The minimum salinity occurred some two months after the maximum river

Table 6.4. The Composition of Natural Waters in Parts Per Million (Wedepohl, 1971, p. 107).

Component	Rain Water; Mainly after Sugawara (1963)	River Water; Mainly after Clarke (1924)	Sea Water	Rain Water to Sea Water Ratio Normalized to Cl Ratio $\equiv 1$
Na	1.1	5.8 [2.5]	10,560	1.8
K	0.26	2.1	380	12
Ca	0.97	20	400	42
Mg	0.36	3.4	1,270	4.9
CL	1.1	5.7 [0.6]	18,980	$\equiv 1$
SO_4^{2-}	4.2	12 [11]	2,650	27
$HCO_3^- + H_2CO_3$	1.2	35	140	157
Si	0.83	8.1	0.05–2 (surface waters) 2.5–5 (deep waters)	4,700

Values in brackets: "cyclic salts" subtracted according to the method of Conway (1942).

discharge and suggested that the lag was due to the time required for the system to attain a new salinity level following seasonal variation in the volume of water in the system. Gibbs (1972) also noted that some elements (i.e., silicon) behaved differently from others during the annual cycle. This example, and others like it, focuses attention on both spatial and temporal variation which are of fundamental interest in landscape geochemistry.

In conclusion, the abundance of elements in the oceans (which constitute over 97% of the hydrosphere) is of relatively minor importance in landscape geochemistry compared with the abundance of elements in subsurface, surface, and atmo-

Table 6.5. General Estimates for the Average Content of Chemical Elements and Ions in Groundwaters (Ward, 1975 after Davis and DeWiest, 1966).

Major constituents (range of concentration 1.0 to 1000 ppm)	
Sodium	Bicarbonate
Calcium	Sulphate
Magnesium	Chloride
Silica	

Secondary constituents (range of concentration 0.01 to 10.0 ppm)	
Iron	Carbonate
Strontium	Nitrate
Potassium	Fluoride
Boron	

Minor constituents (range of concentration 0.00001 to 0.1 ppm)	
Antimony	Lithium
Aluminium	Manganese
Arsenic	Molybdenum
Barium	Nickel
Bromide	Phosphate
Cadmium	Rubidium
Chromium	Selenium
Cobalt	Titanium
Copper	Uranium
Germanium	Vanadium
Iodine	Zinc
Lead	

Trace constituents (range of concentration generally less than 0.001 ppm)	
Beryllium	Ruthenium
Bismuth	Scandium
Cerium	Silver
Cesium	Thallium
Gallium	Thorium
Gold	Tin
Indium	Tungsten
Lanthanum	Ytterbium
Niobium	Yttrium
Platinum	Zirconium
Radium	

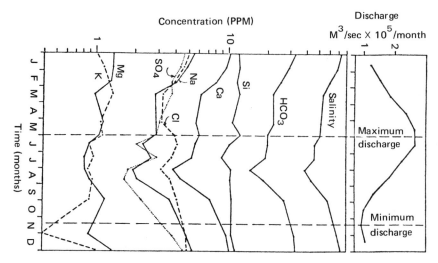

Figure 6.2. Seasonal variation of Amazon River discharge and concentration of dissolved salts [Oltman, 1966 (discharge data), Gibbs, 1972 (chemical data)].

spheric waters. A difficulty with the study of the geochemistry of such waters is that they are most variable in chemical composition—for example within a given season at a particular locality and locally from place to place. Consequently, at present there is no single, universally accepted absolute abundance estimate for the content of elements in fresh waters that can be used as a basis for global comparisons. Until such an estimate becomes available, landscape geochemists will probably still use the Clarke as a standard of referance for elements in waters as described by Polynov in Chapter 2.

The Abundance of Elements in the Atmosphere

Because the atmosphere covers the earth it is a part of every landscape and, as such, is a potential source for fallout of particulate matter of all kinds. Although the absolute abundance of the principal component gasses of the atmosphere hasbeen known to geochemists for a long time (Table 6.6a) it is only within the past twenty-five years that particulate matter, and man-made chemicals in the atmosphere have been studied in detail (Giddings, 1973). Giddings drew a useful distinction between those gasses of the atmosphere that are now safe from disturbance by mankind and those that are vulnerable to manipulation by man's activities (Table 6.6b).

Let us now consider some of the problems of modern atmospheric geochemistry that need to be discussed when the absolute abundance of chemical entities in this geosphere is considered. The dispersion of solid or liquid matter in air is called an aerosol. The natural aerosol of the atmosphere is supplemented by pollutants that man has introduced into it as a by-product of his activities. Spedding (1974) has provided a list of terms generally used to describe the atmospheric aerosol (Table 6.7) and has provided information that indicates that the

Table 6.6.(a) The Average Composition of the Atmosphere on a Weight Percent and Volume Percent Basis (Mason, 1966).

Gas	Composition by Volume (ppm)	Composition by Weight (ppm)	Total Mass (X 10^{20}g)
N_2	780,900	755,100	38,648
O_2	209,500	231,500	11,841
Ar	9,300	12,800	0.655
CO_2	300	460	0.0233[a]
Ne	18	12.5	0.000636
He	5.2	0.72	0.000037
CH_4	1.5	0.94	0.000043
Kr	1	2.9	0.000146
N_2O	0.5	0.8	0.000040
H_2	0.5	0.035	0.000002
O_3[b]	0.4	0.7	0.000035
Xe	0.08	0.36	0.000018

[a] By 1975 this estimate for CO_2 was revised upward to 0.0257 because of man's activity (Baes et al., 1977)
[b] Variable, increases with height.

Table 6.6(b). The Abundance of Major and Trace Gasses and Vapors in the Atmosphere (Giddings, 1973).

	Gas or Vapor	Trillions of Metric Tons in Atmosphere
Gases Now Safe from Disturbance by Mankind	Nitrogen (N_2)	3900
	Oxygen (O_2)	1200
	Argon (Ar)	67
Gases Vulnerable to Disturbance by Mankind	Water vapor (H_2O)	14
	Carbon dioxide (CO_2)	2.5
	Neon (Ne)	0.065
	Krypton (Kr)	0.017
	Methane (CH_4)	0.004
	Helium (He)	0.004
	Ozone (O_3)	0.003
	Xenon (Xe)	0.002
	Dinitrogen oxide (N_2O)	0.002
	Carbon monoxide (CO)	0.0006
	Hydrogen (H_2)	0.0002
	Ammonia (NH_3)	0.00002
	Nitrogen dioxide (NO_2)	0.000013
	Nitric oxide (NO)	0.000005
	Sulfur dioxide (SO_2)	0.000002
	Hydrogen sulfide (H_2S)	0.000001

Table 6.7. Terms Used to Describe the Atmospheric Aerosol (Spedding, 1974). Smoke also Includes Ash, Soot and Other Solids Carried Bodily in the Air.

Aitken particles	Particles of less than 0.1 μm radius
Large particles	Particles of radii in the range 0.1 to 1 μm
Giant particles	Particles of radii greater than 1 μm
Dust	Solid particles broken down from solid material and dispersed by air currents.
Fume	Solid or liquid particles formed by condensation in the vapor phase
Smoke	A fume formed as the result of a combustion process.

chemical composition of large and giant particles is often quite different (Table 6.8). For example, some large particles may be ammonium sulphate (Spedding, 1974). An idea of local variations in the chemical composition of the atmospheric aerosol may be obtained from Table 6.9, which lists data for the content of three ions and seventeen elements in the atmosphere from these districts of different geography. Lake Windermere is a rural site in Great Britain 30 km from the nearest industry; Stockton is an industrial site in Great Britain, and Pasadena is located in an area of California noted for its high incidence of automobile smog. Automobile activity is reflected in the concentration of Pb, Br, and NO_3^-, which are highest at Pasadena and intermediate at Stockton. Vanadium is an element introduced into the atmosphere largely as a result of man's activities. The concentration of this element is highest at Stockton and intermediate at Pasadena

Table 6.8. Distribution of Solution Salts Between Large and Giant Particles in the Atmosphere (μg m^{-3}) (Spedding, 1974).

	Large		Giant	
	Junge	Novakow et al.	Junge	Novakov et al.
Cl^-	0.03	—	1.2	—
Na^+	0.02	—	1.2	—
SO_3^{2-}	4.6	2.86	—	0.42
SO_4^{2-}	—	1.430	1.2	0.65
NO_3^-	0.06 0.8	0.5	0.7	0.14
NH_4^+	—	1.6	0.2	0.14
Amino N	—	2.0	—	0.25
Pyridino N	—	—	—	0.41

Table 6.9. Concentration of Some Elements and Salts from Districts of Different Geography (Spedding, 1974).

	Windermere (μg kg^{-1})	Stockton (μg m^{-3})	Pasadena (μg m^{-3})
Total	20	—	101·5
Al	0·26	0·8	0·8
Br	0·027	0·09	0·6
Ca	0·52	1·3	0·99
Cl	1·75	2·8	0·07
Cu	0·026	—	0·03
Cr	0·002	0·008	—
Fe	0·23	1·7	3·2
I	0·002	—	0·006
K	—	0·3	0·32
Mg	—	0·5	1·1
Mn	0·01	0·1	0·03
Na	2·3	0·8	1·0
Pb	0·09	0·4	3·3
V	0·008	0·02	0·01
Zn	0·08	—	0·18
NO$_3^-$	—	7·2	9·5
NH$_4^+$	—	4·4	0·6
SO$_4^{2-}$	—	11·4	12·1

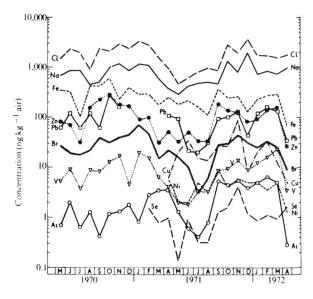

Figure 6.3. Abundance of elements in air near the ground at Wraymires, Westmorland, England (1970–1972) (Peirson et al., 1973)

centration of this element is highest at Stockton and intermediate at Pasadena (Table 6.9). Spedding (1974) noted that the total concentration of particulate matter was three or four times that of the sum of the concentrations of the elements listed on Table 6.9. He attributed this deficit to noncarbonate carbon which can form over 90% of the large particles.

The elemental content of the atmospheric particulate matter varies with season. For example, variation in elements in air particulates collected near the ground at Wraymires, Westmorland, England is listed in Figure 6.3. Clearly a higher content of trace elements was found in the winter months as compared with the summer months (Peirson et al., 1973). These writers also provided average values for the content of twenty-eight elements in rain and in air over a 12-

Table 6.10. Average Content of Elements in Rain and Air at Wraymires, Westmorland, January–December, 1971 (Peirson et al., 1973).

	Concentration in Rain ($\mu g l^{-1}$)	Concentration in Air ($ng\ kg^{-1}$)
Na	2,300	800
Al	160	260
Cl	4,100	1,750
Ca	< 1,100	520
Sc	0.042	0.059
V	4.1	8.0
Cr	2.9	2.4
Mn	8.1	10.6
Fe	200	230
Co	0.25	0.32
Ni*	< 6	5.5
Cu*	23	26
Zn	85	80
As	1.6	2.5
Se*	0.34	0.90
Br	17.0	27.1
Rb*	0.67	1.04
Cd*	< 17.7	< 3
In*	< 0.59	0.3
Sb	1.8	1.6
I	< 2.5	< 2.2
Cs*	0.17	0.16
La*	< 0.3	< 0.5
Ce*	0.42	0.30
W*	0.35	< 0.86
Au	0.01	< 0.01
Hg	< 0.2	< 0.17
Pb*	39	87
Th*	0.12	0.13

* Incomplete data. Annual rainfall 142.7 cm.

month period at the same site (Table 6.10). These data indicate that some elements occur in greater amount in air than in rain and contrast to others where the opposite is the case (Table 6.10). Because the element composition of the atmosphere is quite variable even at a single point on the Earth's surface, the input rate to landscapes from the atmosphere varies with respect to both space and time.

The Abundance of Chemical Elements in the Biosphere

The estimation of the global abundance of elements in the biosphere is much more complicated than is the case with the preparation of absolute abundance estimates for the other three geospheres. According to Bowen (1966) the mass of living organisms that comprise the living "biosphere" is estimated to be 1.2×10^{15} kg on a dry weight basis. Wigglesworth (1964) wrote there are some 3 million species of organisms living on Earth today, all of which require a number of nutrient elements for the perpetuation of the species, plus other elements that are absorbed from the environment but have no known nutrient function. Some of the difficulties in the preparation of estimates for the abundance of elements in the biosphere were described by Bowen (1966) as follows:

> Different workers use different ways of expressing their results, which sometimes makes intercomparison impossible. The majority of workers quote values as a fraction of dry weight, but fresh weight is also used as a reference. A number of workers report concentrations in dry ash. . . . or less often in treated ash. . . . Other modifications include concentrations in terms of fat-free dry weight or nitrogen content. . . . It is common practice to quote the concentration of the major elements in tissues in equivalents or moles per unit weight. (p. 65)

Because of these difficulties and because relatively few species or organisms have been examined for many of the nonnutrients (particularly elements at the minor, or trace, level of concentration), the direct approach to the estimation of the abundance of elements in living organisms is of limited value at present and must be considered a first approximation (Table 6.2).

Fortunately, there is another approach to the estimation of the abundance of elements in the biosphere that may be used for landscapes at the local, regional, or global level of intensity. This is based on the concept of the uniformity of ecosystem types or plant cover communities with respect to the level of concentration of elements within them. These major units of the biosphere are used to obtain general estimates for the abundance of elements in the same way as the different shells (Figure 6.1) of the lithosphere are used when calculating the Clarke of an element.

Two fluxes control the mass of the living biosphere. One involves income rate (e.g., the net primary productivity)· and element uptake of ecosystems on an annual basis; the other is based on loss (e.g., the annual return of organic matter and its elements, e.g., with litter fall). The primary productivity of the biosphere is the creation (by photosynthetic plants) of organic matter incorporating sun light energy. The net primary productivity is that part of the total or gross primary productivity of photosynthetic plants that remains after subtraction of the mass used in the respiration of the plants (Lieth and Whittaker, 1975).

These writers define related productivity units as follows:

> The net productivity is available for harvest by animals and for reduction by saprobes. Net primary productivity provides the energetic and material basis for the life of all organisms besides the plants themselves. Net primary productivity is most commonly measured as dry organic matter synthesized per unit area of the Earth's surface per unit time, and is expressed grams per square metre per year (g/m²/year × 8.92 = lb/acre/year). Net production of ecosystem types in the world is expressed as metric tons (t 10^6g) of dry matter per year (metric tons × 1.1023 = English Short Tons.) Biomass is the dry matter of living organisms present at a given time per unit of the Earth's surface, and may be expressed as kilograms per square meter (kg/m² × 10 = t/ha × 8922 = lb/acre). Productivity may also be expressed as grams of carbon or calories of energy in dry matter formed per unit area and time. (p. 4)

Table 6.11. The Net Primary Production for the World (Lieth, 1975, p. 205).

Vegetation unit	Area (10^6 km²)	Net Primary Productivity		Total production (10^9 t)
		Range (g/m²/year)	Approx. mean	
Forest	50.0		1290	64.5
Tropical rain forest	17.0	1000–3500	2000	34.0
Raingreen forest	7.5	600–3500	1500	11.3
Summergreen forest	7.0	400–2500	1000	7.0
Chaparral	1.5	250–1500	800	1.2
Warm temperate mixed forest	5.0	600–2500	1000	5.0
Boreal forest	12.0	200–1500	500	6.0
Woodland	7.0	200–1000	600	4.2
Dwarf and open scrub	26.0		90	2.4
Tundra	8.0	100–400	140	1.1
Desert scrub	18.0	10–250	70	1.3
Grassland	24.0		600	15.0
Tropical grassland	15.0	200–2000	700	10.5
Temperate grassland	9.0	100–1500	500	4.5
Desert (extreme)	24.0		1	—
Dry desert	8.5	0–10	3	—
Ice desert	15.5	0–1	0	—
Cultivated Land	14.0	100–4000	650	9.1
Freshwater	4.0		1250	5.0
Swamp and marsh	2.0	800–4000	2000	4.0
Lake and stream	2.0	100–1500	500	1.0
Total for continents:	149.0		669	100.2

Leith (1975) provided estimates for the net primary production for the world as shown in Table 6.11. Unfortunately these data were not accompanied by documentation, and the factors for area and amounts per unit area may be too high (Olson, 1970). It is to be expected that more detailed biogeochemical information will become available in the future. This can then be used to calculate estimates of the absolute abundance of all nutrients and other common elements in the living biosphere on the basis of a global consideration of ecosystem data.

Some idea of the distribution of particular nutrient elements in the different vegetation units may be obtained from estimates for the annual return of mineral elements with litter fall as estimated by Rodin and Bazilevich (1967) (Table 6.12, 6.13). Several points of interest in landscape goechemistry may be noted on Table 6.12. For example, with respect to nitrogen, differences in the annual rate of return are most marked between the forest types and the desert types. One surprising observation is that the litter fall in "subtropical semi-shrub desert with lichen" is only a quarter that of "Arctic tundra." The high turnover of sodium and chlorine in the "semi-shrub desert with annuals" indicates that these elements are significantly accumulated in the soils in these areas. The high silica, as opposed to aluminium, in grasses should also be noted in Table 6.13.

So far we have considered general estimates for the productivity and element composition of the living biosphere. Let us now consider an example of the abundance of elements in different plants growing together at one locality. Horovitz et al. (1974) described the content of nine elements in each of thirty-five plant species growing in the Botanical Garden at Tübingen University and the surrounding countryside. This example was chosen for inclusion here because it refers to both common and rare, nutrient and nonnutrient elements in the plants. General estimates for the abundance of elements in land plants taken from Bowen's (1966) data (Table 6.15) are provided for comparation purposes at the foot of Table 6.14. Horovitz et al. (1974) summarized their findings as follows:

> Our data evidenced relatively great variations in the content of some chemical elements among the investigated plant species. We found that the difference between the minimal and maximal content in the plant species studied is greater for scandium and silver (1300 and 1600 times respectively) and smaller for iron and rubidium (less than 100 times). For angiosperms and gymnosperms this difference is very small (less than 30 times for all studied elements), except for silver, for which variation between maximal and minimal values is 400 fold. (p. 402)

Such problems are common in plant biogeochemistry. They will be solved only when test data of this type can be related to general models for the uptake of elements from different substrates by the same plant species. For the present we have no way to predict the absolute abundance of elements in particular plant species or particular cover types. Instead, the relationships between the element composition of plants and the element composition of the living biosphere will be worked out using very general absolute abundance estimates for elements in plants such as those of Tables 6.12 through 6.15.

Table 6.12. Annual Return of Mineral Elements with Litter Fall in the Principal Zonal Types of Plant Communities (kg/ha) (Rodin and Bazilevich, 1967, p. 250–51).

Type of plant community	N	Si	Ca	K	Mg	P	Al	Fe	Mn	S	Na	Cl
Arctic tundra	20	2	4	4	3	1	0.7	0.9	0.2	0.8	0.2	0.5
Shrub tundra	52	6	10	21	5	5	3	2	1	2	2	1
Northern taiga pine forest	24	5	25	18	4	2	2	1	2	3	not det.	not det.
Southern taiga pine forest	16	3	15	10	3	3	2	1	3	4	″	″
Northern taiga spruce forest	13–79	5–7	10–22	4–23	2–12	1–8	1–5	0.5–1	1–5	1–3	0.5	1
Central taiga spruce forest	41–64	8–17	40–50	12–19	5–7	2–4	3–5	4–6	1–4	3–5	not det.	not det.
Southern taiga spruce forest	20–47	8–17	17–41	11–34	2–6	2–5	3–6	1	2–3	2–5	″	″
Birch	80–90	18–20	50–70	14–35	16–35	7–10	4–5	1–2	4–6	5–7	1	″
Aspen forest	60–80	22–32	80–150	70–94	10–16	8–14	8–9	1–2	0.5–1.5	5–10	not det.	″
Lime-tree forest	66–94	10–13	74–108	60–70	11–17	6–11	3–4	0.05	1	5–7	0.5	″
High oak forest (dubrava)	45–72	2–27	80–107	47–65	13–17	3–13	4	1	1.5–2	2–8	not det.	″
Beech forest	83	61	101	35	13	9	29	7	8	4	2	1
Meadow steppe and steppe meadowland	130–161	170–243	25–130	56–116	14–31	9–16	16–31	4–9	not det.	5–8	2–10	7–9
Moderately dry steppe	93–122	138–180	41–62	23–49	7–10	6–7	20–25	11–14	″	3–5	1–3	4–10
Arid steppe	45–109	51–143	16–50	11–45	2–12	5–8	3–11	2–30	1	3–13	1–3	4–8
Semishrub*desert	14–18	2	14–22	3–7	2	1	1	0.5	not det.	1	6–8	4
Semishrub desert with annuals	77–108	71–142	22–65	52–75	10–35	11–17	13–28	6–11	1–4	6–9	3–12	15–23
Subtropical semishrub desert with annuals	26–29	30–56	37–65	6–8	8–10	2	6–12	5–10	not det.	1–3	2	2
Subtropical semishrub desert with lichen	5–6	10	10	2–3	2	0.5	2	2	″	0.5	0.5	0.5
Subtropical deciduous forest	226	144	213	92	31	19	18	16	7	21	5	5
Dry savanna	80	115	55	11	27	4	5	7	not det.	6	2	1
Tropical rain forest	260	770	181–307	53–84	58–72	12–18	56	63	25	33	6	2

Table 6.13. Annual Return of Mineral Elements with Litter Fall in Certain Intrazonal Plant Communities (kg/ha) (Rodin and Bazilevich, 1967, p. 250–51).

Type of plant community	N	Si	Ca	K	Mg	P	Al	Fe	Mn	S	Na	Cl
Forest sphagnum bogs	20–25	6–13	10	4–7	2–5	1–2	2–3	1–2	0.5–1	2	1	1
Solonetzic steppe meadow	193–228	129–221	51–70	54–98	20–32	24–28	21–35	11–26	2	12–35	6–24	23–39
Solonetzic steppe	60–159	33–184	40–99	12–85	1–23	4–21	3–35	1–8	3	6–14	4–15	14–26
Solonetzic arid steppe	18–65	3–184	5–99	12–48	1–10	1–8	2–7	0.5–38	3	1–10	4–20	11–26
Black saxaul woodland	84–205	3–7	66–134	27–52	8–35	4–5	5–7	2	notdet.	8–26	28–130	24–93
Desert solonchak vegetation	9–14	1–5	4–9	5–7	1–4	0.5	0.5–2	0.5–1	″″	2–3	14–27	16–33
Takyr algae communities	1.5–9	0.5	1–2	1–5	0.5	0.5	0.3	0.2	″″	0.5–1	0.5–2	1–3
Grass tugai	450–1300	840–2500	75–270	400–500	30–90	50–80	35–65	5–8	″″	40–110	40–60	300–450
Wooded tugai	46–183	9–19	95–146	92–159	15–23	9–14	10–16	3–7	0.3	11–24	3–12	12–44
Tamarisk tugai	34–71	9–13	59–170	27–43	15–48	6–9	4–5	2–3	notdet.	65–139	39	30–39
Mangroves	82	5	77	45	15	6	6	3	2	19	47	63

Table 6.14. The Abundance of Nine Trace Elements in Each of Thirty-Five Plant Species Growing in the Botanical Garden at Tübingen and the Surrounding Countryside (ppm Oven Dry Weight Basis) (Horovitz et al., 1974).

Species	Order	Origin*	Sc	Th	Cr	Co	Rb	Cs	Fe	Zn	Ag
1 Aspergillus microcysticus Sappe, Tü 502	Eurotiales	E	0.03	0.10	27	0.07	4.0	0.30	460	595	0.02
2 Hypoxylon fragiforme (Pers ex Fr.) Kirkx	Sphaeriales	C	0.003	0.02	0.65	0.06	48	0.65	77	17	0.02
3 Aleuria aurantia Fr. (Fuckel)	Pezizales	C	0.40	—	2.7	0.84	9	0.19	654	170	0.03
4 Bulgaria inquinans Fr.	,,	C	0.03	0.16	1.1	0.27	27	0.27	130	39	1.8
5 Elaphomyces granulatus Fr.	Tuberales	C	0.05	0.07	0.43	2.8	31	2.5	78	51	0.64
6 Clavulina cinerea Schroet.	Agaricales	C	0.19	0.52	1.1	0.29	106	0.53	450	35	16
7 Stereum hirsutum (Wild. ex Fr.) Fr.	,,	C	0.18	0.20	3.3	1.0	38	0.58	790	45	0.01
8 Lycoperdon pyriforme Schf.	Lycoperdales	C	0.48	—	3.0	0.79	11	0.33	1420	53	0.18
9 Scleroderma verrucosa Pers.	Sclerodermatales	C	1.3	1.6	9	2.0	115	0.93	460	440	4.3
10 Streptomyces arenaea Waksm. a. Heur	Actinomycetales	E	0.02	—	—	—	—	—	935	490	7
11 Cladonia retipora (Lab.) Fr.	Lichens	C	0.25	0.20	4.6	0.72	7	0.40	940	28	4.3
12 Anacystis nidulas Men-gh	Chroococcales	E	0.009	0.14	1.2	0.04	2.3	0.28	280	180	0.02
13 Aphanizomenon flos-aquae (L.) Ralfs	Oscillatoriales	E	0.04	0.22	9	2.1	6	0.65	3370	39	0.02
14 Laminaria saccharina (L.) Lamour	Laminariales	D	0.07	—	6	0.19	47	0.13	200	46	0.18
15 Ahnpheltia plicata (Huds.) Fries	Gigartinales	D	0.45	—	2.9	1.4	9	0.39	1840	38	0.35
16 Caulerpa prolifera (Forsk) Lam.	Caulerpales	A	0.07	—	11	10	2.2	0.03	458	230	10
17 Chara fragilis H.a.J. Groves	Charales	A	0.17	—	0.82	0.92	7	0.26	787	1970	0.04
18 Marchantia polymorpha L.	Marchantiales	A	2.0–0.11**	1.9	14–1.0	2.8–0.29	44	2.1–0.45	2510–390	250–50	1.6–0.03
19 Sphagnum acutifolium Ehrh.	Sphagnidae	C	0.30	—	7	0.91	25	0.70	984	840	2.7
20 Polytrichum commune Hedw.	Polytrichidae	C	0.13	—	2.5	0.69	8	0.25	456	42	0.07
21 Hypnum cupressiforme L. ap. Hedw.	Bryidae	B	0.99	1.57	8	2.0	28	0.63	2420	80	0.03
22 Psilotum triquetrum Sw.	Psilotales	B	0.009–0.01	0.76	0.97–0.27	0.09–0.04	35–11	0.50–0.13	190–53	23–120	0.31–0.02
23 Selaginella willdenowii Back	Selaginaellales	A	0.001	—	0.43	0.05	69	1.49	74	12	0.01
24 Lycopodium circinatum L.	Lycopodiales	B	0.01–0.004	0.05	0.59–0.33	0.06–0.03	43–14	0.55–0.12	39–31	26–256	1.0–0.02
25 Equisetum giganteum L.	Equisetales	B	0.01	—	0.45	0.10	49	0.12	93	120	0.05
26 Ophioglossum pedunculosum L.	Ophioglossales	B	0.03	—	0.45	0.13	62	0.11	149	116	2.3
27 Salvinia auriculata Aubl.	Salviniales	B	0.35–0.006	0.42	4.4–1.0	1.7–0.13	70–23	0.67–0.55	750–50	880–240	0.10–0.04
28 Encephalartos lehmanii (Eckl et Zeih) Eckl	Cycadales	B	0.006–0.003	0.06	0.53–0.05	0.02–0.01	7–0.7	0.06–0.01	41–21	8–19	0.15–0.02
29 Ginkgo biloba L.	Ginkgoales	A	0.01	—	0.20	0.10	4.4	0.01	59	17	2.0
30 Juniperus communis L.	Coniferales	A	0.03–0.02	0.06	0.64–0.37	0.09–0.12	5–3.6	0.08–0.03	130–66	13–22	1.5–0.02
31 Ephedra gerardiana Wall. ex. Stoph	Ephedrales	A	0.01–0.04	0.007	0.56–0.45	0.05–0.14	12–9	0.27–0.10	63–79	10–36	8–0.02
32 Liriodendron tulipifera L.	Magnoliales	A	0.007	—	0.37	0.04	2.0	0.01	50	19	0.02
33 Pulmonaria saccharata Mill.	Boraginales	A	0.001	—	0.43	0.20	20	0.11	98	28	0.01
34 Elodea canadensis Michx	Butomales	A	0.03	—	0.43	0.39	14	0.12	430	120	0.02
35 Carex pendula Huds.	Cyperales	A	0.01	—	0.12	0.03	3.8	0.01	61	15	1.0
x̄	—	—	0.19	0.45	3.45	0.89	26.4	0.43	640	193	1.7

* Source of the plant material: A-exposed land of the Botanical Garden; B-greenhouse of the Botanical Garden; C-forest; D-Helgoland, North Sea; E-synthetic media
** All double values indicate analyses for the same plant species, taken in May and September, respectively

Table 6.15. Estimates for the Distribution of Elements in Animals and Plants on an Oven Dry Weight Basis (Bowen, 1966).

Element	Marine Plants	Land Plants	Marine Animals	Land Animals
O	470,000	410,000	400,000	186,000
Si	1,500–20,000	200–5,000	70–1,000	120–6,000
Al	60	500	10–50	4–100
Fe	700	140	400	160
Ca	10,000–300,000	18,000	1,500–20,000	200–85,000
Mg	5,200	3,200	5,000	1,000
Na	33,000	1,200	4,000–48,000	4,000
K	52,000	14,000	5,000–30,000	7,400
Ti	12–80	1.0	0.2–20	< 0.2
H	41,000	55,000	52,000	70,000
P	3,500	2,300	4,000–18,000	17,000–44,000
Mn	53	630	1–60	0.2
F	4.5	0.5–40	2	150–500
Ba	30	14	0.2–3	0.75
Sr	260–1,400	26	20–500	14
S	12,000	3,400	5,000–19,000	5,000
C	345,000	454,000	400,000	465,000
Zr	20	0.64	0.1–1	< 0.3
V	2	1.6	0.14–2	0.15
Cl	4,700	2,000	5,000–90,000	2,800
Cr	1	0.23	0.2–1	0.075
Ni	3	3	0.4–25	0.8
Rb	7.4	20	20	17
Zn	150	100	6–1,500	160
Cu	11	14	4–50	2.4
Ce	—	(320)	—	< 0.03
Nd	5	(460)	0.5	—
La	10	0.085	0.1	0.0001
Y	—	< 0.6	0.1–0.2	0.04
Co	0.7	0.5	0.5–5	0.03
Sc	—	0.008	—	0.00006
Nb	—	0.3	0.001	—
N	15,000	30,000	75,000	100,000
Ga	0.5	0.06	0.5	0.006
Li	5.0	0.1	1.0	< 0.02
Pb	8.4	2.7	0.5	2.0
B	120	50	20–50	0.5
Th	—	—	0.003–0.03	0.003–0.1
Sm	—	.0055	0.04–0.08	0.01

General Summary Statement on the Absolute Abundance of Elements in the Four Geospheres

In lithogeochemistry the Clarke may be used with confidence as a standard of reference for the distribution of elements in different shells (major components) of the Earth's crust. Similarly, on the regional or local scale, the scale of the Clarke may be used to relate the chemical composition of rocks or minerals to that of the Earth's crust as a whole. Such relationships are discussed in the section that follows dealing with the relative abundance of elements.

The absolute abundance of elements in the hydrosphere is usually related to the chemical composition of sea water (Table 6.2) which is quite different from that of various fresh waters. In landscape geochemistry a reliable estimate (similar to the Clarke, as used for rocks) for the chemical composition of the freshwaters of the hydrosphere might be useful as a standard of reference. Difficulties involved in the establishment of such estimates have already been discussed, and it is concluded that the establishment of such a reference datum is not realistic at present.

The absolute abundance of the principal gaseous components of the atmosphere is quite well known (Table 6.6). But before a standard datum for the composition of the atmosphere can be established on a weight percent (or volume percent) basis, two problems must be solved. One concerns the level of pollutant gasses and vapors in the atmosphere which, like the abundance of particular cations dissolved in waters, is very variable in time and space; the other concerns the content of particulate matter in the atmosphere and how it relates to its chemical composition. At present there is no generally accepted estimate for the gases, vapors and particulate content of the atmosphere described by geochemists for comparative geochemistry.

The estimation of the absolute abundance of nutrient and nonnutrient elements in the biosphere is even more difficult than in the case of the hydrosphere or atmosphere. Although first approximation estimates for the abundance of elements in the biosphere have been known for many years (e.g., Vinogradov in Perel'man, 1966; Bowen, 1966) and some attempt has been made to estimate the composition of vegetation in particular vegetation cover types (e.g., Rodin and Bazilivich, 1967), as yet there is not a single comprehensive estimate of the abundance of elements in the biosphere as a whole.

The Relative Abundance of Elements in the Four Geospheres

Outside Russia, little attention has been paid to the formalization of the concept of relative abundance of elements in natural materials. Although the Clarke of Concentration has been known about for many years (Mason, 1958) little attempt has been made to apply it systematically either in relation to the tenor of ores in mineral deposits or in environmental geochemistry where it is potentially of considerable importance.

In Russia the Clarke of Concentration has been used quite commonly: for example, on Perel'man's (1972) diagrams of landscape components, one for chlorine and the other for bromine in Figure 6.4. These two elements have similar chemical properties, although the Clarke value for chlorine (137 ppm) is much higher than that for bromine (2.5 ppm). The Clarke of Concentration for the two elements are also quite similar in different landscape components even though the absolute abundance of the two elements has nearly a hundred-fold difference. The broadly similar behavior of these two elements in different landscape com-

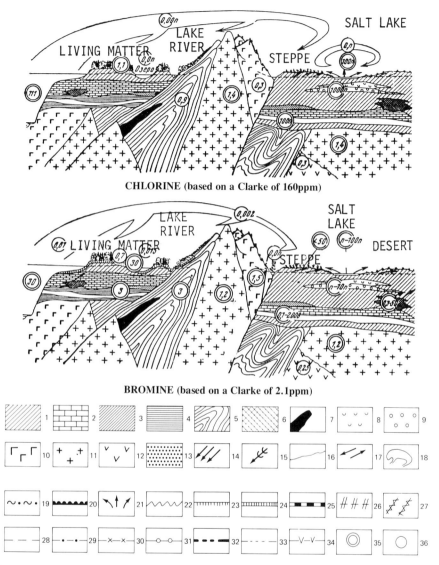

Figure 6.4. Diagrams drawn to show the relative abundance of Chlorine and Bromine in the environment. **Landscape Components:** (1)Buried sedimentary rocks, (2)Limestones, (3)Brown coal, (4)Clays, (5)Clays and schists, (6)Petroleum, (7)Antharacite, (8)Gypsoliths, (9)Halloliths, (10)Basic igneous rocks, (11)Acid igneous rocks, (12)Ultrabasic rocks, (13)Sapropel and peat, (14)Acid leaching, (15)Sulphuric acid leaching, (16)Oxygen boundary, (17)Direction of water flow, (18)Air migration. **Geochemical Barriers:** (19)Sorptive, (20)Thermodynamic, (21)Evaporative, (22)Reducing, (23)Reducing and sorptive, (24)Biogeochemical, (25)Biogeochemical and sorptive, (26)Carbonatic, (27)Sulfatic, (28)Alkaline, (29)Acid, (30)Hydrogenic, (31)Oxygenic, (32)Calcic, (33)Arenaceous, (34)Sulphidic. **Abundance Ratios:** (35)equal to or more than 1.0, (36)less than 1.0. This is a general legend provided by Perel'man (1972) for a series of 15 diagrams each of which involves the relative abundance of a different element. The list is provided in full to indicate the potential scope of such diagrams.

ponents reflects their similar chemical properties. The difference in concentration levels involved reflects their geochemical behavior in nature. One way in which the simultaneous study of many elements in specific natural materials may be facilitated is by comparisons between their Clarkes of Concentration in the natural materials of interest and the Clarke of Concentration for the same element in the same natural material quoted in the literature (e.g., in Figure 6.4). If this screening is done automatically by the computer on large amounts of semi-quantitative geochemical data, it should be possible to separate out sample landscape components with abnormal Clarkes of Concentration for particular elements and work on them later in greater detail. This approach could be used to locate subtle variations in the geochemistry of the environment related to the incidence of disease in plants, animals, or man, and for the delineation of biogeochemical provinces.

The Partial Abundance of Elements

The partial abundance of elements in natural materials can be considered in two ways. First, the partial abundance of element x in material y can be measured using a specified technique; second, the amount of an element in the environment is related to a particular chemical species, or substance, that may be isolated from the environmental sample by one or more of a number of techniques of chemical analysis. Let us consider examples of each of these approaches.

A popular technique employed by exploration geochemists is stream sediment sampling which, under suitable landscape conditions, is used to locate the presence of mineral deposits within a given region or locality. This approach involves the collection of samples of a complex natural material (stream sediment) at regular intervals along streams that occur within the area prospected. The samples are usually dried, sieved (to obtain a sample material of a suitable grain size and the mix of the sample material) weighed, and subjected to an extraction procedure of elements of interest by shaking with a solution of a specific composition for a specific time. The amount of a metal or metals present in the extract is then determined by colorimetric or instrumental (i.e., atomic absorption) techniques. During the past twenty years, millions of samples of stream sediment material have been treated by these techniques and many mineral deposits have been located as a result of information obtained from them (Levinson, 1974). For example, Lang (1970) described a survey carried out in the Yukon Territory of Canada by R. W. Boyle that involved the collection and analysis of stream and spring sediments as well as stream and spring waters (Figure 6.5). Both kinds of sample material were analyzed by a colorimetric method for heavy metals that included zinc and, under certain conditions, copper, and lead. The data for the content of heavy metals in spring and stream sediments and waters listed in Figure 6.5 is interpreted to indicate the presence of mineralization in the vicinity of the north and middle forks of Parent Creek. It is interesting that the high metal values in the stream sediments were found on the middle fork of the creek in contrast to the waters where high values were known in the north fork. In either

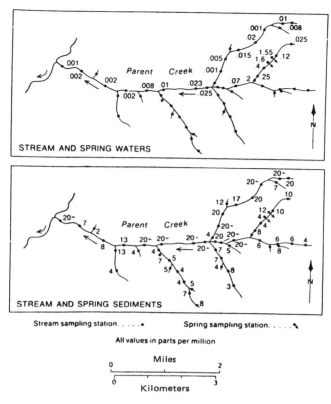

Figure 6.5. The heavy metal content in waters from streams and springs (upper) and from stream and spring sediments (lower). Based upon data described by Levinson (1974) after Lang (1970).

case attention of the geochemist was drawn directly to the upper part of Parent Creek where geochemical anomalies were discovered.

This is an example of a successful environmental investigation based on partial abundance information of the first kind described above (i.e., the partial abundance of heavy metals was determined by a standard technique in stream and spring sediments). It is a good example because the end (i.e., the detection of the geochemical anomalies) was achieved even though the material on which the surveys were based (i.e., stream sediments and waters) were complex and variable in chemical composition (sediment material varies in composition from place to place, stream waters vary in chemical composition with time and space), and the technique of chemical analysis used was not specific for a single element. Thus, in this case, positive results were obtained even though the landscape components studied were variable in chemical composition and the exact amount of the elements of interest extracted from them was not determined (i.e., this survey was a first approximation, only). Environmental geochemical investigations that are based on partial extractions are also carried out in order to solve similar problems in environmental geochemistry; and although they may not pro-

Table 6.16. Chemical Properties of Some Representative Noncalcic Brown Soils from California (Harradine, 1967).

Horizon	Depth	pH of Soil Paste	Extractable Cations				Cation Exchange Capacity	Base Saturation	Organic Carbon
			Ca	Mg	Na	K			
	in			*me/100 g soil*				%	%
GREENFIELD fine sandy loam—Minimal development									
Ap1	0–10	6.5	7.4	1.7	0.16	0.69	13.0	77	1.36
A12	10–20	6.4	6.6	2.2	0.19	0.24	11.0	85	0.57
B2	20–30	6.3	8.9	3.9	0.24	0.16	15.3	87	0.13
B3	30–36	7.0	8.2	4.8	0.21	0.11	14.3	94	0.23
C	36–60	7.2	6.0	3.9	0.18	0.09	11.2	91	0.11
TEHAMA loam—Medial development									
Ap1	0–8	5.5	3.8	2.3	0.1	0.1	9.5	66	0.86
A3	8–19	5.9	5.3	4.4	0.1	0.1	13.4	74	0.33
B2	19–31	6.5	6.8	11.9	0.1	0.2	23.0	83	0.25
B31	31–42	7.0	6.1	10.7	0.1	0.2	23.0	74	0.12
B32	42–56	7.1	6.2	10.6	0.1	0.1	23.5	72	0.14
KIMBALL Loam—Maximal development									
Ap1	0–4	6.4	6.0	2.7	0.1	0.3	11.4	80	1.01
A3	4–10	6.0	3.7	2.2	0.1	0.2	8.7	71	0.38
B1	10–17	6.1	4.6	3.4	0.1	0.1	11.8	69	0.20
B2	17–34	6.6	11.2	8.6	0.2	0.2	22.0	92	0.19
B31	34–46	7.1	11.6	8.4	0.2	0.2	21.0	97	0.09
B32	46–64	7.0	10.6	7.0	0.2	0.1	18.8	95	0.07
SAN JOAQUIN loam—Maximal development with hardpan									
A1	0–6	5.6	3.6	1.6	0.08	0.54	8.3	70	1.41
A3	6–15	5.7	3.3	1.6	0.36	0.35	6.7	84	1.02
B1	15–24	6.1	3.6	2.4	0.14	0.16	7.3	86	0.76
B2	24–30	6.0	4.0	11.6	0.45	0.18	16.5	99	0.25
Cm	30–38	—	—	*a*	*a*	*a*	—	—	—
C	38–72	6.8	8.1	4.3	0.20	0.20	13.7	93	0.22

a Iron silica-cemented hardpan.

duce quantitative information, the patterns in the data are of considerable local significance. Such data are called appraisal level data and, during the past twenty years, much misunderstanding has occurred when scientists have confused this appraisal level geochemical data with absolute abundance data that requires much more rigorous techniques to produce.

The second type of partial abundance data is more specific with respect to chemical elements although the principle involved is similar to that described in relation to the Parent Creek investigation. For example, Harradine (1967) studied

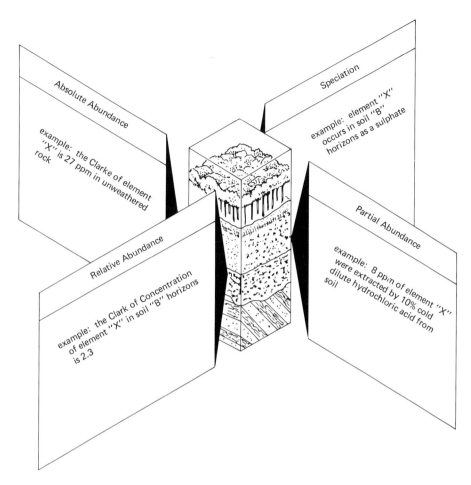

Figure 6.6. Different kinds of abundance of the same element within a single landscape prism.

the partial abundance of elements in the noncalcic brown soils of California using standard methods of chemical analysis that are published in the soil science literature. He described in some detail the vertical distribution patterns for partially extractable calcium, magnesium, sodium, and potassium in soil profiles together with the pH of the soils, their cation exchange capacity, and organic carbon content (Table 6.16). In this case the partial abundance data for the particular elements in the soils were used as a basis for a discussion of the genesis of the soils (Table 6.16). Although Harradine's partial abundance data were more exact than data described in the previous example, they have significance only in relation to the morphology of soil profiles under discussion with unknown geochemical significance on a global scale. These two examples indicate some of the advantages and disadvantages of collecting partial abundance geochemical data and provide reasons why such data are usually only of local significance.

The Concept of Element Abundance and Holism

The fact that landscape geochemistry is concerned with the circulation of all elements (plus other chemical entities) in the environment focuses attention on the relationship between chemicals in the environment at the local, regional, or global level of study. The absolute abundance of elements in the environment is often a starting point for more complex studies in landscape geochemistry, and the relative abundance ratio is a convenient way of making first approximation synthesis and analysis of geochemical data. This procedure also facilitates the comparison of local with regional or global information. When the details of the circulation of chemical substances and elements are studied within a particular locality partial abundance data are often of considerable importance. Such investigations contribute to a holistic view of the details of the behavior of such entities within a particular component or components of landscape. (see Figure 6.6.)

Summary and Conclusions

In landscape geochemistry, element abundance can be considered on a local, regional, or global level, and relationships on these different scales are often of importance. Similarly, relationships between absolute, relative, or partial abundance of elements and the hierarchy of chemical complexity are of considerable interest. This chapter describes some of the problems encountered in the preparation of reliable general estimates in geospheres with particular reference to the atmosphere and biosphere.

Discussion Topics

1) The advantages and disadvantages of using volume weight rather than a weight/weight basis for geochemical data collected in any of the geospheres.

2) The development of Perel'man diagrams for migrant elements at the local, regional, and global scale of environmental geochemistry.

7. Element Migration in Landscapes

"The continuous influx of solar energy onto the surface of the Earth promotes numerous supergene processes and increases their complexities, energies, and intensities. Although temperatures and pressures are low, the migration of chemical elements in this zone is pronounced. Substances may become highly dispersed or segregated to an extent unknown in hypogene environments."

A. I. Perel'man, *The Geochemistry of Epigenesis*. Translated by N. N. Kohanowski (New York: Plenum Press, 1967), p. 1.

Introduction

Like the concept of abundance, the concept of the migration or mobility of elements in landscapes can be discussed on an absolute or relative basis. Until now the tendency, outside Russia, has been to stress the study of absolute mobility of elements in landscapes. But in Russia, ever since the pioneer studies of Polynov (1937), both kinds of mobility have been considered by landscape geochemists. Consequently, as the study of environmental geochemistry becomes more sophisticated, the need for the description of relative mobilities of elements in particular landscapes will increase. For example, using simulation models in certain types of landscapes it should be possible to predict accurately the migration patterns for most mobile elements based on data and information obtained for the behavior of relatively few mobile elements.

In this chapter, examples are given of studies that involve both the absolute and relative mobility of elements in landscapes. With respect to relative mobility, Perel'man (1966, 1967, 1972) has provided a series of general, empirical, parameters to describe the behavior of migrant elements on a local, regional, or a global scale. He has also used the values obtained for these parameters to describe a classification of migrant elements that is of considerable importance in landscape geochemistry.

A Study of the Absolute Mobility of Elements in a Landscape

A good example of a descriptive (empirical) study of the absolute mobility of elements (and ions) in a landscape was described by Likens and Borman (1972). Over a three-year period they studied the gross export, by streams, of eleven kinds of ions from each of two catchment areas within the Hubbard Brook

Table 7.1. Comparative Gross Losses of Dissolved Substances in Stream Water from Undisturbed (W6) and Deforested (W2) Catchment Areas, Hubbard Brook Experimental Forest (Likens and Borman, 1972).

	Metric tons/km³/yr					
	1966–67		1967–68		1968–69	
Element	W2	W6	W2	W6	W2	W6
Ca^{++}	7.7	1.1	9.3	1.2	7.0	1.2
K^+	2.3	0.2	3.6	0.2	3.3	0.2
Al^{+++}	1.8	0.3	2.5	0.3	2.1	0.3
Mg^{++}	1.6	0.3	1.9	0.3	1.3	0.3
Na^+	1.8	0.7	1.9	0.9	1.3	0.7
NH_4^+	0.09	0.04	0.07	0.02	0.06	0.01
NO_3^-	46.0	0.6	65.2	1.2	46.9	1.2
$SO_4^=$	4.6	5.1	4.5	5.8	5.0	5.1
HCO^-3	0.1	0.2	0	0.3	0	0.1
Cl^-	1.1	0.5	0.9	0.5	0.7	0.5
$SiO_2 aq$	6.7	3.7	7.0	3.6	5.9	2.9
TOTAL	73.8	12.7	96.9	14.3	73.6	12.5
THREE-YEAR AVERAGE					81.4	13.2

Experimental Forest (Table 7.1). They found that the export of dissolved substances (exclusive of organic matter) from an undisturbed catchment areas was 13m tons/ha/year compared with 82 metric tons/ha/year from a similar cutover catchment area in which arboricides were used to suppress uptake by secondary plant growth. Coupled with this increased export of dissolved matter was a ninefold increase in the output of organic particulate matter from the deforested catchment area. One of the reasons for this increase was the physical disturbance of the landscape resulting in the reduction of the fallen leaf cover of the soil causing a reduction in the soil binding action of small roots. Although the stream water from the two catchment areas appeared clear and potable, the nitrate content of the water from the deforested catchment area was continuously higher after the removal of the forest cover and at times almost doubled the maximum amount recommended for drinking water (Likens and Borman, 1972). This effect, combined with the increased amount of solar radiation (due to the absence of the forest canopy), also resulted in the eutrophication of the stream. In this case changes in the rate of absolute mobility of particular ions due to logging resulted in secondary effects that were of considerable interest in environmental geochemistry.

This study is a good example of a description of the effects of man's disturbance on the migration rate of particular ions from similar catchment areas. The rate of absolute migration of elements from particular landscapes is not constant but varies with local conditions during the year or as a result of man's activity.

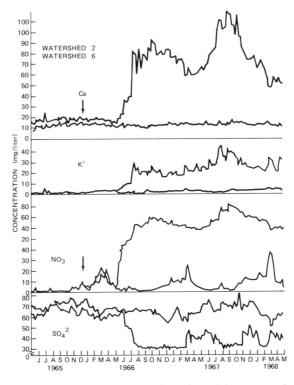

Figure 7.1. Measured streamwater concentrations for calcium, potassium, nitrate, and sulphate in catchment areas (2) deforested and (6) control Hubbard Brook Experimental Forest. Arrow indicates completion of cutting in Catchment (2) (Likens and Bormann, 1972).

A graphic representation of the effects of the logging on the content of Ca^{++}, $K^+ NO_3^-$ and $SO_4^=$ in the disturbed catchment area and in the control catchment area during the period 1965–1968 appears as Figure 7.1. These graphs show clearly that the effects on the migration of elements of man's activities are greater during the four-year period than the natural seasonal variations within the control catchment area. One advantage of descriptive investigations of this type is that they measure directly the migration rate of elements in the catchment area particularly when the natural circulation patterns for elements are disturbed by man's activities. A disadvantage of this approach is the amount of time and effort required to produce the required information and hence the limited range and selection of factors that can be covered.

The Migration Rate Equation

The Russian landscape geochemists, notably Polynov (1937) and Perel'man (1966, 1967) have considered the problem of the migration rate of elements in landscapes using a general conceptual (and mathematical) approach. Perel'man (1966) defined the migration rate equation to describe the rate at which an element is brought into solution within a given landscape. This was the first step towards the establishment of a number of general parameters to describe the relative mobility of chemical elements in landscapes.

Perel'man (1967) defined the parameter P_x as follows:

> Assuming the quantity of the element "X" in a given natural system to be b_x, its amount in the mobile state at any given moment of time is:
>
> $$\Delta b_x/b_x \cdot 1/\Delta t$$
>
> In its application to the weathered crust, this quantity, P_x, expresses the velocity of leaching of 1 g of a substance containing the given element. Rewriting it in the differential form and noting that $db/b = d \ln b$, we get:
>
> $$P_x = d \ln b_x/dt$$
>
> Unfortunately, the equation has no numerical solution, since the rate of change of the quantity of the given element in rocks with time is unknown [i.e., $b = f(t)$ is unknown]. (Perel'man, 1967, p. 149)

Because of the experimental difficulty of finding solutions to this equation by directly using experimental techniques (such as that used by Likens and Bormann, 1972, 1975; or Elwood and Henderson, 1975), Perel'man (1967) discussed an alternative solution that also expressed the idea of relative migration of elements in landscapes. He wrote:

> Since the determination of P_x is quite difficult in most cases, the author has proposed the use of the coefficient of aqueous migration K_x which is obtained by dividing the element's content in the rock into its content in waters draining from that rock.* The greater the value for K_x, the greater will be the migrational ability of a given element:
>
> $$K_x = m_x 100/an_x$$
>
> where m_x is the X element's content in water, mg/liter, n_x is its content in the rock, %, and a is the mineral residue contained in the water, %. The K_x factors afford a convenient means of comparing the intensities of migration of the principal and accessory elements. (Perel'man, 1967, p. 150)

Table 7.2 is an example of the relationship between the abundance of five elements in the lithosphere and the coefficient of aqueous migration. The K_x values may conveniently be used as a basis for the comparison of migration intensities of both common and accessory elements. For example, the element zinc (Table 7.2) is seen to be 17 times more mobile than silicon although the Clarke values for the two elements are 83 ppm and 296,000 ppm, respectively (Perel'man,

---

* The following relationship exists between the values P_x and K_x:

$$P_x - P_y = K_x/K_y$$

Table 7.2. Calculation of the Coefficient of Aqueous Migration K_x from Streamwater Abundance Data and Clarke Values (Perel'man, 1967).

Abundance	Units	Si	Ca	Zn	Cu	Fe
Stream waters, m_x	mg/liter	10	50	$5 \cdot 10^{-2}$	$3 \cdot 10^{-3}$	1
Lithosphere, n_x	%	29.5	2.96	$8.3 \cdot 10^{-3}$	$4.7 \cdot 10^{-3}$	4.65
Coefficient, K_x	—	0.07	3.3	1.2	0.12	0.04

Note: The mineral residue of water is assumed to weigh 500 mg/liter; the content of the elements in rocks is taken at their abundance values.

1966). It should be noted that the coefficient of aqueous migration allows for the determination of the mobility of elements in rivers on a relative scale regardless of the actual concentration at the time of sampling.

Perel'man (1967) calculated the magnitude of the K_x values for migrant elements of silicious rocks in temperate climates and ranked them as shown in Table 7.3. This table also included reference to the behavior of elements in oxidizing or reducing environments. He noted that "very strong migrants" have K_x values in excess of 20 and strong migrants have values between 20 and 1. Elements such as iron, titanium, and aluminium which are weak migrants were calculated to have K_x values below 0.1.

The K_x parameter is a first approximation only and is subject to variation due to the environmental conditions found within a given locality. For example, the variation of the organic matter content of streamwaters, the input of cyclic salts in coastal areas, or elements derived from air pollution may influence the K_x parameter value in a particular locality. The fundamental importance of the K_x

Table 7.3. Migrational Series for Elements Released from Silicious Rocks in Temperate Climates (Perel'man, 1967).

Migrational Intensity	Oxidizing Environment K_x (1000 100 10 1 01 001 0001)	Contrast of Migration (Weak ← → Strong)	Strongly Reducing Environments K_x (1000 100 10 1 01 001 0001)
Very strong	Cl,I Br,S	Cl,Br,I	Cl,2 Br
Strong	Ca,Mg Na,F Sr,Zn U	Ca,Mg,Na,F,Sr / Zn,U	Ca,Mg, Na,F Br
Medium	Co,Si, P,Cu Ni,Mn K	Sl,P,K / Cu,Nt,Co	St,P, K
Weak and very weak	Fe,Al,Ti,Y, Th,Zr,Hf, Nb,Ta,Ru, Rh,Pd,Os, Pt,Sn	Al,Ti,Zr,Hf,Nb,Ta,Pt,TR,Sn	Al,Ti,Se,V,Cu,Nt, Co,Ma,Th,Zr,HF,Nb, Ta,Ru,RH,Pd,Os,Zn, U. Pt

factor in landscape geochemistry is that it provides a first approximation for the comparison of the relative mobility of the same elements in different landscapes.

Perel'man (1967) also defined a second parameter, the migrational contrast coefficient, to express the change in rate of migration of the same element in the same environment as a result of the chemical species in which it migrates. For example, he noted that a migrational contrast coefficient around 100 was obtained by dividing the K_xZn (as sulphide) into K_xZn (as oxide) for these ions migrating in the same landscape. The field of migrational contrast determinations for different chemical species of elements in the landscapes is currently of considerable interest, particularly in exploration geochemistry.

A Classification Scheme for Migrant Elements

Perel'man (1967) used the concept of the aqueous migration coefficient (K_x) as a basis for a classification of migrant elements in landscapes. After distinguishing between air and water migrants he divided the latter on the basis of the empirical rate of migration and, in some cases, used pH/Eh relationships as well (Figure 7.2). This classification is of particular interest to the beginner in environmental geochemistry who requires a guide to the likely migration behavior of specific elements in the same landscape. The classification scheme is less satisfactory for air migrants because it does not allow for elements in particulate matter which, as was noted in Chap. 6, are of considerable importance in atmogeochemistry.

The Perel'man classification scheme for water migrant elements is based on descriptive and abundance criteria. A chemical classification for the migration of elements during the minor geochemical cycle had been worked out prior to this by Goldschmidt (Mason, 1966). Goldschmidt considered six general chemical products of the weathering process. They are:

1) Resistates: minerals that are resistant to chemical and physical weathering
2) Hydroslates: secondary minerals consisting of Si, Al and K, eg. clays
3) Oxidates: including iron precipitated as ferric hydroxide
4) Carbonates: including the precipitation (inorganically or organically) of carbonates
5) Evaporates: including sodium chloride and sulphates of calcium and magnesium
6) Redusates: including organic sedimentary materials, sedimentary sulphides, and sedimentary sulphur that accumulates under reducing conditions

Goldschmidt also considered the chemical speciation of elements during the transport stage of the minor geochemical cycle. He noted that, in general, the geochemical behavior of elements in landscapes was directly related to the ionic potential of the ions involved. He defined the ionic potential as the ionic charge of an ion divided by its ionic radius (Z/r). This parameter may be used to explain why elements with markedly different chemical properties behave in a similar

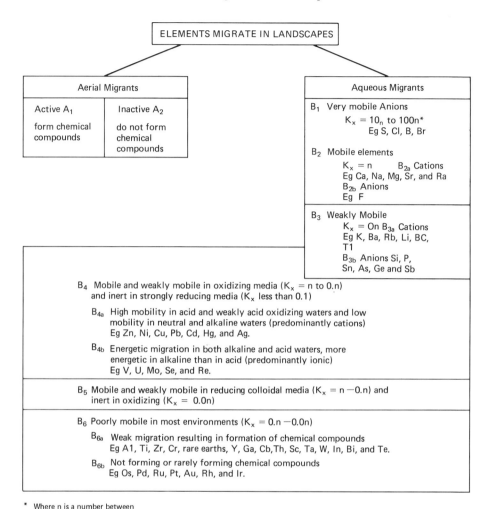

Figure 7.2. A classification of migrant elements in landscapes (Perel'man, 1967).

way geochemically during migration in landscapes. Thus, using the ionic potential as a criteria for classification, elements fall into three groups: (1) those that have a low ionic potential that migrate in solution; (2) those with intermediate ionic potential that precipitate out as hydrolysates; and (3) those with high ionic

pointed out, the ionic potential provides a chemical explanation for the tendency of bivalent beryllium, trivalent aluminium, and quadrivalent titanium to precipetate together during sedimentation, although this behavior may be due in part to the resistance of the minerals containing these elements to chemical weathering processes. Unexpected behavior of elements during the migration process such as this is of considerable interest in landscape geochemistry—particularly in

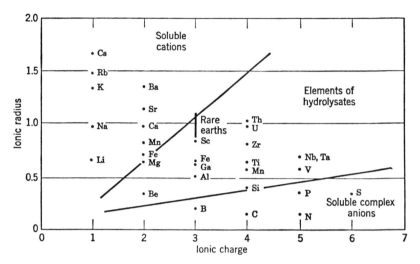

Figure 7.3. The geochemical separation of some important elements on the basis of their ionic potential (Mason, 1966, p. 163).

relation to the incidence of diseases that may be related to the presence or absence of specific trace elements in waters or foods.

Other chemical considerations that are taken into account in relation to the migration of elements in landscapes include the pH of environments and Eh relationships. This is a well-known and complex field of classic environmental geochemistry which is mentioned here only briefly with reference to two examples. Figure 7.4 shows the behavior of silicon and aluminium with respect to change in pH in an inorganic system. This suggests that silicon would increase in mobility in tropical landscapes with increasing alkalinity of soils and that aluminium

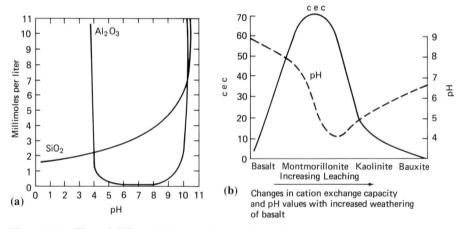

Figure 7.4. The solubility of silicon and aluminium with respect to pH in the laboratory (a). Changes in the cation exchange capacity (cec) and pH with increasing weathering of basalt (b) [(Mason, 1966 (a), Loughnan, 1969 (b)].

would be insoluble under very acid or very alkaline soil conditions. It should be noted that from pH 5 to pH 9 the solubility of silicon is relatively high but aluminium is relatively insoluble. In the field, these inorganic chemical facts account for the accumulation of aluminium in bauxites under favorable conditions (Mason, 1966). On the basis of data presented in Figure 7.4a, Mason (1966) predicted ground water pH offered an explanation for the formation of particular clay minerals from the same parent rock under different environmental conditions. He predicted that under alkaline conditions (pH 8–9) the relative abundance of silica in solution with respect to aluminium would favor the formation of montmorillionite clay whereas under more acid conditions (where there was less silica in solution) kaolinite would form. This supposition is supported by information quoted by Loughman (1969) (Figure 7.4b) concerning the weathering basalt from the New England area of New South Wales, Australia. Similarly, ph/Eh diagrams of the type listed as Figure 7.5 can also be used to predict the behavior of particular ions in other natural environments, although expected relationships may be modified by the presence of living organic matter in some areas.

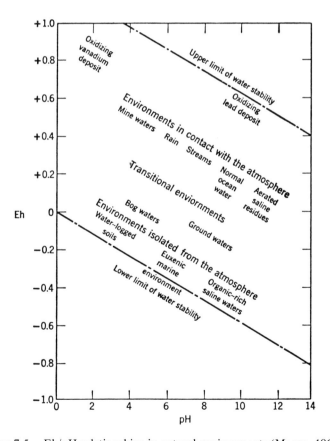

Figure 7.5. Eh/pH relationships in natural environments (Mason, 1966).

		Coefficients of biologic accumulation					
		$100 \cdot n$	$10 \cdot n$	n	$0 \cdot n$	$0.0n$	$0.00n$
Elements of biologic capture	Very strong	P, S, Cl					
	Strong		Ca, K, Mg, Na, Sr, B, Zn, As, Mo, F				
Elements of biologic accumulation	Moderate			Si, Fe, Ba, Rb, Cu, Ge, Ni, Co, Li, Y, Cs, Ra, Se, Hg			
	Weak					Al, Ti, V, Cr, Pb, Sn, U	
	Very weak						Sc, Zr, Nb, Ta, Ru, Rh, Pd, Os, Ir, Pt, Hf, W

Figure 7.6. The BAC of elements (Perel'man, 1967, p. 9).

Both descriptive/empirical and chemical classification schemes for the study of the mobility of elements in landscapes have their place in landscape geochemistry.

The Migration of Elements in the Biosphere

The living biosphere plays some part in the migration of elements in almost all landscapes. Because of this the Russian landscape geochemists have defined a number of descriptive parameters for use in biogeochemistry. Three of these are discussed here: (1) the biological absorption coefficient (BAC); (2) the relative absorption coefficient (RAC); and (3) the acropetal coefficient (AC) (Brooks, 1973). Perel'man (1966) defined the BAC as follows:

$$BAC = 1_x/n_x$$

where 1_x is the content of element x in the ash of organisms taken from a landscape (usually plants), n_x is the content of the element in soil, rock, or the Earth's crust.

It should be noted that the greater the value for the BAC of an element the more readily is it absorbed from the substrate by a plant (or plants). Perel'man (1967) noted that the most intensively absorbed elements have BAC values of over 100n (where n is a number between 1 and 10) (Figure 7.6) and weakly absorbed elements have a factor as low as 0.00n. The level of concentration of an element in the Earth's crust seldom relates directly to its uptake by the biosphere. For example, the BAC for common anions P, S, Cl, which are strongly absorbed by plants (BAC = 100n), is in contrast to the low BAC for common cations Si, Al, Fe, which are only weakly absorbed.

Brooks (1973) listed BAC values for 26 elements and rare earths and noted that the BAC values for nutrient elements was usually higher than for other elements (Table 7.4). So-called accumulator plants may have exceptionally high BAC values for particular elements. For example, Brooks (1973) noted that the BAC for nickel in *hybanthus floribundus* was found to be over 500. The BAC for a given plant species may vary from element to element and with environmental conditions. A relatively simple example of this phenomena was described by Loneragan (1975) who provided data for the uptake of copper and molybdenum by three plant species growing in the same soil at different pH values (Figure 7.7). Under these experimental conditions the uptake of molybdenum is *medicago denticulata* varied from 1.6 ppm to 12 ppm as the pH of the soil varied from

Table 7.4. BAC Values for Essential and Nonessential Elements (Brooks, 1973).

Perel'man Classification	Element	Physiological Role	BAC
Strong absorption	Boron	Essential	1 · 70
Intermediate absorption	Sulphur	Essential	0 · 96
	Zinc	Essential	0 · 90
	Phosphorus	Essential	0 · 88
	Manganese	Essential	0 · 40
	Silver	Non-essential	0 · 25
	Calcium	Essential	0 · 14
	Strontium	Non-essential	0 · 13
	Copper	Essential	0 · 13
	Potassium	Essential	0 · 12
	Barium	Non-essential	0 · 12
	Selenium	Essential	0 · 10
Weak absorption	Molybdenum	Essential	0 · 040
	Magnesium	Essential	0 · 034
	Nickel	Non-essential	0 · 030
	Cobalt	Non-essential	0 · 020
	Uranium	Non-essential	0 · 020
	Iron	Essential	0 · 012
Very weak absorption	Sodium	Non-essential	0 · 007
	Rubidium	Non-essential	0 · 007
	Rare earths	Non-essential	0 · 003
	Chromium	Non-essential	0 · 003
	Lithium	Non-essential	0 · 0015
	Silicon	Non-essential	0 · 0006
	Vanadium	Non-essential	0 · 0006
	Titanium	Non-essential	0 · 0003
	Alumimum	Non-essential	0 · 0003

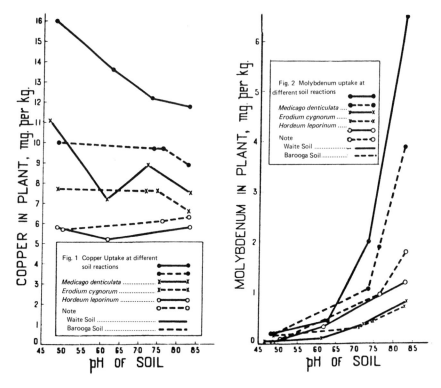

Figure 7.7. Relationship between the pH of the soil and content of copper and molybdenum in several plant species (Longeragan, 1975 after Piper and Beckwith).

4.5 to 8.5. Although this may be an extreme example, it indicates some of the complexities involved when the uptake of trace elements is related directly to changes in environmental conditions. It also provides an example of how the increase in content of one element, owing to change in envrionmental conditions, may be accompanied by a decrease in the content of another element (Figure 7.7).

The second biological absorption coefficient listed by Brooks (1973) was defined by Kozlovskiy (1969) as follows:

$$\text{Relative Absorption Coefficient (RAC)} = C_p/C_r$$

where C_p is the concentration of an element in the plant ash of one species and C_r is the concentration of the same element in the ash of another plant used as a reference standard. The RAC was devised for use in biogeochemical surveying and is used when one wishes to change the sampling species on which a survey is based. According to Brooks (1973) this parameter may be used to compare the uptake of two plants of similar growth one from each of the two species when they are located within 3m of each other (Brooks, 1973).

A third coefficient was defined by Sabine (1955) as follows:

$$AC = C_o/C_x$$

where C_o is the concentration of an element in an organ of a plant and C_x is the concentration of the same element in a reference, or standard organ of the same plant species (Brooks, 1973).

Values for AC are often similar for different individuals of the same species. Brooks (1973) noted that a variation in AC value may indicate that a plant is under environmental stress. It is evident from these examples that biogeochemical methods for the assessment of the migration patterns of elements in plant cover types are now at an early stage of development. Descriptive biogeochemistry has an important role to play in landscape geochemistry as a link between purely descriptive information regarding the content of elements in plants (e.g., the data shown on Table 6.14) and information regarding biogeochemical cycling (e.g., Figure 4.1).

Element Migration and Holism

A major difference between studies of the major and minor geochemical cycles for elements is in the rate of formation of chemical substances and the mode of study of samples taken from them. Field investigations, which involve the products of the major geochemical cycle, are undertaken by geologists who study the details of rock specimens and, as a result of such studies, estimate the nature and rate of chemical processes that occurred when the crystalline rocks were formed. Because of the long time span involved, details of rates of synthesis and decomposition of such rocks cannot be directly studied in the field. In the case of diagenesis of sedimentary rocks, the migration of elements and waters within the rocks—as a result of depth, and/or heat gradients—is also carried out using geological methods.

When one considers the behavior of elements and other chemical entities in landscapes, the picture is very different, because we are dealing with dynamic systems. In such systems chemical entities can migrate at rates and over distances that are not possible during the major goechemical cycle. This is evidenced by the migration of radioactive isotopes (resulting from nuclear tests in the atmosphere) that are now found in detectable quantities in all landscapes. Thus, in the study of the minor geochemical cycle, the direct study of migration rates is not only desirable, it is mandatory. (Figure 7.8.)

Let us consider a point source of pollution of the environment such as a smelter operation. The chemicals from the smelter enter the atmosphere and migrate relatively short distances within it until they fall out or rain out on the landscape. When this occurs the living biosphere, the pedosphere, the fresh water hydrosphere and, in some cases, the lithosphere are all affected by the chemicals from the smelter. Thus one can say that the whole environment is

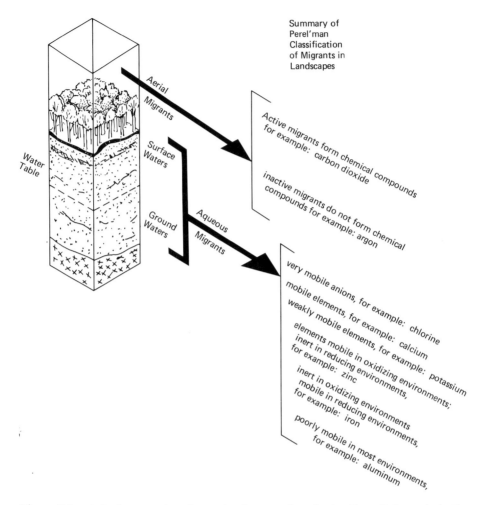

Figure 7.8. A landscape prism drawn to show modes of migration of elements in the atmosphere and waters together with a summary of Perel'man's classification of migrant ions.

affected, although the migration rate of the chemicals from the smelter and their effects on the landscape vary from component to component. In order to detect the effects of such migrations a holistic approach to the study of landscape is required and the study of connected, rather than unconnected, data bases in both time and space is needed.

The concept of element migration and the approaches to the study of the rate problem in different kinds of landscapes described in this chapter are part of such a holistic approach. Such studies are in their infancy in landscape geochemistry and the absolute and relative migration rates of particular chemical entities in particular types of landscapes cannot be predicted on the basis of simulation models.

Summary and Conclusions

Like the abundance of elements, the migration rate of elements in landscapes can be discussed on an absolute or relative basis. Both approaches have their place in landscape geochemistry; examples of both were outlined in this chapter. The behavior of elements, and other chemical entities, in landscapes may be described by means of empirical parameters as favored by Polynov, Perel'man and the other Russian workers or on the basis of the chemical and geochemical properties of the elements or the entities concerned. The empirical approach may be considered as a first approximation leading to more intensive chemical studies that offer a second approximation. In landscapes a combination of the empirical and chemical aspects of element migration is desirable. As quantitative, multielement, geochemical data becomes available in large amounts for landscapes of all kinds studies of the usual and unusual rates of migration of particular entities in landscapes will become recognizable within raw data using computerized scanning techniques. Migration rate anomalies of this type will then require explanation and the environmental effects of such anomalies in the short and long term must then be estimated.

Discussion Topics

1) Should methods for the estimation of the absolute migration rate of elements in landscapes be standardized?

2) Relationships between the BAC and the problem of making general estimates for the abundance of chemical elements in the biosphere.

3) Are the six general chemical products of weathering as described by Goldschmidt all of equal importance as bases for geochemical surveys designed to measure the migration rate of elements in landscapes in which they are found?

8. Geochemical Flows in Landscapes

"I think one could imagine some process which may be effective in that manner, a process connected with plant life, for instance in a forest. The soil solutions dissolve the organic constituents of the subsoil according to their solubility as we normally only have a small amount of solvent compared with the bulk of the mineral matter. These solutions enter the plant organism through the roots of the plants, and at the place of strongest evaporation, especially in the leaves the greater part of the mineral matter is deposited. The dead leaves accumulate at the surface of the ground, decaying into humus substances. The most soluble mineral constituents of the leaves and their products of decay, such as carbonates and sulphates of lime, magnesium, sodium, potassium, and humates of iron are swept downwards by circulating rainwater, insoluble or sparingly soluble compounds are filtered in the humus layer, perhaps in some cases being precipitated as organic complex compounds. By such processes many different chemical elements have been found to become concentrated in the uppermost layer of forest soils:"

V. M. Goldschmidt, The Principles of Distribution of Chemical Elements in Minerals and Rocks, Journal Chemical Society of London, 1937, p. 670.

Introduction

This classic description by Goldschmidt of biogeochemical cycling of elements in a forest ecosystem is an example of the Holistic geochemical approach to the study of the circulation of elements in landscapes. More recently Kozlovskiy (1972) provided a series of conceptual models and principles, from the viewpoint of theoretical landscape geochemistry, that together summarize neatly the principal flow patterns of elements in landscapes. His paper provides a broad conceptual basis for the discussion of the cycling of elements in landscapes together with the inflow and outflow of elements to and from landscapes in groundwaters and the atmosphere.

The Flow of Substances through Landscapes

Kozlovskiy (1972) commenced his paper with a description of "the imaginary streams of substances existing in landscapes over a period of time." He divided these streams into two kinds called (1) independent migrants that create a moving phase (or stream) in the landscape and (2) dependent migrants that are transported within the flows of independent migrants. For example, flowing groundwater is an independent migrant, and the salts dissolved in it are dependent migrants. According to Kozlovskiy (1972), independent migrants flow along geochemical channels in the landscape. His conceptual model for the flow of elements in landscapes is based on the assumption that the geochemical structure of the landscape (in the broad sense) is determined by its migrational structure (in the narrow sense) acting during geological and pedological time, and, more specifically, by mechanical, water, air, and biological migration of chemical elements and substances. From Kozlovskiy's viewpoint the migration of elements and substances in most landscapes is governed by (1) the chemical conservativism of the lithological-geomorphological "framework" which, under a given water and air flow regime, controls the paths and intensity of the intralandscape water-air flows, and (2) by the long-term stability of the biocenotic (ecosystem) structure. Kozlovskiy's paper deals largely with hypothetical landscapes where these systems are relatively stable. Unstable landscapes include those where, for example, frequent forest fires are common or where salt accumulation has taken place under arid conditions. It is relatively simple to add complexity to a landscape flow pattern once the simple situation is known, but it is often impossible to extract the simple pattern from a complex landscape situation.

When considering the flows of elements through landscapes, Kozlovskiy (1972) noted that the most difficult problem was the definition of the lower boundary of the landscape. He solved this problem by reference to what he called the principle of maximum migrational interaction, which he defined as "the migrational relationship within some elementary area of a geochemical landscape must be stronger than its relationships with zones of the Earth's crust beyond its boundaries." (p. 229). For example, in glaciated areas where a layer of coarse textured glaciofluvial material lies on a glaciated pavement of silicious rocks, the lower limit of the landscape is the contact between the bedrock surface and the overlying material.

Geochemical Flow Patterns within Landscapes

Kozlovskiy (1972) described three principal flow patterns for elements in landscapes that occur concurrently within the same landscape. The first type is similar to that during biogeochemical cycles (see Figure 4.1) which approach a steady-state condition in mature landscapes. During such flows the movement of matter is predominantly vertical upward from soil to plants and animals and then downward from plant and animal back to soil. This flow type, which involves the nutrient and nonnutrient elements (or chemical substances), was called the main migrational cycle (MMC) by Kozlovskiy (1972) (Figure 8.1).

The second flow pattern is the landscape geochemical flow (LGF) which involves a progressive movement of substances parallel to the Earth's surface. The LGF within a landscape may occur in the atmosphere, the pedosphere, or the lithosphere (Figure 8.1). An example of a chemically inactive air migrant in the LGF is argon. (Incidentally if it were to be dissolved in groundwater, it would also be an example of an inactive element in the soil LGF.) The air and water LGF's within a landscape prism may flow in different directions in response to seasonal change or diurnal change in local weather conditions.

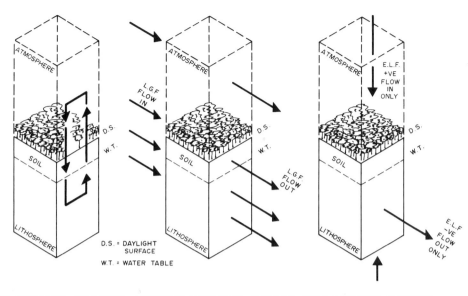

Figure 8.1. Landscape prisms drawn to illustrate the three flow patterns which occur simultaneously within landscapes (Fortescue, 1974a; terminology for flows from Kozlovskiy, 1972).

The third flow type is the extra landscape flow (ELF), which may be either positive or negative. If positive, the ELF involves the addition of chemical elements or substances to the landscape—usually via groundwater or the atmosphere. A negative ELF involves the removal of elements or substances from the landscape (1) in groundwaters, (2) in surface waters, (3) by gravity, or (4) in the atmosphere. Estimates for the removal of material from landscapes by erosion are examples of negative ELF's relating to pedological or geological time.

The advantage of combining the ideas on geochemical flows described by Kozlovskiy (1972) and the landscape prism concept (Figure 8.1) is that it allows one to envisage the migration rate of elements through a point in space located in any landscape at a particular instant in time. The migration of elements through a landscape prism may be described on either a volume percent or a weight percent basis. For example, flows of elements ions or chemical substances may be described with reference to prisms $1m^2$, $10m^2$ or $100m^2$. For descriptive purposes, prisms of 1m (or less) in cross section are most convenient although flow rates for elements are frequently estimated on a kg/ha basis. For example, Poole (1974) described the rate of cycling of calcium through a British pine forest as summarized by an arrow diagram (Figure 8.2); the same data has been included in Figure 8.2 using the landscape prism concept which separates the abundance data (left side) from the flow data (right side), which is mainly related to the MMC (right side) and ELF (top and bottom). Provided the information was available, a series of such diagrams could be drawn to show the

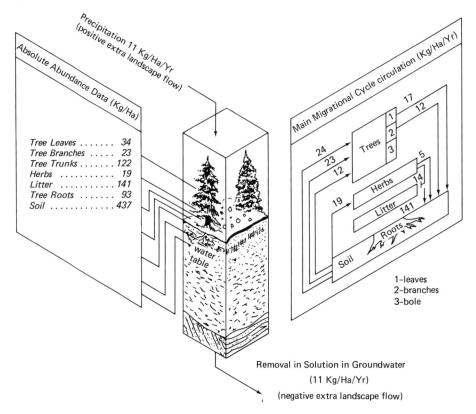

Figure 8.2. Landscape prism diagram drawn to illustrate relationships between absolute abundance, extra landscape flows, and main migrational flow rates for calcium in a pine forest ecosystem (Poole, 1974)

increase in the MMC and the abundance of the calcium in the standing crop during different stages in the life cycle of the pine trees. A series of such diagrams could then be used for purposes of comparison with similar data obtained for calcium in other forest stands. The input of calcium from the atmosphere indicated in Figure 8.2 is, using Kozlovskiy's terminology, a positive ELF derived from the LGF of the atmosphere.

Elementary Landscape Cells

Kozlovskiy (1972) called areas of country where MMC predominates over the other flow patterns Elementary Landscape Cells (ELC's). In Chapter 2, the relationship between Jenny's concept of the tessera and the concept of the landscape prisms was mentioned, as was the relationship between the concept of the polypedon and the pedon which is used in soil science. Although the ELC is defined on geochemical rather than ecological criteria, the relationship between it and the landscape prism is similar to that between the tessera and the landscape

Geochemical Flows in Landscapes 99

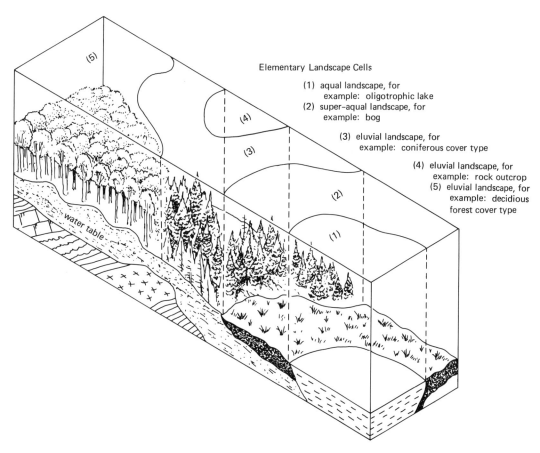

Figure 8.3. Idealized hypothetical landscape section drawn to show examples of elementary landscapes (as defined by Polynov) and elementary landscape cells (as defined by Kozlovskiy)

prism. Because of the predominance of the vertical movement of chemical substances in an ELC during the MMC, such a cell is recognized by the presence of horizontal geochemical barriers such as those that result in the formation of soil horizons. If ELC is accompanied by horizontal movement of substances—for example, on a slope where material tends to move downhill—then the point at which the change occurs is taken to be the center of the ELC (Kozlovskiy, 1972).

Kozlovskiy noted that the concept of ELC is not the same as that of the elementary landscape type as described by Polynov many years before. For example, in a complex landscape of the kind illustrated in Figure 8.3 there are according to Polynov's terminology two elementary landscapes: a transeluvial landscape on the hillside, and a superaqual landscape in the bog area (see Figure 5.1). However, from the viewpoint of geochemical flows in the landscape as described to Kozlovskiy, there are four ELCs.

Three of these have centers on the tops of sand dunes, and the other occurs in the bog area (Figure 8.3). In general, ELCs tend to be smaller than elementary

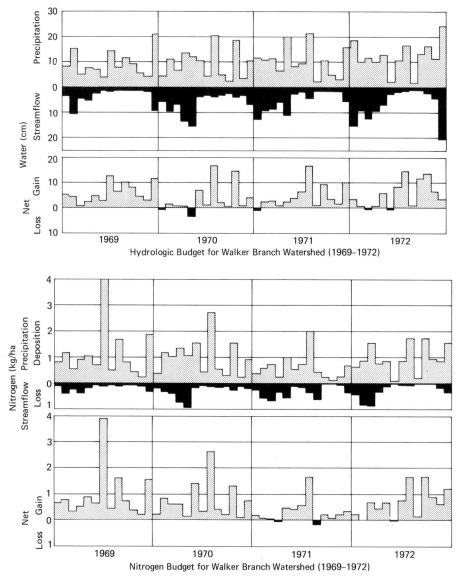

Figure 8.4. Seasonal distribution by four week periods (13/year) of precipitation, streamflow and net gain or loss (precipitation-streamflow) and (bottom) nitrogen deposition in precipitation, loss in streamflow and net gain or loss (deposition-loss) for the Walker Branch Watershed from 1969 through 1972. In values are expressed as cm of water over total area of catchment (97.5 ha) and annual net gain represents evaportranspiration (Henderson and Harris, 1975).

landscape types in complex landscapes and larger than elementary landscapes types in areas of level, or nearly level, ground. Kozlovskiy (1972) also distinguished between "open" and "closed" Landscape Geochemical Cells (LGC). An open LGC is one in which the migrants derived from it are removed from it by natural processes, in contrast to a closed ELC in which they accumulate. In general, the movement of material within a landscape is usually horizontal with relatively minor amount of matter moving in (or out) via the upper or lower boundaries.

Like Perel'man (1966) (see Chapter 2) Kozlovskiy (1972) recognized the problem of unconnected data bases with respect to space, time, and chemical complexity in landscapes. He recognized that MMC ordinarily consists of oppositively directed migrational flows (e.g., from soil to tree root to tree leaf and then back to soil via the latter layer) in forest and to-and-fro horizontal movements in flood plains at different seasons of the year. He pointed out that some of the processes that occur in landscapes can be described adequately on the basis of annual cycles although in other cases a longer period of time (such as a climatic cycle) may be required to describe adequately the movement patterns of chemical substances in a landscape. For practical purposes, the systematic study of the behavior of chemical elements in an ELC should be carried out on the basis of the minimum space and time unit that applies within a particular study area. He concluded that the ELC is a space/time functional geochemical unit. According to Kozlovskiy (1972) the time interval chosen includes the following flows:

(1) external inflow of substances and their outflow beyond the limits of a particular ELC,

(2) mobilization of migrants and their immobilization (i.e., transition from motionless to mobile forms and vice versa),

(3) the redistribution of migrants within an ELC (p. 232).

Kozlovskiy also noted that, from the theoretical viewpoint, "the capacity of an ELC with respect to a migrant usually exceeds many times the number of migrant particles that pass from one ELC to another within an elementary time interval" (p. 233). Consequently, the average residence time of a migrating particle within an ELC greatly exceeds the elementary time interval and, during this time, the particle is circulated within the ELC in conformity with the internal migrational pattern of the cell. This leads to an equilibrium between the amount of the particle material in the ELC as a whole and the part in the LFG channel. For example, if excess nitrogen fertilizer is applied to an ELC located in a farmer's field, some of this nitrogen will be retained by the crop and recycled in the organic material (i.e., circulated within the MMC) but the excess nitrogen fertilizer which is not absorbed by the plants may be washed out of the field in the groundwater LGF or escape into the atmosphere as N_2 and be lost to the ELC. On the basis of this conceptual scheme, Kozlovskiy (1972) devised a mathematical model for the migration of elements within certain kinds of simple landscapes. (Further details of this interesting approach can be obtained from the paper by Kozlovskiy, 1972).

General Principles for the Flow of Elements through Landscapes

Kozlovskiy (1972) described four general principles for the flow of chemicals through landscapes that are of fundamental importance to landscape geochemistry. Let us consider them briefly.

Principle I. The Macrostructure of Landscapes

The space time periodicity of landscape geochemical processes, based on the essential periodicity of migration, is probably universal. The temporal periodicity of these processes is well known. . . . The uneven distribution of migrants in space is also well known and is frequently rythmic (complexity, spottiness). This aspect of landscape geochemistry is not given the attention it deserves, since its methological importance is still underestimated. Yet there are reasons to believe that the spatial periodicity of the distribution of migrants is as universal as temporal periodicity inasmuch as it is largely controlled by the latter. Both periodicities and especially the spatial one are one of the manifestations of a more general landscape (biogeocoenitic) development process, the differentiation of its subunits and their integration into a macrosystem in particular. (Kozlovskiy, 1972, p. 240)

The principle is of considerable importance in both theoretical and practical landscape geochemistry. From the theoretical viewpoint it focuses attention on the need to pay very careful attention to the size of units for the description of the migration patterns of elements through landscapes. If the units chosen are too small, the data for migration of elements through them may not be representative because of "spottiness" within the landscape component; but if they are too large then the investigation may include more than one ELC. Consequently, the choice of the nature and extent of field sites for fundamental studies in the circulation of elements in landscapes is of crucial importance. In practical applications of landscape geochemistry (particularly in agriculture and forestry) attention to the choice of units for investigations in relation to the flow pattern in landscape may also be very important. For example, the selection of areas for fertilizer trials should be done with reference to the uniformity of the landscape conditions within which the trial plots are laid out. In theoretical and fundamental landscape geochemistry, the description of flow patterns in relation to landscape prisms located at the center of ELCs is to be recommended in order to minimize "edge effects," or variations on flow rates due to transitions between the different horizontal layers within the ELCs studied.

Principle II. The Specific Geochemistry of Migration Processes

The diversity and "structural" organization of the physiochemical and biological processes affecting migration exclude the possibility of describing them reliably from the quantitative aspect by summing up the effects of particular mass-exchange processes (potential flows, convection, diffusion, sorption, etc.) computed from a single measurement of the corresponding parameters. A description of this kind would lead unavoidably to an accumulation of errors as soon as we reached beyond the limits of a single, or a few, cycles.

The physical meaning of the parameters in the equations of geochemical migration cannot completely be identified with the physical meaning of particular mechanical, or physiochemical forms of motion, even if we know that migration in a landscape is associated primarily with one of them. The parameters of migration must be regarded specifically as geochemical quantities. This must be taken into account when comparing them with related physical quantities of other structural levels of matter. (Kozlovskiy 1972, p. 241)

This principle relates to the distinction between the "piecemeal" approach to the study of the geochemistry of landscapes adopted by many environmental scientists and the "holistic" approach favored by landscape geochemists. Limitations of the piecemeal approach have already been referred to in relation to the principle of unconnected data bases described by Perel'man (see Chapter 2). Because of the principle of the specific geochemistry of migration processes, one cannot extrapolate the descriptive chemical data from a landscape obtained as an "isolated incident" in space and time to the general circulation of the elements concerned during ecological, pedological, or geological time in the same landscape.

Once this is clearly understood one is able to identify a limitation of the descriptive approach to the cycling of elements in particular landscapes. Thus, if the input/output of precipitation/streamflow for a particular catchment area is studied over a period of time, the effects of the spottiness within the ELCs involved is minimized, and we are mainly concerned with the temporal variations in the water balance equations. For example, Henderson and Harris (1975) provided data for the seasonal distribution of precipitation, streamflow, and net gain or loss (precipitation-streamflow) and nitrogen deposition in precipitation loss from streamflow and net gain or loss (deposition-loss) for the Walker Branch Watershed located near the Oak Ridge National Laboratory in Tennessee (Figure 8.4). Suppose we assume that the sampling and the time interval hydrological and chemical methodology of this experiement are accurate in all respects. Then, in the ideal case, the histograms for the four year-period would be identical. In reality, they are quite different from season to season and from year to year because of variations in the weather each year. Using Kozlovskiy's terminology, this experiment focuses attention on the difficulty, if not the impossiblity, of describing (the physical, chemical and biological processes) reliably from the quantitative aspect by summing effects of particular mass exchange processes computed from a single measurement of the corresponding parameters.

Instead, Kozlovskiy (1972) suggested that some other approach to the solution, based on geochemical quantities, should be used to describe the migration patterns of elements in landscapes. Presumably, if this alternative approach were applied to suitable data obtained from the field as shown in Figure 8.4, identical information would be provided regardless of the annual cycle selected for study. Although this ideal of landscape geochemistry is still far from realization, solutions to problems of this type (based on procedures of systems simulation analysis) have already been obtained in certain ecosystems by ecologists (e.g., at the Walker Branch Watershed at Oak Ridge).

Principle III. The Stability of Geochemical Parameters of Migrational Macroprocesses

This principle is expressed in the presumed stability in time of the parameters of migrational macrostructures (MMC and LGF) and the mobility parameters of dependent migrants for a given type of migrational structure. We have spoken earlier of the prerequisites for the stability of the migrational structure of a landscape. Although

> we fully admit the possibility of reciprocal actions of geochemical processes, such as that of salt accumulation on the migrational structure; for example, we assume that these effects do not change the general plan of the latter. . . . It can be assumed that landscapes of the same genetic type have similar migrational macrostructures. (Kozlovskiy, p. 241)

This principle is concerned with the relationship between the dynamic processes (i.e., the MMC and the LFG) that occur continuously in landscapes from season to season in ecological time and the slow processes that take place during pedological and geological time. For example, Likens and Bormann (1972) summarized the cycling of calcium during a period of one year in a catchment area at the Hubbard Brook Experimental Forest in New Hampshire:

> Some 570 kg calcium/ha are held by the vegetation and 1,740 kg calcium/ha in the organic debris; some 690 kg/ha are available in soil water and on exchange surfaces; and the soil and rock minerals contain about 28,550 kg calcium/ha. Annual uptake by the vegetation slightly exceeds 49 kg/ha, while about 49 kg/ha are released by decomposition and leaching. Thus about $1/14$th of the available calcium pool is cycled through the vegetation each year. The average annual input (1963–1969) of calcium is 2.6 kg/ha and the output is 12.0 kg/ha, of which 11.7 kg/ha is lost in dissolved form in streamwater. These data applied to the model, show that weathering generated on the average ecosystem is efficient in retaining and circulating calcium. The annual net loss (9.1 kg/ha) represents only about 1.3 percent of the total available calcium (690 kg/ha)

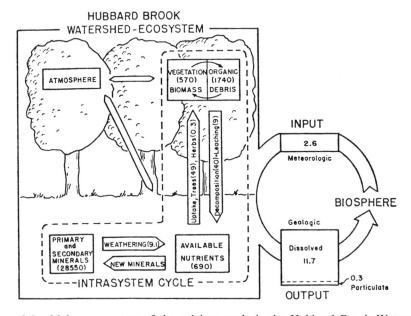

Figure 8.5. Major parameters of the calcium cycle in the Hubbard Brook Watershed ecosystems. All the data in kg/ha and kg/ha/yr. Data on organic debris, leaching and decomposition from Gosz and Eaton (unpubl.); vegetation biomass and uptake from Whittaker (unpubl.); primary and secondary minerals, available nutrients and organic debris (all to a depth of 61 cm) from Lunt (1932) (Bormann and Likens, 1972).

in the ecosystem, and about 18 percent of the amount circulated annually by the vegetation through uptake and release." (p. 53)

Kozlovskiy's (1972) terminology of the "cycling" of calcium, as shown in Figure 8.5, between the "available nutrients," the "vegetation biomass," and the "organic debris" would constitute the MMC; the "weathering" and "formation of new minerals" would be positive and negative ELF's respectively. The "meterological input" and the "geological output" would be positive and negative ELF's in relation to the system. These data are the circulation of calcium in the landscape as carried out by the migrational structure of the landscape. The migrational macrostructure of calcium in this example would be the primary and secondary calcium minerals that involved an estimated 28,550 kg of this element in the soil and rocks. The principle of the stability of geochemical parameters of migrational macroprocesses assumes that landscapes of a similar genetic type and history will have similar migrational structures. If the general migrational structures can be modeled effectively for one area of a landscape in a given locality then, with proper modifications, the same model should be applicable for similar areas elsewhere. Using the terminology discussed in Chapter 2, tactical information on the geochemical flows as measured at specific points in a landscape may be synthesized into a strategic model that may then be used as a guide to the behavior of elements in all similar ELC's in that locality. Once significant progress along these lines has taken place, many practical applications of this principle in environmental conservation and land use planning will follow.

Principle IV. The Specific Geochemistry of Migrants

Inasmuch as migration as a whole is the result of the effect on the migrating substance of various forces acting together or in a definite sequence, no substance can be regarded a priori to be a complete geochemical analogue of another, be it a dependent or independent migrant. Hence, the migration of each element in a landscape of given category is described, in principle, by parameters obtained from a study of the given element in a given group of landscapes, rather than from its "witness" except for isotopes, perhaps." (Kozlovskiy, 1972, p. 241)

The substance of this principle has been mentioned already in Chapter 7. As Kozlovskiy (1972) noted, one way in which the behavior pattern of a specific element may be studied in detail within a particular landscape (or ecosystem) is on the basis of radioactive isotopes. Suppose a known amount of a radioactive isotope of a macronutrient element is added to a particular ecosystem at a particular phenological instant in time then, after a suitable time period has elapsed, if the amounts of the isotope in the "sinks" of the biogeochemical cycle (see Figure 4.2) are determined, the migration pattern for that element can be established in detail for the time period involved. For example, Thomas (1969) studied the flow rates of the macronutrient element calcium in Dogwood trees using this approach, using ^{45}Ca as the radioactive tracer (Figure 8.6). The trees were tagged on May 4, 1966 and harvested on November 4, 1966. One hundred units of ^{45}Ca

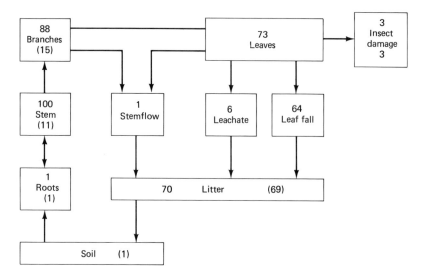

Figure 8.6. Distribution of ^{45}Ca after the stem innoculation of dogwood trees. Numbers not in parentheses represent maximum percentages of innoculum in components of the tree soil system during the growing season. Numbers in parentheses are percentages of innoculum found after leaf abscission. Trees were tagged on May 4th and harvested on November 4th, 1966 (Thomas, 1969).

were injected into the stem of the tree and then the number of units of the tracer found in each of the ecosystem components was estimated at the end of the experiment. Information of this type is of considerable interest with respect to the flow patterns of elements in landscapes as well as being useful in studies of migration rates of particular nutrients in ecosystems of different kinds.

Landscape Geochemical Flows and Holism

It is evident from this chapter that Kozlovskiy's approach to the migration of chemical elements in landscapes is holistic with respect to both space and time. If the field study of the major geochemical cycle by geologists and geochemists is likened to the study of a still picture, then Kozlovskiy's approach to the study of the minor geochemical cycle is a moving picture. One conclusion from Kozlovskiy's work is that you cannot study the geochemistry of landscapes in detail based on information collected at a single instant in time. Thus, in order to study the flows of elements through landscapes realistically one requires a period of several seasons in time, and the data bases involved must all be connected to each other so that generalizations can be made regarding the flow of elements regardless of seasonal variations.

The holistic study of landscapes by geochemists is still in a very early stage of development and much can be learned from the experience of systems ecologists and geomorphologists by geochemists involved in this type of landscape study. Clearly, the systematic study of geochemical flows in natural and man-made landscapes presents a major challenge to students of landscape geochemistry.

Kozlovskiy's Approach to the Description of Flow Patterns in Landscapes

Types of Flow Patterns[a]	General Principles for the Migration of Elements through Landscapes
1) The Main Migrational Cycle flow (MMC) includes the biogeochemical cycling of nutrient and non nutrient elements.	a) The principle of the macrostructure of landscapes stresses the importance of the periodicity of geochemical processes in landscapes with respect to both space and time.
2) The Landscape Geochemical Flow (LGF) involves the flow of fluids (i. e., the atmosphere or waters) through landscapes with no addition or subtraction of matter.	b) The principle of the specific geochemistry of migration processes draws a distinction between the measurable migration cycles and flows in landscapes and the general geochemical flows and cycles of elements for landscapes of a particular type. The descriptions of these geochemical parameters presents problems at this time.
3) The Positive Extra Landscape Flow (+ve ELF) which involves the addition of matter to the landscape.	
4) The Negative Extra Landscape Flow (−ve ELF) involves the removal of matter from the landscape.	c) The principle of the stability of the geochemical parameters of migrational macroprocesses deals with the relationships between the lithological and active processes of biological circulation that occur in landscapes.
These flows are all considered to occur simultaneously in Elementary Landscape Cells (ELC's) that occur in areas where the MMC predominates over the other flows. A convenient conceptual model for the description of flows within an ELC is a Landscape PRISM, as shown on Fig. 8.1.	d) The principle of the specific geochemistry of migrants states that each element behaves in a characteristic way in landscapes. Such patterns for element circulation may conveniently be studied using radioactive isotopes.

Figure 8.7. Concepts and principles for the description of the flow patterns of elements in landscapes (Kozlovskiy, 1972)

Summary and Conclusions

This chapter has dealt with concepts and principles described by Kozlovskiy (1972) and others for the flows of elements and substances through landscapes (Figure 8.7). The concepts and principles of Kozlovskiy (1972) provide a starting point for the discussion of the circulation of elements in landscapes at the local, regional, and perhaps the global levels of study. Systematic comparisons of flow rates and abundance data can provide clues to variations in the behavior of numerous elements in different localities, which may be difficult to discover by other approaches.

The concepts and principles described by Kozlovskiy (1972) form a theoretical foundation for the detailed discussion of the migration of chemical elements and substances through landscapes during geological, pedological, ecological, and technological time periods. As such they should be of interest to every environmental geochemist regardless of discipline as a means of placing specific projects in geochemical perspective and as an incentive to the description of new ideas and approaches to the solutions to problems of environmental geochemistry.

Discussion Topics

(1) What landscape types are best suited for the discussion of the geochemical flows described by Kozlovskiy and for what reasons?

(2) Which of the flow patterns described by Kozlovskiy is of greatest importance in relation to man's activities in the landscape?

(3) Should systems modeling and systems simulation modeling be based on the concept of the ecosystem, on the concept of the biogencoenose, or the concept of the elementary landscape cell in closed basins in arid areas?

9. Geochemistry Gradients

"Catena: A sequence of soils of about the same geological age, derived from similar parent materials and occurring under similar climatic conditions but having different characteristics due to variations in relief and in natural drainage."

    R. L. Donahue, J. C. Shickluna, and L. S. Robertson, *Soils: An Introduction to Soils and Plant Growth*, 3rd ed. (Englewood Cliffs, N.J.: Prentice Hall, 1971), p. 508.

Introduction

Gradient is an important concept in ecology and landscape study. For example, in ecology gradients are frequently studied in relation to the change in plant cover types or in response to some environmental factor; in soil science the concept is often used in relation to relief or natural drainage. Two kinds of gradient are commonly distinguished: (1) continuous gradients within which the rate of change is gradual and relatively uniform; and (2) discontinuous gradients along which abrupt changes occur.

Both continuous and discontinuous geochemical gradients are important in landscape geochemistry. A continuous geochemical gradient of a pollutant is found in the environment in the downwind direction from a smelter stack or in a river downstream from a source of pollution. Discontinuous geochemical gradients often result from changes in the chemical composition of bedrock that underlie landscapes. For example, a discontinuous gradient for the content of an element may occur in a soil in the downhill direction on a slope underlain by bands of rocks containing different concentration levels of the element that strike at right angles to the slope.

In order to describe geochemical gradients, two other terms are of importance. Geochemical gradients may be either visible or invisible within a landscape. A visible gradient is one associated with a visible effect on the landscape such as the modification of growth of plant cover by smelter fumes. An invisible geochemical gradient is one that cannot readily be detected by the eye although it may be detected readily on the basis of the chemical analysis of samples drawn from a suitable landscape component sampled along it. For example, variations in the copper content of branches from trees growing near a copper mineral deposit in a temperate coniferous forest cannot be detected by direct observation of the sample material. They are part of an invisible biogeochemical gradient for copper that can be detected by chemical analysis of branches or needles.

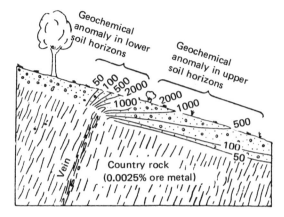

Figure 9.1 A conceptual model of an ideal geochemical anomaly in soil derived from a mineral vein (Malyuga, 1964; after Sergeev et al. 1954). [Sic: presumably contours for 1000 and 2000 may be reversed.]

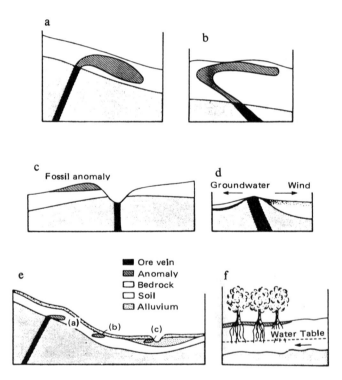

Figure 9.2. Secondary dispersion patterns developed in soils in the vicinity of mineral deposits. a) an anomaly in residual soil with no surface expression, b) anomaly developed in surface soil with surface expression, c) fossil soil anomaly before erosion formed the valley, d) anomalies formed uphill due to wind and groundwater, e) multiple disrupted anomalies in soils and alluvium originating from the same deposit, f) false anomaly resulting from the uptake by vegetation of trace metals from groundwater (Levinson, 1974, p. 353).

Geochemical Gradients

The classic case of a geochemical gradient is in the vertical and horizontal distribution patterns for elements in landscape components situated downslope from a mineral deposit located on a hillside. An idealized conceptual model for such a gradient was described by Sergeev et al. (1941) over thirty years ago (Figure 9.1). In this model the mineral deposit is a vein striking at right angles to a uniform slope covered with a deep soil and subsoil. The geochemical anomaly is superimposed on the soil from below and decreases downslope from the source. In this example, the "background values" for the element in the soil are 50 ppm and the maximum anomaly is over 40 times as strong. Conceptual models of more complicated geochemical gradients associated with other types of mineral deposits in more complex landscape conditions were described by Levinson (1974) as illustrated in Figure 9.2. In this example A, B, and D have geochemical anomalies with continuous gradients and C and E have discontinuous gradients.

Another simple example of a geochemical gradient results from a plume downwind from a smelter or coal-fired power station. For example, Bolter et al. (1972) described the content of lead, zinc, and sulphur at various distances from a lead smelter as it affected the chemical composition of surface soil material (Table 9.1). In this case the decrease in content of lead was from 5220 ppm to 355 ppm over a 2 mile distance. Corresponding values for sulphur were from 1088 ppm to 582 ppm and for zinc from 635 ppm to 79 ppm. It should be noted that the concentration level of the geochemical gradient over the two mile gradient varies from element to element. In the case of lead, the high to low contrast is near 17 to 1, compared with 8 to 1 for zinc. This is an example of the complex relationship between geochemical gradients and relative mobility of elements in landscapes affected by smelter fumes.

The Russian geochemists have described several conceptual models in relation to geochemical gradients on the regional or global scale. For example, Strakhov (in Malyuga, 1964) provided a generalized conceptual model for the formation of the weathered crust in technically inactive areas along a gradient from the north pole to the equator (Figure 9.3). The purpose of this idealistic, global-level model was to indicate relationships between climatic and geochemical effects of weathering over periods of geological time. Variation in both of the principal

Table 9.1. Averaged Contents of Lead, Zinc and Sulphur in the 0-1 Inch Layer of Soil Material in the Vicinity of a Smelter in Missouri (Bolter et al., 1972).

Distance from Smelter	Lead	Zinc	Sulphur
0.5 miles	5220(32)*	635(37)	1088(42)
0.5-1 miles	1285(65)	210(64)	627(70)
1.0–1.5 miles	540(65)	113(64)	593(70)
1.5–2.0 miles	355(33)	79(37)	582(33)
over 2 miles	188(52)	62(62)	408(60)

* Number of analyses

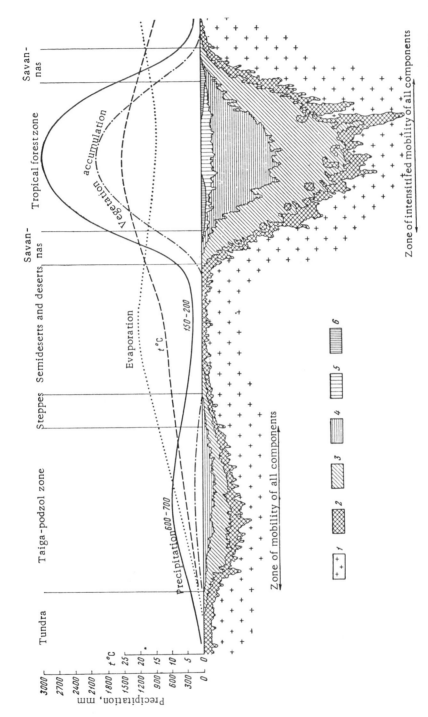

Figure 9.3. Idealized conceptual model for the formation of the weathered crust in tectonically inactive areas 1) New Crust, 2) Grus Zone; chemically inert little modified, 3) Hydromica-montmorrillonite-beidellite zone, 4) Kaolinite zone, 5) Ochers, Al_2O_3, 6) Hard layer $Fe_2O_3 + Al_2O_3$. (Strakhov, 1960; from Malyuga, 1964).

aspects of climate (i.e., temperature and precipitation) with latitude are indicated on the model as well as the evaporation rate in the different climatic zones. The diagram lists two zones of energetic chemical weathering, one in the "taiga-podzol zone" and the other in the "tropical forest zone." As Malyuga (1964) pointed out, the depth of the former is around two metres compared with the latter where weathering may be hundreds of metres deep.

Perel'man (1966) carried this line of thinking a stage further and defined the two types of geochemical gradient that are of fundamental importance in landscape geochemistry on the global, regional, or local scale. One of these he termed water series and the other thermal series. A thermal series of landscapes is characterized by an optimal supply of water along it where the solar heat inflow varies. As an example of such a series Perel'man (1966) listed the following: (1) wet equatorical forest, (2) wet subtropical forest, (3) wet broadleafed forest, (4) taiga landscape, and (5) oceanic tundra landscape. In the case of a water series where the heat influx is relatively uniform but the water supply is limited, he listed the following example: (1) wet equatorial forest, (2) high grass savanna and woodland, (3) typical savanna, (4) Savanna tending to desert, (5) tropical semi-desert, and (6) tropical desert.

One of the interesting aspects of landscape geochemistry, which has yet to be explored in detail, is the formal study of variations in migration rates and relative mobilities of elements that occur along either a thermal series or a water series. Although such investigations would be difficult to plan in the field (due to the difficulty in locating suitable localities for sampling), they would certainly provide general geochemical information of considerable importance particularly if studies were carried out systematically on a global scale.

Gradients in climatic conditions during geological time may give rise to very complex landscapes. Similarly, stable geological conditions may result in the evolution of similar plant cover types for the same climate, but as Perel'man (1966) put it: "Even with identical climatic conditions the landscape in early Palaeozoic time differed considerably from recent ones."

Kovda and Perel'man (Lukaschev, 1970) considered changes in landscape conditions that occur along the thermal series: tundra, forest, steppe, and semi-desert (Figure 9.4). Using a bar graph diagram approach these writers provided an interesting overview of the changes in the landscape parameters that occur in eluvial landscapes along a major climatic gradient. Note that the maxima for the majority of the parameters does not occur at either end of the series but somewhere between the two extremes. For example, the "total living matter" is at a maximum in "forests" and the "exchangeable hydrogen and aluminium" reach a maximum and then decrease with increasing heat flux. This focuses attention on a common characteristic of gradients in natural systems. They are seldom uniform for the range of a parameter either with respect to chemical elements (or substances) or environmental variables.

Thus gradients are of considerable importance in landscape geochemistry both with respect to the behavior or particular chemical elements and with respect to environmental conditions. Geochemical gradients relate closely to Kozlovskiy's ideas for geochemical flows in landscapes and to the concepts of element migration in landscapes that were described in Chapter 7.

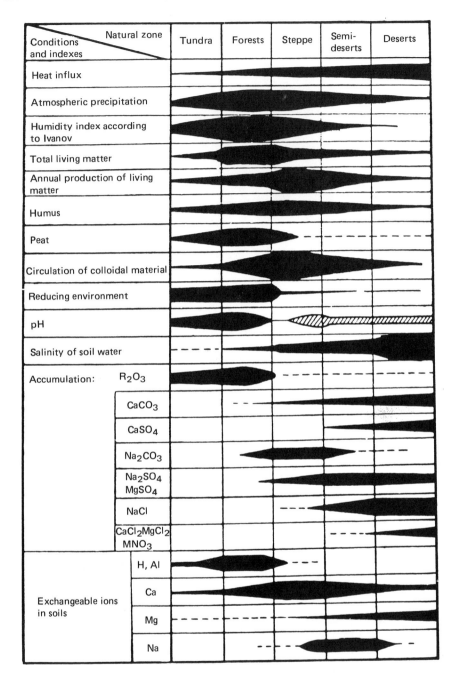

Figure 9.4. Geochemical features of landscapes along a thermal series according to Kovda and Perel'man (Lukashev, 1970, p. 113).

Ecological Gradients

Although geochemical gradients may be visible in relation to plant and/or soil cover types within a particular landscape, a very close relationship often occurs between geochemical gradients and gradients in the living biosphere. Consequently, the study of ecological gradients is intimately related to that of geochemical gradients. Some examples follow that will illustrate common types of ecological gradients occuring in landscapes. As an example of a regional ecological gradient, let us consider a climatic and topographical gradient described for Venezuela by Walter (1973):

> .... large climatic contrasts can be observed distinctly within a very small area in Venezuela (Figure 9.5). Venezuela lies between the equator and 12°N and provides examples of every altitudinal belt from sea level up to the glaciated Pico Bolivar (5,007m). The northern part of the country is exposed to trade winds from November until March which, however, bring rain only to the mountainous districts. The lowlands, therefore, have a distinct dry season lasting five months and a rainy season of seven months. Only in the south, in the Amazon basin, does no month have rainfall less than 200mm. In Venezuela the annual rainfall rises steadily towards the south, from 150mm on the island of La Orchila to more than 3,500mm. In mountainous regions precipitation increases rapidly on the windward side up to cloud level, but decreases again above this. At the same time the mean annual temperature falls by 0.57°C per 100m increase in altitude. The inner valleys of the Andes, situated in the rain shadow, are extremely dry. [Figure 9.5] shows schematically the variations in vegetation from north to south with increasing rainfall, as well as the altitudinal belts. (pp. 33 and 34)

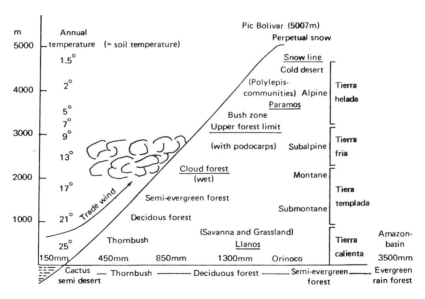

Figure 9.5. Schematic representation of vegetation cover types and zones in Venezuela from north to south with annual temperatures in C. (Walter, 1973, p. 34).

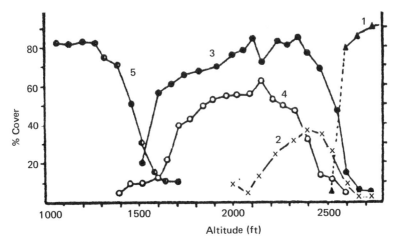

Figure 9.6. A single altitudinal ecological gradient transect at Corserine Mt. Rhinns of Kells Range, S. W. Scotland: 1) *Salix herbacea,* 2) *Carex bigelowii,* 3) *Festuca rubra,* 4) *Nardus stricta,* and 5) *Molinia caerulea* (Shimwell, 1972).

On the local scale, ecological gradients may be the result of variations in topography. For example, Shimwell (1972) described a simple altitudinal transect in Scotland where samples were collected from 10^2 m plots spaced at 50 ft intervals (Figure 9.6). He found that *festuca rubra* and *nardus stricta* had a range between 1,500 ft and 2,500 ft in contrast to *molinia caerulea,* which disappeared at the 1,700 ft level, and *Salix herbacea,* which makes an appearance at 2,250 ft

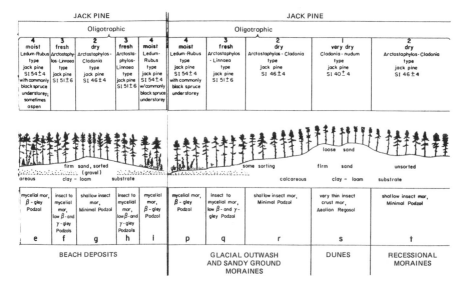

Figure 9.7. Oligotrophic (i.e., nutrient poor) forest habitat types from Southeast Manitoba (Mueller-Dombois and Ellenberg, 1974).

and rapidly assumes dominance in terms of cover. It is likely that this gradient is also associated with variations in the element composition of humus (associated with the different plant cover types) although this was not mentioned by Shimwell (1972).

A more complicated, and holistic, ecological gradient was described by Mueller-Dombois and Ellenberg (1974) who described the concept of an ecological series which is considered in detail here because it relates directly to the concept of geochemical gradients and provides an introduction to the description of land-

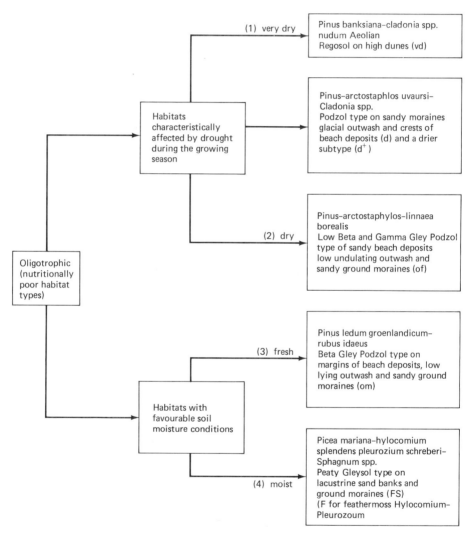

Figure 9.8. Oligotrophic forest habitat types in Southeastern Manitoba and their major ecological characteristics (Mueller-Dombois and Ellenberg, 1974).

scapes to be discussed in Chapter 12. Mueller-Dombois and Ellenberg (1974) defined their regional concept for an ecological series as follows:

> An ecological series is here defined as a group of two or more habitats along a transect which differ from one another by a different stage of intensity of a major environmental control factor. The concept is similar to the well known catena concept in use in soil science. This refers to soils comprised of the same parent material that differs in soil water relations along a transect. The concept is also similar to the environment gradient concept, except that the latter does not require recognition of habitats or communities. (p. 321)

HABITAT TYPE	DRY			MOIST		
	1 vd	(2) d+	2 d	3 f	4 m	5 vm
W1^b Restricted to dry						
L Cladonia silvatica	■	+	•	•		
M Polytrichum piliferum	■	+	•			
F Solidago nemoralis	■	■	+	+		
G Koeleria cristata	•	■	•	•		
W2 Mainly dry						
S Prunus pumila	■	■	■	•	•	
LS Arctostaphylos uva-ursi	■	■	■	■	•	
G Oryzopsis pungens	■	■	■	■	•	
W3 Intermediate						
F Campanula rotundifolia	•	■	■	+	•	
LS Gaultheria procumbens	+	+	■	■	+	•
LS Chimaphila umbellata	•	•	■	■	•	•
G Oryzopsis asperifolia	•	+	■	■	■	•
F Fragaria virginiana		•	■	■	■	•
W4 Mainly moist						
LS Linnaea borealis		•	•	■	■	■
F Rubus pubescens			•	•	■	■
S Rubus idaeus			•	•	■	•
F Cornus canadensis				+	■	■
S Cornus stolonifera		•	•	•	■	■
S Ledum groenlandicum				•	■	+
F Trientalis borealis				•	■	+
W5 Restricted to moist						
F Petasites palmatus				•	+	■
LS Vaccinium vitis-idaea				•	•	■
F Coptis trifolia				•	•	■
W6 Indifferent						
S Vaccinium angustifolium	+	■	■	■	■	■
S Rosa acicularis	•	■	■	■	■	■
F Maianthemum canadense	•	■	■	■	■	•

Figure 9.9. The distribution of indicator plant species across the Oligotrophic groundwater series in Southeastern Manitoba (modified slightly, Mueller-Dombois and Ellenberg, 1974).

Geochemical Gradients

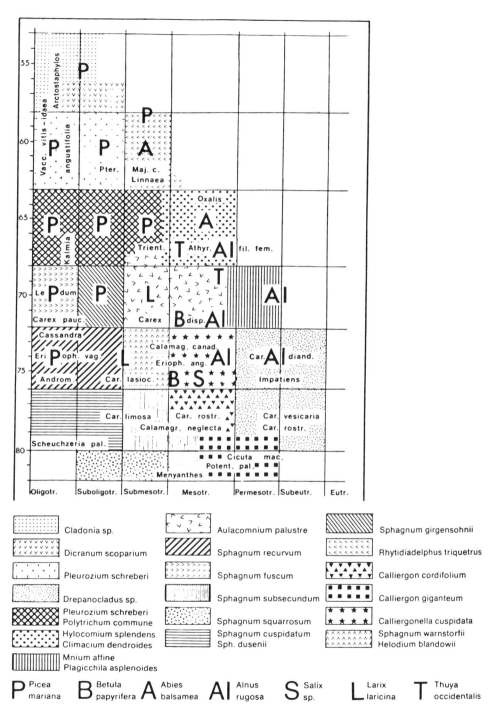

Figure 9.10. A classification of muskeg based on nutrient and moisture gradients (Stanek, 1977; after Ramenskij, 1962).

As an example of this concept these writers described a complex conceptual model for the southeast of the province of Manitoba, Canada. A part of the model is reproduced as Figure 9.7.

The conceptual model (Figure 9.7) indicates relationships between topography, geological substrate, soil type, and ground vegetation under the overstory tree cover of jack pine *(pinus banksiana)*. Other parts of the conceptual model (not shown) are concerned with other types of overstory cover. A gradient in moisture conditions from 1 (very dry) to 4 (moist) is indicated in Figure 9.8 which also lists the names of plant species used to identify the different vegetation types associated with either dry or moist conditions. Each habitat type is associated with both overstory and soil cover types. In the case of moist sites, black spruce *(picea mariana)* is also included as an overstory tree species. A more detailed breakdown of the plant cover type, indicator species across the oligotrophic groundwater series in southeast Manitoba are listed on Figure 9.9; the composition of the understory and the ground flora is used as a guide to the depth of the water table, and not all the listed plant species have the same life form. This ensures that the classification is based on species that sample different levels of moisture in the soil and do not all draw water from the same depth (Mueller-Dombois and Ellenberg, 1974). Consequently, the indicator species chosen are complementary to each other in the plant communities they represent and do not directly compete with each other in their utilization of the habitat.

Superaqual landscapes may be classified on the basis of nutrient availability and moisture gradients. For example, Stanek (1977) provided a classification for Canadian organic terrain based on relationships between the availability of nutrients (ologtrophic through eutrophic) against moisture that involved the principal ground and overstory plant cover associated with this relationship. He noted that, according to Ramenskij (1962), the richest muskegs in the overall nutrient series are subeutrophic (Figure 9.10).

Summary and Conclusions

Gradients often relate directly to both plant cover and soil cover. In this chapter simple and complex geochemical gradients were described and general relationships between geochemical gradients, climate, and vegetation cover have been outlined. The close relationship between geochemical gradients and ecological gradients has been stressed and the use of simple or complex ecological gradients in plant ecology has been introduced.

The systematic description of geochemical gradients is thus an important objective of landscape geochemistry both with respect to fundamental studies (which may relate directly to ecological investigations) and in applied geochemistry where natural or man made geochemical gradients are of considerable interest in relation to air or water pollution of the environment.

Discussion Topics

1) Is the distinction between continuous and discontinuous gradients of importance in the explanation and interpretation of geochemical anomaly patterns due to all types of mineral deposits?

2) Are water series and thermal series of equal importance at the local, regional, and global levels of landscape investigation?

3) Should an ecological gradient be described before or after a geochemical gradient?

10. Geochemical Barriers

"Lithologic boundaries in the supergene zone where conditions of migration change drastically and concentrations of chemical elements begin to rise are called geochemical barriers."

A. I. Perel'man, *Geochemistry of Epigenesis* (New York: Plenum Press, 1967), P. 213.

Introduction

Perel'man (1966, 1967) defined and illustrated the concept of a geochemical barrier which is of considerable importance in landscape geochemistry and in geology where certain kinds of mineral deposits are known to form as a result of geochemical barriers. Perel'man (1967) defined two general kinds of geochemical barriers of interest in landscape geochemistry: (1) areal barriers which may extend over relatively large areas of country (e.g., in grasslands, forests, and bogs) that are usually isotropic and isometric in plan, and (2) linear barriers which are due to processes which are anisotrophic and cover relatively small areas (e.g., slope discontinuities, fault zones, and spring lines). Perel'man (1966) noted that most environmental scientists tend to concern themselves with areal parts of landscapes (e.g., uniform ecosystem types); however, from the viewpoint of landscape geochemistry, linear features (e.g., transitions between ecosystem types) may often be of greater interest.

Kinds of Geochemical Barriers

According to Perel'man (1967) geochemical barriers are predominantly mechanical, physicochemical, or biological in nature although they usually result from a mix of all three types. An example of a predominantly mechanical barrier is a placer stream deposit in which heavy minerals are separated mechanically from the stream waters. Physicochemical barriers result from sharp changes in the pH or pH/Eh relationships (e.g., in superaqual landscapes). Indeed, the study of pH/Eh relationships in relation to natural environments is very closely related to the study of the geochemical barriers which result from such changes. Perhaps the classic example of a predominantly biological barrier was described by Goldschmidt (1937) (Figure 10.1). In this case the humus layer of forest soil acts as a horizontal barrier for certain trace elements which accumulate there. These ele-

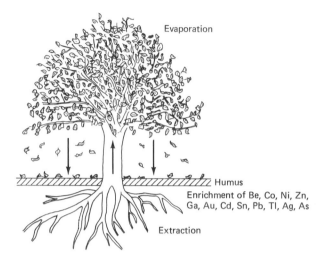

Figure 10.1 Humus acting as a biogeochemical barrier in a deciduous forest (Goldschmidt, 1937).

ments are taken up by the trees from soil and then returned to the soil in litter. Not all elements are enriched in the humus layer, some pass through it into the mineral soil. Thus, geochemical barriers are often selective of elements, or chemical substances, accumulated within them.

Geochemical barriers may also form within landscapes below the daylight surface. For example, geochemical barriers result in perched water tables that are of considerable interest in landscape geochemistry because they modify the migration behavior of elements flowing through the landscapes containing them. The size of a geochemical barrier may vary from a few centimeters long in soils to several kilometers in certain geological formations.

Effects of Geochemical Barriers

The surface of soil in eluvial landscapes acts as a physical and geochemical barrier for fallout from the atmosphere. An extreme example here is the deposition of loess on soil formation as described by Ruhe (1969) who summarized relations between the characteristics in the Loveland loess along a gradient from Bently to Loveland in southwestern Iowa (Figure 10.2). The source of this loess material is in the Missouri river valley some 13 miles west of cut 50 (Figure 10.2). As the thickness of the loess decreased from cut 50 to cut 4 the texture of the material along the gradient also changed. The change in texture of the soil (which subsequently developed upon the loess) was due to a decrease in coarse and fine silt from the loess source accompanied by an increase in the clay-sized material. In terms of geochemical barriers this example is important because it shows how the same mechanical geochemical barrier may accumulate different materials

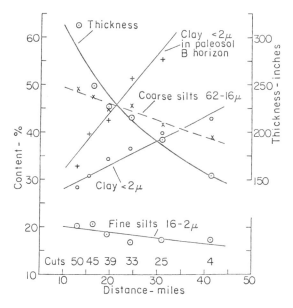

Figure 10.2 Relations of thickness and texture of the Loveland loess along a traverse from Bentley to Atlantic in southwestern Iowa (Ruhe, 1969).

along its length. In general, mechanical barriers develop where the velocity of the landscape geochemical flow (LGF) of water (or air) changes abruptly. Other mechanical barriers are developed in glaciated areas when the flow of ice affects the normal migration patterns of minerals, rocks, or waters. Below the daylight surface mechanical barriers often occur in aquifers owing to changes in the texture and permeability characteristics of rocks.

Perel'man (1967) listed common types of physicochemical barriers (Table 10.1). The precipitation of iron and manganese where oxygenated waters mix with gley waters in superaqual landscapes in an excellent example of an oxygen barrier. Perel'man (1967) noted that the effects of oxygenated waters may be more subtle; he recorded the effects of such waters over 500 m below the day-

Table 10.1. Types of Physicochemical Barriers Which Occur in Landscapes.[a]

1. The oxygen type for Fe, Mn, Co, and S
2. The hydrogen sulfide, reducing type for V, Fe, Cu, Co, As, Se, Ag, Ni, Zn, Cd, Hg, Pb, and U
3. The sulfatic and carbonatic type for Ca, Sr, and Ba
4. The alkaline type for Ca, Mg, Sr, V, Cr, Mn, Fe, Co, Ni, Cu, Zn, Cd, and Pb
5. The acid type for SiO_2
6. The evaporate type for Li, N, F, Na, Mg, S, Cl, K, Ca, Zn, Sr, Rb, Mo, I, and U
7. The sorptive type for Mg, P, S, K, Ca, V, Cr, Co, Ni, Cu, Rb, Mo, Zn, As, Hg, Pb, Ra, and U

[a] From Perel'man (1967).

light surface (Perel'man, 1967). A hydrogen sulfide barrier may occur at the bottom of eutrophic lakes. This is associated with the formation of black shales which, like their marine equivalents, are enriched in minor elements (Mason, 1966). Sulfate and carbonate barriers occur where waters rich in these ions mix with others containing calcium, barium, and strontium. Thus geochemical barriers in the rocks below landscapes may influence the chemical composition of springs and surface waters.

Let us consider conceptual models provided by Perel'man (1967) for two geological environments which include geochemical barriers. The first involves the localization of uranium ore in sedimentary strata (Figure 10.3).

In this case descending oxygenated waters at the periphery of artesian basins are reduced in strata of grey unconsolidated permeable sandstones (Perel'man, 1967, p. 220):

> The ore is restricted to strata of grey unconsolidated, permeable sandstones. The oxygenated ground waters contain $n.10^{-5}$ to 1.10^{-4}g/liter uranium, and have Eh values of over 0.25V. These waters oxidize the grey sandstones, turning them brown due to limonitization. In such cases a local oxydation zone develops. Below it, sandstones resume their gray color, the uranium content drops to 1 to s $.10^{-6}$g/liter and the Eh becomes negative (-0.05 to -2V).

As a result of this process "manto"-type ore bodies are formed by the precipitation of uranium from the oxygenated waters at the barrier where the oxidation zone ends. This barrier type is recognized in the field by the change from brown (limonitized) to gray (pyritic) sandstones (Figure 10.3).

The second example involves both physicochemical and biological activity. In this case hydrogen sulfide is presumed to be generated as a result of bacterial

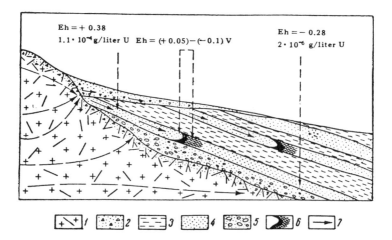

Figure 10.3 A geochemical barrier resulting in the localization of uranium ore in sedimentary strata: (1) granites, (2) deluvial–proluvial sediments, (3) impermeable argillaceous rocks, (4) sandy water-bearing strata, (5) conglomerates, (6) ore bodies, and (7) the direction of water flow (Perel'man, 1967; after Evseeva, 1962).

activity. As this gas migrates upward to the surface it is oxidized by other kinds of bacteria and a deposit of native sulfur is formed, (Figure 10.4). The reaction is $2H_2S + O_2 = 2H_2O + 2S$.

Other biological barriers may occur on the local, regional, or global scale. For example, the whole biosphere may be considered as a geochemical barrier for certain elements (such as carbon). This barrier is complicated by the behavior of living matter which results in the circulation of carbon from biosphere to atmosphere and back to biosphere by respiration and photosynthesis as well as by the slower carbon cycle of living (and dead) matter within the biosphere. In geological time, the accumulation of organic matter (which eventually results in the formation of oil and gas and coal) is also due to geochemical barriers in the carbon cycle. Perhaps the best example of a biological barrier at the Earth's surface is the formation of a domed peat bog in a humid climate. In this case, because of the coincidence of the water table with the daylight surface and the nature of the peaty material, an oxygen barrier forms just below the growing moss cover. This does not allow for the circulation of oxygen-carrying groundwaters and consequently remains acid and free from oxygen, which, in turn, preserves the underlying organic matter from oxydation and decompostion. Under these environmental conditions peat thicknesses of over 5 m are not uncommon (Radforth and Brawner, 1977).

Typomorphic Elements

Closely related to the concept of geochemical barriers is that of typomorphic elements. This concept was defined by Perel'man (1967, p. 143) as follows:

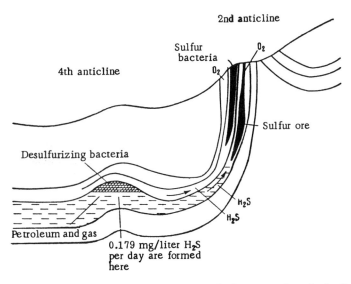

Figure 10.4 The participation by different groups of micro-organisms in the formation of the Shor-Suh sulfur deposit (revised after Kuznetsov et al., 1963; Perel'man, 1967).

Chemical elements, ions and compounds are said to be typomorphic if their migration characterizes a given epigenetic process. In turn, typomorphism depends on the abundance and the migrational ability of an element. With regard to their abundance, chemical elements are either principal or accessory. To be principal, the elements should be abundant enough to constitute the bulk of rocks Si, Al, Fe, Ca, Mg, Na, K, P, Cl, Ti, etc. Being present in significant amounts, such elements control the migration of other elements. Some of the principal elements might be typomorphic.

Elements which occur only in minor amounts were called "accessory" by Perel'man (1967) The alkali elements potassium and sodium are, from time to time, typomorphic in landscapes while cesium, lithium, and rubidium are accessory because they have low Clarkes. Thus the Clarke of the element is often a guide to the degree to which it will be typomorphic in landscapes. Elements such as calcium, sulfur, and iron may or may not be typomorphic in landscapes according to local conditions in contrast to other elements—such as radium, uranium, and cobalt which are never typomorphic. Typomorphism is of particular interest in areas where evaporation exceeds precipitation because, under such conditions, typomorphic elements tend to accumulate in soils and salt pans.

Epigenetic Processes

Perel'man (1967) discussed relationships between air and water migrants and listed the common types of geochemical processes which occur at or near the daylight surface of the Earth (Table 10.2). He noted that several of the processes

Table 10.2. Basic Geochemical Types of Epigenetic Processes.[a]

	Aerial Migrants			
	O_2 Oxidizing		CO_2, CH_4 Reducing, without H_2S	H_2S Reducing, with H_2S
Aqueous Migrants	In Rocks Containing Reducers	In Rocks Not Containing Reducers		
H^+, SO_4^{2-}, Fe^{2+}, Zn^{2+}, Cu^{2+}	Sulfatic	—	—	Sulfatic, sulfidic
H^+, HCO_3^-, organic acids	Oxidation by acid waters	Acid	Carbonate-free colloidal	—
Ca^{2+}, HCO_3^-, Mg^{2+}	Oxidation by neutral weakly mineralized waters	Neutral, calcareous	Carbonatic colloidal	—
Cl^-, SO_4^{2-}, Na^+	Oxidation by neutral strongly mineralized waters	Chlor-sulfatic	Oxysalt-bearing colloidal	Oxysalt-sulfide-bearing
Ca^{2+}, SO_4^{2-}	—	Gypsiferous	Oxysalt-bearing colloidal	
Na^+, HCO_3^-, OH^-, SiO_2	Oxidation by soda waters	Sodic	Sodic, colloidal	Soda and hydrogen sulfide

[a] From Perel'man (1967).

may occur side by side within the same landscape depending upon local conditions. The importance of a classification of epigenetic processes (based on water and air migrants) is that it focuses attention upon variation in landscape conditions which may be expected to occur either vertically (or horizontally) within particular landscapes types. Some environmental conditions are associated with geochemical gradients and other with geochemical barriers. In the latter case, the elements accumulated may become typomorphic.

Perel'man (1967) also attempted to relate the basic types of epigenetic processes to soil, the weathered crust, continental deposits, and aquifers (Table 10.3). Although the information on this table is very general and is related to the global scale of epigenetic processes the approach which it embodies is of considerable significance in landscape geochemistry on both the regional and local scales. Relationships between geology, climate, and soil formation provide the geochemical environments in which the living biosphere occurs. An important aspect of geochemical barriers is that they focus attention upon volumes of the landscape in which chemical environments are variable in contrast to geochemical gradients which usually relate to gradual abundance changes in a particular direction. The importance of the basic types of epigenetic processes also varies along water and thermal geochemical gradients. Consequently, relationships between soil profiles, the types of water and air migrants, and the influence of waters from aquifers all contribute to an understanding of the migration patterns of elements in landscapes.

Geochemical Barriers and Holism

Geochemical barriers and gradients are of fundamental and practical importance. Their study is now at a descriptive stage in most landscapes but formal mathematical models describing the behavior of independent and dependent migrant elements will be used to an increasing extent in landscape geochemistry. Hydrologists and geochemists have very similar problems with groundwater flow and the flow of soil water. For example in a discussion of soil water behavior Klute (1973) summarized the problems as follows:

> The soil water flow theory based on the Darcy equation provides a first approximation to the description and prediction of soil water behavior. Solute-water and heat-water interactions may in some situations produce significant deviations from the Darcy-based flow theory. Methods of application of flow concepts range from their use as general, qualitative background, through various approximate uses of flow equations, to full-scale detailed prediction of the behavior of a given field flow situation. The latter is not generally feasible because of the cost and lack of the detailed knowledge of the pertinent hydraulic properties of the soil. The lack of rapid, reliable, routine methods of assessing the water flow properties of soils, and the difficulties of coping with the spatial and temporal variability of these properties are important barriers to the quantitative application of the flow theory.

In landscape geochemistry we are concerned with geochemical barriers which occur above, at, or below the daylight surface in landscape. We are also con-

Table 10.3. The Occurrence of Basic Types of Epigenetic Processes.[a]

Vertical Subzones	Epigenetic process						
	Sulfatic	Acidic	Neutral Carbonatic	Chlor-sulfatic	Gypsiferous	Sodic
Soil	Over sulfide ore bodies	Meadow podzols; red, gray, brown soils of forested steppes; salt flats	In steppes: chernozems, chestnut-brown, or gray soils	Upper layers of some salt marshes in steppes and deserts	Ancient salt marshes on terraces	Salt marshes
Weathered crust	Oxidation zone of sulfur and sulfide ore deposits; over pyritic shales and clays	Carbonate-free rocks in temperate and tropical humid climates	In steppes and deserts	Over salt-bearing rocks and salt deposits	Gypsum cappings over salt bodies	—
Continental deposits	—	Deluvium in humid climates	Deluvium and proluvium in arid climates	Saline sediments in deserts	—	Red sediments
Subzone of catagenesis (aquifers)	Sulfide-bearing rocks, flushed by oxygenated waters	Carbonate-free rocks in humid climates	Carbonate-rich rocks (limestones, marls, loesses)	Salt-bearing rocks; aquifers in deserts	Places where deep chlor-calcic waters mix with surficial sulfatic waters	Wastes of steppes: deep-lying sand with artesian waters

[a] From Perel'man (1967).

Table 10.3 (Cont.)

Vertical Subzones	Epigenetic process							
	Carbonate-free colloidal	Carbonate-rich colloidal	Saline colloidal	Gypsiferous colloidal	Sodic colloidal	Oxysalt-sulfidic	Sodic with H$_2$S	
Soil	In swamps of taiga, tundra, or tropics	Meadow and swamp soils of northern steppes; carbonate-rich on meadows and in swamps of forest and tundra zones	Salt marshes with weakly-reducing quality	Gypsum horizons of meadows	Soda-salt marshes on meadows	Lower horizons of salt marshes	Salt marshes	
Weathered crust	Plains of northern taiga and tundra	—	—	—	—	—	—	
Continental deposits	Alluvium in humid climates	Alluvium in forested and northern steppes	Muds in saline lakes with weak reducing media	—	Muds of soda lakes	Muds of saline lakes with strong reducing media	—	
Subzones of catagenesis (aquifers)	Carbonate-free rocks in predominantly humid climates	Carbonate-rich and sulfate-poor rocks	Rocks in arid climates	Aquifers in gypsum strata	—	Deep oil deposits	Oil deposits, bituminous limestones and shales in process of destruction; flushed by weakly mineralized waters	

cerned with the behavior of all chemical entities which circulate in landscapes at such barriers. This leads to the study of landscapes from the viewpoint of (1) geochemical flows, (2) geochemical gradients, and (3) geochemical barriers with less emphasis upon the ecosystems, or plant or soil cover types present. When one considers the circulation of chemicals in landscapes using these concepts the importance of the origin and chemistry of the inert material in the landscape descreases leading to a separation of the study of migration from the study of landscape morphology. This in turn leads to the study of landscapes on the basis of migration rates (e.g., in prisms) regardless of man's disturbance of the landscape materials. Thus, the migration of chemicals in landscapes is to landscape geochemistry what the living biosphere is to ecology—regardless of the modifications which man has made to particular areas of country.

Summary and Conclusions

Like the sister concept of a geochemical gradient the concept of geochemical barriers is of fundamental importance in landscape geochemistry. Geochemical barriers are important in exploration geochemistry because their existence may relate directly to the shape and size of geochemical anomalies due to mineral deposits which result from the current, or a previous, cycle of weathering. Geochemical barriers are also of practical importance to man especially in the solution of problems which relate to the disposal of wastes.

The study of geochemical gradients (and geochemical barriers) leads to a consideration of the dynamics of the migration of elements in general due to epigenetic processes. It is to be expected that as the relative importance of such processes in particular landscapes becomes better understood systems and simulation models may be used to separate out the roles played by individual processes in landscapes where they can be studied readily.

Discussion Topics

1) Are laboratory studies of the formation of geochemical barriers and gradients in model landscape systems likely to contribute significantly to landscape geochemistry?

2) Rank the types of physicochemical barriers listed in Table 10.1 in order of importance in a humid tundra landscape.

3) Are geological geochemical barriers in unweathered rocks likely to affect the geochemistry of the landscapes developed above them to a significant degree?

11. Historical Geochemistry

"We must not underestimate the importance of an understanding of landscape evolution and its strong effect on the time factor in soil formation. Although a given landscape may appear quite uniform and simple from a casual glance, chances are that it has a complex geomorphic history and that some soils differ on this landscape because of differences in their time zero of soil formation related to the landform they occupy."

S. W. Buol, F. D. Hole, R. J. McCracken, *Soil Genesis and Classification* (Ames, Iowa: Iowa State University Press, 1973), p. 167.

Introduction

The concepts of geochemical barriers and epigenetic processes focus attention on the time factor in the evolution of landscapes. The purpose of this chapter is to describe aspects of the development of landscapes over time. In order to do this landscapes are divided into two groups; (1) those which have evolved from a substrate of uniform chemical and physical composition during a single time period (during which relatively uniform environmental conditions are obtained)—isogeochemical landscapes, and (2) landscapes which have evolved during more than one set of climatic and ecological conditions. Unlike common landscapes, isogeochemical landscapes do not include relict structures within them. The historical development of landscapes is normally formally studied in relation to eluvial landscapes although, as we shall see, under favorable conditions superaqual and aqual landscapes may also be of considerable importance.

Polynov's Concept of the Weathering Process

In his classic book *The Cycle of Weathering,* Polynov (1937, pp. 7, 10–11) discussed a general conceptual model for the behavior of elements in landscapes during the long-term weathering process:

> We thus see that in the Earth's crust there takes place diverse transformations of material which form, as it were, an undying cycle—a cyclic process, in which after a series of changes matter invariably returns to its original form. Such cycles may be followed not only for the great mass of rock-forming material, but also for many individual substances.
>
> [and]. . . . it is incorrect to express the cyclic processes of the Earth's crust as closed circles. These "cyclic" processes have at the same time a progressive movement,

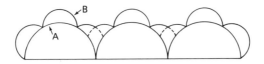

Figure 11.1. A complex cycloid conceptual model (Polynov, 1937).

and would be more correctly expressed as a curve traced by a point on the circumference of a circle rolling in a straight line. Such a curve is known as a cycloid. A complex cycloid may be taken as the symbol for the cyclic processes of the whole earth's crust, with little constituent cycles representing transformations restricted to individual sheaths or their parts.

The subject of our study is not the whole Earth's crust, but that part only which bears the name crust, or zone of weathering. In considering that alone we limit ourselves more or less artifically; we separate from the whole process certain parts of it—from the unbroken curve of the complex cycloid we select only a small piece. But the study of the nature of that piece by itself is possible only provided its place on the general curve is recognized. That condition does not allow us to leave out of consideration entirely other parts of the earth's crust. We shall have to consider especially frequently the sheaths that cover the lithosphere, these being very closely connected with the phenomena and transformations on its surface.

Thus landscapes can be envisaged with general weathering cycles (A in Figure 11.1) occurring simultaneously with shorter cycles (B in Figure 11.1). The smaller cycles (e.g., annual biogeochemical cycles in forests) relate to longer cycles (e.g., the cycle of the whole tree in a forest) which in turn relate to pedological and geological cycles of even longer duration. The model of the complex cycle relates directly to Kozlovskiy's ideas (Chap. 8) where the macrostructure of a landscape is a large (or geological cycle) and the migrational structure of the landscape is a small (or ecological/pedological cycle) within it.

McFarlane (1976, p. 106) described a series of conceptual models which illustrate this point very well; he described how laterite is formed in a tropical climate as a result of a number of stages (Figure 11.2):

Stage A	Iron is segregated into pisoliths, within the narrow range of oscillation of the groundwater table in saprolite during a late stage of landsurface reduction.
Stage B&C	Zone of precipitate formation lowers as land surface is reduced and the precipitates accumulate as a sheet at the base of the soil.
Stage D&E	Landsurface reduction ceases and the accumulated sheet of pisoliths becomes altered and hydrated to form a massive sheet of planation surface laterite.
Stage F	Water-table once again begins to lower upon the initiation of the succeeding weathering cycle and pisolith formation resumes.
Stage G, H&I	With continued water table lowering, the zone of pisolith formation lowers leaving above it a spread of pisoliths.
Stage J	The water-table is ultimately lowered beyond which the pisoliths can form.
Stage K L&M	Leaching through the carapace depletes the saprolite underlying the spread of pisoliths, forming a palid zone.

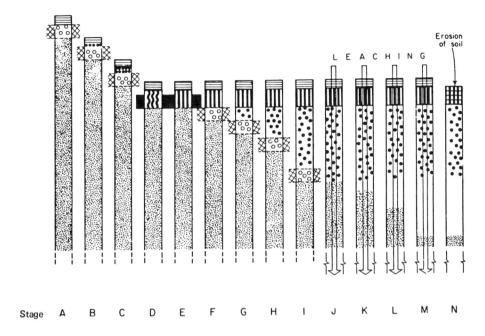

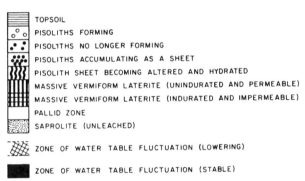

Figure 11.2. Stages in the formation of laterite (McFarlane, 1976, p. 106).

Stage N   Deforestation leads to induration of the laterite and its loss of permeability. Pallid zone formation ceases.

If the indurated layer (Stage N, Figure 11.2) is eroded mechanically to the water table then the whole cycle of weathering will start again until another layer of indurated laterite is formed. McFarlane (1976) also provided a series of conceptual models for the evolution of the relief in soil in a tropical climate (Figure 11.3), in which there is a relief inversion as the process of laterization proceeds. Similar cycles of evolution of landscapes occur in other climates where the importance of physical and mechanical processes is often more important than is chemical weathering in tropical climates.

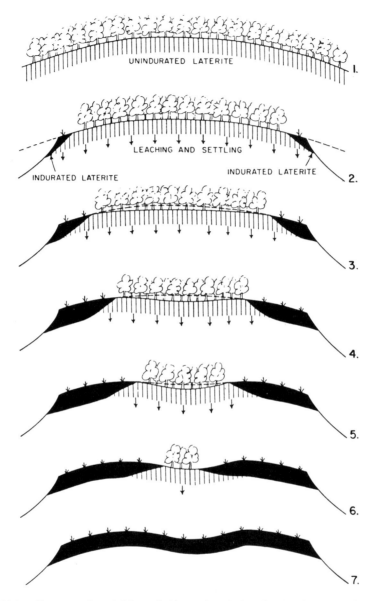

Figure 11.3. Conceptual model for relief inversion during the development of the "soup-plate" form of the Buganda Surface Mesas (McFarlane, 1976, p.99).

Stage 1. A permeable laterite sheet develops under forest cover.

Stage 2. Incision and erosion exposes the laterite at the margins of the original interfluve. It becomes indurated and impermeable and the forest vegetation deteriorates. Where permeability is maintained, there is leaching through the carapace and settling of the surface.

Stages 3, 4, 5, and 6. Induration of the laterite and vegetation deterioration extend toward the center of the mesa gradually reducing the area in which settling occurs.

Stage 7. Ultimately the entire mesa is deforested and the laterite indurated and impermeable. The central area, subject to the longest period of leaching and settling, now lies relatively lower than the periphery.

These two examples have described how sequences of weathering occur in an ideal situation. It should be remembered that conditions are seldom ideal and, consequently, sequences of events which result in the formation of landscapes during pedological and geological time are often very complicated and difficult to trace. This aspect of landscape evolution has been very actively researched by modern geomorphology (see Leopold et al., 1964).

It is evident that the weathering sequence described by McFarlane (1976) is close to the ultimate in chemical decomposition of rocks. In many climates this ultimate is not reached. Consequently, soil scientists have provided a classification which indicates the degree or stage at which a soil has developed (Table 11.1).

From the viewpoint of landscape goechemistry the soil is a cumulate of the more resistant primary and secondary minerals plus organic matter which remains after the more mobile chemical substances have been removed from the original parent materials of the soil. Some idea of the relationship between the chemical and mineralogical composition of fresh rocks and senile soils in tropical areas may be obtained from data from Harrison (1934) who studied the weathering products from granite and dolerite and related them to the chemical and mineralogical composition of fresh rock nearby. Details of some of his findings and observations are listed on Tables 11.2 and 11.3. He summarized his conclusions as follows (from Harrison, 1934; quoted by Mohr and Van Baren, 1959):

a) Under either topical or temperate conditions the superficial weathering of igneous rocks results in the katametamorphism of the minerals of such rocks by processes of hydration and oxydation. The weathering produces hydrated silicates of aluminia, iron, residual quartz and other resistant minerals.

b) Where tropical conditions prevail, the katametamorphism of *basic* and *intermediate* rock at, or close to, the water table, under more or less ideal conditions of drainage is accompanied by almost complete removal of silica, and of calcium, mag-

Table 11.1. Stages in the Development of Soils.[a]

Stage of Soil Formation	Characteristics of Soil Formation
Young soils	Azonal soils
Immature soils	Intrazonal soils the development of which has been dominated by excessive water, salts, or carbonates
Mature soils	Normal zonal soils in equilibrium with the environment
Senile soils	Pedogenic accumulations of inert materials: sesquioxides and heavy minerals

[a] From Buol et al (1973, p. 155).

Table 11.2. Chemical and Calculated Mineralogical Composition of Granite and Its Decomposition Products Mazaruni Quarry, British Guiana.[a,b]

	I	II	III	IV	V	VI
Quartz	30.77	41.67	34.51	37.06	32.12	29.66
SiO_2	41.38	28.50	32.12	29.04	30.84	31.36
Al_2O_3	14.75	18.84	21.00	22.07	24.13	25.26
Fe_2O_3	2.25	1.78	1.55	1.75	1.68	1.22
FeO	0.31	0.25	0.29	0.33	0.36	0.05
H_2O	0.52	5.17	6.25	6.76	8.01	10.32
TiO_2	0.71	0.68	0.61	0.75	0.75	1.10
MnO	0.22	—	—	—	—	—
MgO	0.70	0.17	0.47	0.21	0.26	0.11
CaO	1.08	0.03	0.13	0.03	0.03	0.01
K_2O	4.70	2.70	2.75	1.97	1.67	0.48
Na_2O	2.74	0.21	0.27	0.21	0.11	0.14
P_2O_5	—	—	—	—	—	—
Quartz	31	42	35	37	32	30
Orthoclase	17	9	9	6	6	1
Plagioclase	28	2	3	2	1	1
Muscovite	18	8	9	7	4	1
Ilmenite	1	1	1	1	1	2
Kaolinite	3	37	42	44	51	63
Hematite	2	—	—	1	—	—
Goethite	—	1	1	—	2	—
Gibbsite	—	—	—	2	3	—
Water	—	—	—	—	—	2

[a] From Mohr and Van Baren (1959) after Harrison (1934).
[b] I, Granite mass in quarry just beneath lowest stained layer, depth 5 m; II, granite, thin, grey-colored layers, resembling soft shale and crumbling to granite-sand; depth 3.65 m; III, granite-sand, still showing structure of granite; depth 2.00 m; IV, argillaceous granite-sand, loose and incoherent, orange red in color; depth 1.70 m; V, argillaceous granite-sand, compact, pale brownish-grey, mottled with cream and pale red; depth 0.85 m; VI, argillaceous, greyish, compact subsoil; depth 0.20 m.

nesium, potassium and sodic oxide, leaving an earthy residuum of aluminum trihydrate (in its crystalline form) (i.e. gibbsite), limonite, a few unaltered fragments of feldspar, in some cases secondary quartz, and the various resistant minerals originally present in the rock. The residium is termed *primary laterite*.

c) The process of primary laterization is followed by one of *resilication* and this gradually results in the vast masses of lateritic earths or *argillaceous laterite* which so frequently cover wide wide areas of basaltic and intermediate rocks in the tropics.

d) Where tropical conditions prevail, acidic rocks such as apalites, pegmatites or granite and granitic gneisses, do not undergo primary laterization, but gradually change, through katametamorphic processes, into pipe or pot clays, or into more or less quartziferous and impure kaolines.

e) Where tropical conditions prevail, lateritic earths, and even pot clays, may undergo desilification accompanied by the formation of concretionary and superficial masses of bauxite.

Table 11.3. Chemical and Calculated Minerological Composition of Colerite and Weathering Products Tumatumari British Guiana.[a,b]

	I	II	III	IV	V	VI	VII	VIII	IX
Quartz	1.60	13.24	12.66	11.31	11.64	12.28	47.41	48.09	0.04
SiO_2	49.69	7.94	3.41	4.23	4.46	21.89	3.30	4.72	1.56
Al_2O_3	15.20	26.93	29.11	32.77	26.81	24.94	26.33	24.97	19.08
Fe_2O_3	3.08	27.94	28.62	27.80	33.97	26.93	10.67	11.07	55.63
FeO	11.20	3.14	3.34	3.21	2.68	1.36	0.0	0.0	0.80
H_2O	0.30	17.26	20.20	19.65	18.86	11.29	11.28	10.90	17.39
TiO_3	1.00	1.80	2.03	1.37	1.13	0.38	0.67	0.67	5.50
MnO	nil	nil	nil	nil	nil	nil	nil	nil	nil
MgO	5.63	0.69	0.06	0.09	0.14	0.68	0.21	0.07	nil
CaO	9.58	0.48	0.02	0.02	0.02	nil	0.23	0.02	nil
K_2O	0.60	0.32	0.06	0.07	0.04	0.05	0.21	0.02	nil
Na_2O	2.09	0.78	0.05	0.07	nil	0.02	0.14	0.16	nil
P_2O_5	0.01	nil	nil	trace	trace	0.16	trace	trace	trace
Quartz	2	13	13	12	12	12	47	48	—
Plagioclase	51	10	—	—	—	—	—	—	—
Augite	42	3	—	—	—	—	—	—	—
Magnetite	4	4	5	6	5	4	—	—	—
Ilmenite	1	3	4	2	2	1	1	1	11
Gibbsite	—	38	40	43	35	10	25	26	27
Diaspore	—	—	—	—	—	—	9	5	—
Goethite	—	28	28	28	34	8	—	—	57
Hematite	—	—	—	—	—	13	11	10	—
Kaolinite	—	—	7	9	10	52	7	10	3
Water	—	1	6	—	2	—	—	—	2

[a] From Mohr and Van Baren (1954) after Harrison (1934).
[b] I, Parent rock, dolerite, II, primary laterite; first layer (3 mm); III, primary laterite; second layer (18 mm); IV, primary laterite; third layer (50 mm); VI, lateritic earth (5–6 m); VII, subsoil (1 m); VIII, surface soil (25 cm); IX, surfacial ironstone.

It is clearly evident that during geological time and in a tropical climate chemical weathering may result in the almost complete decomposition of rocks which leads to the formation of senile soils. Crompton (1962, p. 8) summarized the problem of chemical decomposition of rocks:

> Under the variety of climates in the world many combinations of weathering and translocation are possible from those in hot, dry, regions where intense weathering for short periods follows infrequent rains, insufficient to remove the weathered products, to those in cool, wet climates where every ion released is like to be leached unless, like iron and aluminum, it immediately forms a new, insoluble product. . . .
> We therefore arrive at the concept of the *richness of weathering* which is the product of the quantity and the variety of elements in the parent material capable of being brought into the soil solution and those effects of temperature, site, vegetation, etc. which tend to bring them into solution. The richness of weathering is offset by the *intensity of leaching* an outcome of the rainfall/evapo-transpiration balance and soil permeability.

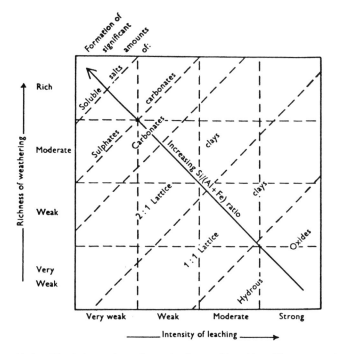

Figure 11.4. The joint action of weathering and leaching (Crompton, 1962).

He also provided a generalized conceptual model diagram for the joint action of weathering and leaching under different climatic conditions (Figure 11.4).

Using the terminology described in this book we can say that the epigenetic processes acting over a period of pedological and geological time result in the richness of weathering of rocks which is offset by the intensity of leaching which relates directly to significant changes in climate. From the viewpoint of the historical geochemical development of landscapes, changes in climate result in changes in the rate of weathering and leaching which in turn have their effect on the formation of soils. In some places this yields to the rapid formation of deep soils. But in others it results in the formation of relict structures in soils which may remain long after the environmental conditions under which they were formed have changed. Consequently, the "cycle of weathering" of Polynov is usually longer in a particular landscape than the time required to develop a single soil type. Thus, Polynov's concept relates to geological time as well as pedological time.

The Role of Groundwaters in Landscape Development

Polynov (1937, p. 21) was also concerned with the role of groundwaters in the weathering process; he distinguished between two types of weathering which occur in landscapes and which result in the solution or precipitation of chemical substances:

> 1) *THE REGION OR ZONE OF WEATHERING,* i.e. the upper part of the Lithosphere which in different places and at different times may consist of various minerals such as igneous and metamorphic masses, and loose sedimentary rocks, *but within which the processes are directed towards the disintegration and comminution of rocks and the formation of the crust of weathering.*
> 2) *THE PRESENT DAY CRUST OF WEATHERING,* i.e. those parts of the surface sheath of the Lithosphere which at this moment already consist of loose disintegrated products of weathering; in other words, any kind of sediment or deposit, and sedimentary rocks which are not yet undergoing metamorphism.

Both zones are included in the idealized landscape prism model (Figure 5.1). The distinction between the crust of weathering and the zone of weathering is important in landscape geochemistry as a whole which includes eluvial, superaqual, and aqual landscapes. Although the crust of weathering is absent in bogs, lakes, and rivers the underlying zone of weathering is usually present. The role of the zone of weathering has not been studied much outside Russia, but in Russia it has been reported that substantial amounts of weathering products are removed annually by groundwaters. For example, Zverev (1972) estimated that subsurface waters in the Soviet Union remove 301,000,000 tons/year of dissolved matter compared with 472,000,000 tons/year discharged as fine silt.

Polynov (1937, p. 21) further described the two weathering zones as follows:

> The thickness of the zone, or region of weathering, is determined by the depth to which the agents of weathering penetrate, i.e. fluctuations of temperature and the action of solutions, oxygen, carbon dioxide of the air and other vadose agents. This depth is considered to be 0.05 km from the surface of the Lithosphere, i.e. from the surface of the dry land or bottoms of oceans. . . . As regards the thickness of the present day crust of weathering it cannot exceed that of the zone of weathering but it may be as little as 1m or less. The thinnest parts occur on outcrops of igneous or metamorphic rocks or recently cooled lavas. When the crust of weathering is formed it is not entirely uniform; we can distinguish an upper, fairly distinct part occupied by the biosphere, i.e. the soil.

Thus, Polynov considered the behavior of waters in relation to the historical development of landscapes in the long-term and distinguished between deep-seated effects and near-surface effects on the chemistry of groundwaters.

The Geological Substrate of Landscapes

Polynov (1937) discussed the general chemical characteristics of the lithosphere which underlies landscapes and noted that the products are of three general types:

1. Orthoeluvium: results from the weathering of crystalline igneous and metamorphic rocks which are products of the Major Geochemical Cycle (Figure 2.1)

and are composed of minerals which may be metastable under the pressure/temperature conditions which are found at the Earth's surface.

2. Paraeluvium: results from the weathering of sedimentary rocks which may be consolidated but which are not affected by metamorphism.

3. Neoeluvium: results from the weathering of unconsolidated materials, for example, the products of Pleistocene glaciation.

These terms are important to the history of landscapes where orthoeluvium is overlain by a thin cover of paraeluvium and/or neoeluvium. Under such conditions the chemical substances derived from the underlying geological formations may have a significant and unexpected influence upon the surface soils and waters.

Details of Landscape Development during Pedological and Ecological Time

Superaqual Landscapes

So far we have been concerned with general principles relating to the geological and pedological history of landscapes and the weathering processes which occur within them. Now let us consider a more detailed approach to the history of landscape evolution which has evolved for superaqual landscapes during ecological time periods. Superaqual landscapes are chosen because the events which relate to the development of such landscapes are simpler to describe than eluvial landscapes.

Useful conceptual models for the evolution of peatlands in Scandinavia were described many years ago by Magnusson et al. (1957) (Figure 11.5). These include ombrogenic bogs which develop in immobile groundwaters, and soligenic bogs which develop in areas of moving groundwaters. Peatlands were divided into two types topographic peatlands which develop as flat bogs by filling in of lake basins and mixed peatlands which result from a combination of other types. The historical development of a peat bog may, under favorable conditions, be worked out in detail on the basis of fossil palynology and ^{14}C dating (used to provide absolute age data). For example, Moore and Bellamy (1974) described a conceptual model for the evolution of a topographic peatland in organic terrain. If a peat core located at A in Figure 11.6 were sampled every few centimeters and its morphology, palynology, and geochemistry described, this would allow for a reconstruction of the events which took place during the formation of the bog. If ^{14}C dates were obtained from the core material the historical time scale, during the formation of the bog, could also be worked out.

Studies have been made relating the vertical distribution of elements in peat to the morphology and palynology of bog core material. For example, Fortescue (1975b) (Figure 11.7) plotted the content of zinc, lead, nickel, copper, calcium, magnesium, iron, and manganese in a peat core from a bog at Dorset, Ontario, Canada. The history of this bog was interpreted as follows. A decrease in the ash

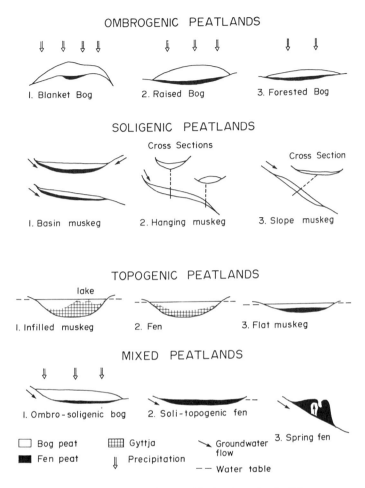

Figure 11.5. Conceptual models for different kinds of peatlands (Magnusson et al., 1957, from Fortescue, 1975b).

content of the core marked the change from laucustrian conditions to bog conditions in the landscape. The palynological data indicates the succession of plant cover types in the Dorset area at the time the bog was being formed and this may also be reflected in the geochemical data. For example, birch *Betula* sp. is known to be an accumulator of zinc and the amount of zinc in the core material is seen to increase at the time the *Betula* increased. It was found that some of the plant cover types corresponded to similar data from other bogs in the Dorset region and other data related to the local conditions in the Dorset bog during its formation.

It is clear from these two examples that, under favorable conditions, the history of the development of organic terrain can be described on the basis of the palynology, morphology, and, in some cases, the abundance of elements in peat

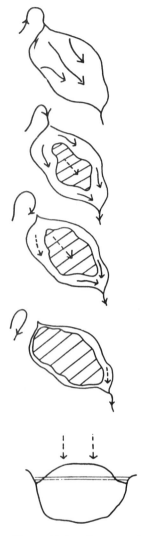

Stage I. Two alternatives when peat commences to build up a lake: (1) a large flow of water brings much allochthonous material into the system-slow development of very heavy peat below the flowing water (2) less water flowing, less allochthonous material brought in, faster rate of growth of peat: floating mat of light peat formed water flow beneath the peat.

Stage II. Except at periods of high water the water is now canalized around the peat area. Two alternatives: (1) the whole peat mass is undated, (2) the peat mass is not undated.

Stage III. Continued peat growth diverts the flow of water from the basin. Peat gets reduced flow from the surrounding basin and water from precipitation on to the bog. Main drainage tracts in the mire get slow continuous flow.

Stage IV. Further growth of peat leaves large areas of the bog unaffected by moving waters but subject to undation during heavy rainfall.

Stage V. Further growth of the peat leaves large areas of the bog unaffected by moving water due to a cupola being formed which is above the water table of the lake. It has its own water table fed by rain falling directly upon it.

Figure 11.6. Conceptual model of the succession of mire types in a hypothetical lake situated in organic terrain (Moore and Bellamy, 1974).

cores. The relative ease with which this can be done is in marked contrast to the situation in most eluvial landscapes which contains soil resulting from both the cumulative and dynamic effects of the weathering processes.

Aqual Landscapes

Under favorable conditions lake or river sediments may be used to describe the history of a landscape. The use of lake sediments to date and describe the history

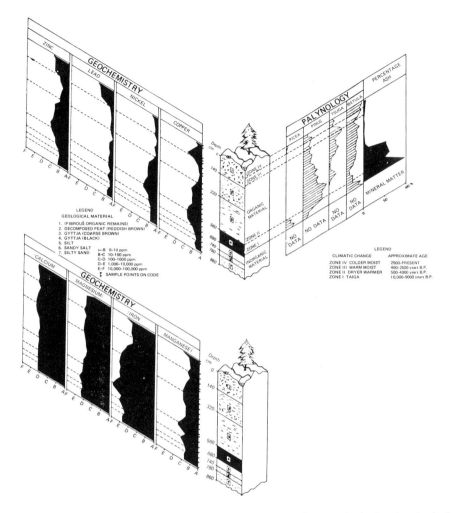

Figure 11.7. A landscape prism drawn to show morphological, geological, palynological and geochemical conditions in a peat bog (Fortescue, 1975b) based on data from Woerns, 1975 (unpubl. B. Sc. thesis Brock University, St. Catharines, Ontario).

of landscapes is now quite common and many interesting studies have been made in merimictic lakes.

In the bottoms of lakes, which are not subject to erosion by currents, layers of sediment material accumulate year by year. If such sediment material is then removed from the lake by a coring device which retains an intact, undisturbed core, this can be examined in the same way as the peat core described previously. Ruttner (1963) described a study of this type carried out many years ago by Nipkow (1920) on a sediment core from Zürichsee in Switzerland (Figure 11.8). In this case the changes which occurred in the biological and geological conditions in the lake over technological time were directly related to the mor-

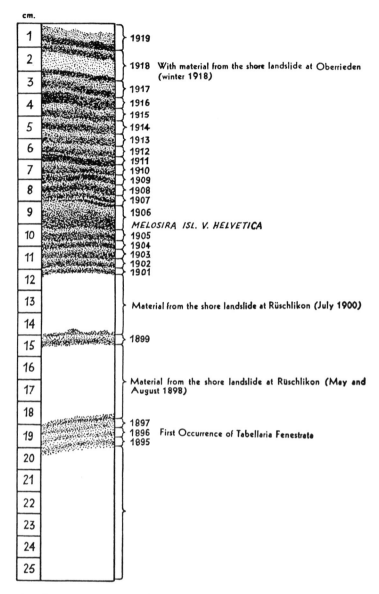

Figure 11.8. Stratified sediments from the Zurichsee (Ruttner, 1963; after Nipkow, 1920).

phology of the core (based on the examination of annual layers of sediment material). The supply of oxygen in the water diminished around 1896 as a result of man's activity around the lake, and *Tabellaria fenestrata* bloomed which indicates incipient eutrophication of the lake. Material from landslides into the lake in 1898 and 1900 was clearly identifiable in the core as well as annual layers of sediment material including a summer portion of brightly colored blooms from *Oscillatoria rubescens* and a winter layer colored black by FeS. It is clear that under favorable conditions, such as those in the Zürichsee, lake sediments may be used to describe accurately events which have taken place in technological time. Similar studies may describe events in ecological time due to changes in climate.

Eluvial Landscapes

Under favorable conditions, the history of superaqual and aqual landscapes can be described in detail on the basis of peat (or lake sediment) cores. Conceptual models may often be used to describe the sequence of events during a weathering cycle in an eluvial landscape in a tropical climate. Unfortunately, it is not often possible under normal conditions to distinguish the sequence of events within an eluvial landscape which resulted in the formation of a soil profile during pedological time. The study of the evolution of soil cover types is called paleopedology (Yaalon, 1971). According to Ruellan (1971) (in Yaalon, 1971, p.74), paleosols are of three types:

> 1) Soils whose characteristics are in balance with the present environment. These soils show no evidence of their historical evolution different from that at present;
> 2) Soils which contain certain features, so called relicts, indicating past pedological conditions different from contemporary ones; and
> 3) Buried soils which are, in fact, a particular case of the previous group.

Using the terminology described in this chapter soils of type (1) are those developed during the current weathering cycle compared with those of type (2) which result from more than one weathering cycle. Soils of type (3) usually result from a disturbance of the landscape which may, or may not, mark the end of a weathering cycle.

The literature of soil science on soil formation is voluminious and an evaluation of it lies outside the scope of this book. However, there is one concept used by soil scientists which is of considerable interest in landscape geochemistry because when it is used in relation to field studies it describes a sequence of events which takes place during soil formation: a chronosequence which is defined as a sequence of related soils which differ from one another in certain properties, primarily as the result of time as a soil forming factor. Vreeken (1975) distinguished three types of chronosequences:

1. Postincisive sequences involve soils forming today which began to form at different times in the past. For example, on different beds of volcanic ash which arrived sequentially during the evolution of a landscape.

2. Preincisive sequences involve a group of soils which commenced to form at the same time but includes areas buried at different times resulting in a series of paleosols (i.e., those buried at an earlier stage of development).

3. Time transgressive sequences include soils in a sequence in which initiation and burial took place at various and different times. Time transgressive chronosequences are of two kinds, those which have a historical overlap and those which do not.

Figure 11.9. Diagrammatic section of the mud flow area near Mt. Shasta, California, showing stratigraphic relationships, buried soils, and vegetation cover (Dickson and Crocker, 1953).

Vreeken (1975) discussed the limitation of the chronosequence concept for the detailed description of events which occurred during the formation of a particular soil in a particular locality. Briefly, the difficulty with postincisive chronosequences is that the climate and the intensity of the soil-forming processes related to it may vary during pedological time resulting in an irregular rate of soil formation. One way around this difficulty is to make parallel studies of peat cores and lakes sediment cores in humid landscapes which would indicate variations in the pollen record related to climatic changes. A difficulty with preincisive soil sequences is that in addition to the problems just mentioned chemical changes may occur in the palaeosol after burial. Time transgressive sequences with historical overlap are even less suited to the reconstruction of soil development owing to the many kinds of variables involved. Consequently, if the geochemical history of landscapes is to be worked out based on chronosequences, great care should be taken to identify the type of chronosequence involved prior to the interpretation of data from field studies.

A relatively simple, and well-documented, example of a postincisive chronosequence occurs near Mt. Shasta in Northern California, which was described in a series of papers by Dickson and Crocker (1953a, b, 1954). They studied details of soil profiles developed on a series of volcanic mud flows of different ages superimposed in a regular stratigraphic sequence. Figure 11.9 shows a landscape section across five of the mudflows (aged 27, 60, 205, 566, and 1200+ years). The age of flows A, B, C, and D was established by dendrochronology and flow E was aged at 1200+ years on the basis of soil morphological data.

Dickson and Crocker (1954) provided interesting chemical and pedological data on the development of the soils [e.g., the data on soil pH and base exchange of the soils (Figures 11.10 and 11.11)]. Dickson and Crocker (1954, p. 175) interpreted the pH data on the soils as follows:

> The pH of the surface 2½ inches rapidly dropped to a minimum of 5.5 at the 60 year point. The pH values of the 2½ and 5-12 inch layers were also reduced but did not

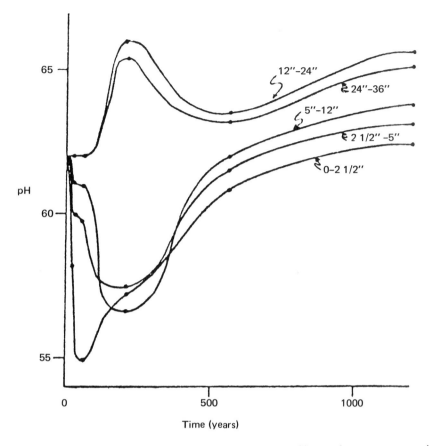

Figure 11.10. Variation of pH of soil samples from the Mt. Shasta chronosequence with depth and age of the soil (Dixon and Crocker, 1954).

show their minima until later. This lag phenomenon appears to be related in some way to the downward movement of organic matter. The pH however, was not related to the quantity of organic matter, nor was it related to any of the other properties measured including the exchangeable cations. The increase in pH of the subsoil at the 1205-year point might have been due to the leaching of bases from the heavy deposit of litter prior to this time.

The patterns of the variation in the exchangeable cations obtained from the 36" deep samples increased rapidly during the 27- to 60-year period, increased slightly during the 60- to 205-year period, and then rose a further 50%+ in the oldest soils. The total exchangeable bases of the oldest soil is only slightly different from the youngest but in the older soils there is a much larger amount of exhangeable hydrogen. Hence the long-term effect is to increase the acidity of the soil. This example is of particular interest in landscape geochemistry because it focuses attention upon the complexity of the chemical changes which occur in

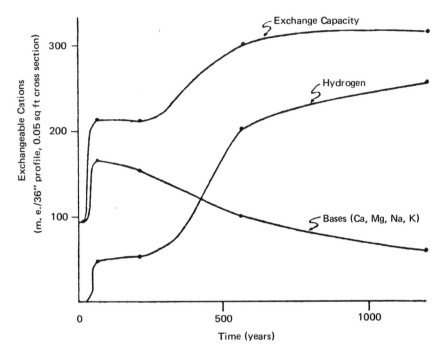

Figure 11.11. Variation in the exchange capacity, exchangeable calcium, magnesium, sodium, and potassium plus hydrogen in the 36 inch profile of soil from Mt. Shasta chronosequence (Dickson and Crocker, 1954).

azonal and immature soils in the Mt. Shasta area. The fact that the soils took some 500 years to become mature is of considerable importance in relation to man's activities in the environment.

Unfortunately, the soils of most landscapes are not as simple as those in the Mt. Shasta chronosequence. An example of a complex eluvial landscape was described by Hunt (1972). Hunt (1972, p. 201) summarized the features of the area as follows:

> .... in northern Pennsylvania. ... Wisconsin drift is parent material for Grey Brown Podzolic Soils and Sol Brun Acide; the drift overlaps red, deeply weathered, older drift in which boulders and cobbles are altered to clay. The pre-Wisconsin soil is parent material for a Red and Yellow Podsol (Sweden Soil): southward from the drift border, this soil grades into yellowish brown, strongly acid friable loam about three feet thick overlying reddish brown, yellowish-red, or red-silt clay 3 to 7 feet thick containing pebbles which have been altered to clay.

In complex eluvial landscapes, such as that in northern Pennsylvania, where several types of soil occur within the same area of country, it is important for the landscape geochemist to have a general idea of the geological and pedological history of the area. But such descriptions are only a first step toward an understanding of the migration of chemical elements through the eluvial parts of the

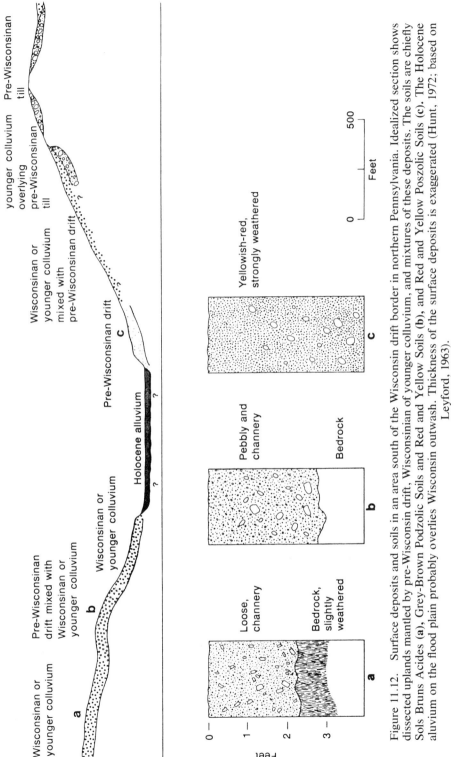

Figure 11.12. Surface deposits and soils in an area south of the Wisconsin drift border in northern Pennsylvania. Idealized section shows dissected uplands mantled by pre-Wisconsin drift, Wisconsinan of younger colluvium, and mixtures of these deposits. The soils are chiefly Sols Bruns Acides (a), Grey-Brown Podzolic Soils and Red and Yellow Soils (b), and Red and Yellow Poszolic Soils (c). The Holocene aluvium on the flood plain probably overlies Wisconsin outwash. Thickness of the surface deposits is exaggerated (Hunt, 1972; based on Leyford, 1963).

landscape. Consequently in complex landscapes better information on the migration patterns of elements during the formation of the landscape can often be obtained from lake, or (in some cases) peat cores taken from associated aqual, or superaqual landscapes rather than from eluvial landscapes.

Historical Geochemistry and Holism

Landscape geochemistry is concerned with the study of the migration rates of all chemical entities in all components of landscapes. Such migration rates are relatively simple to study in the atmosphere and in aqual landscapes. In eluvial landscapes the history of the formation of the soil as a function of the geology, climate drainage, and other variables all affect the migration rate of elements through the soil system. Consequently, it has been found in humid regions that bog cores or lake sediment cores are of greater importance in the description of historical geochemistry of a landscape than the detailed study of eluvial soil cover. One advantage of the landscape geochemistry approach is that, unlike soil science, it is not restricted to the study of eluvial landscapes but includes superaqual and aqual landscapes as well. Another advantage of our approach is that in landscape geochemistry the transitions between the different landscape types which may include marked geochemical gradients, or barriers may provide information on the history of the geochemistry of a landscape which could not be obtained solely from soil science, limnology, or plant ecology.

Summary and Conclusions

From the viewpoint of historical evolution geochemical landscapes are either simple or complex. Simple landscapes result from a single cycle of weathering and include no relict structures in their soils. If a simple landscape is developed upon a geological formation which is uniform mineralogically, chemically, and with respect to texture, such a landscape is called isogeochemical. Although they are rare, isogeochemical landscapes are of considerable potential importance in fundamental landscape geochemistry research.

In complex landscapes the study of the sequence of events during their formation is often very difficult—particularly if only the soils of eluvial landscapes are studied. In other landscapes the parallel study of the historical record in lake and river sediments (i.e., aqual or superaqual landscapes) is of considerable importance because it may provide detailed information on the migration patterns of elements and substances in time. Effects of changes in climate cannot be readily obtained from a study of soils.

From the viewpoint of landscape geochemistry (which is concerned with the general migration patterns of elements in landscapes) the geological substrate of the landscape is often adequately described by rather general broad terms. For this reason Polynov defined the terms orthoeluvium (from crystalline rocks), paraeluvium (from sedimentary rocks), and neoeluvium (from unconsolidated materials) to describe eluvium in relation to its origin. In this way properties of the lithosphere underlying a landscape are described in general terms which relate to the hydrology of landscapes regardless of the climate and the nature of the weathering crust.

Under favorable conditions technological time (e.g., in chronosequences), and geological time (e.g., in lake sediments and certain eluvial landscapes in the tropics) may all be measured on the basis of an examination of landscape components. Within each of these time scales significant changes in landscape conditions may be identified and dated. I conclude that the landscape components hold the keys to the understanding of the historical development of the landscape.

Discussion Topics

1) Are Polynov's theories of weathering outdated?

2) Would the detailed study of the geochemistry of landscapes where there are chronosequences facilitate the simulation modeling of particular landscape types?

3) Are there any landscapes in which the historical development sequence cannot be studied? Why?

12. Geochemical Landscape Classification

"The purpose of any classification is to so organize our knowledge that the properties of the objects may be remembered and their relationships may be understood most easily *for a specific observation.*"

Marlin G. Cline, "Basic Principles of Soil Classification," *Soil Science,* vol. 67, 1949, p. 81.

Introduction

This chapter deals with the most holistic concept of formal landscape geochemistry—the geochemical classification of landscapes—which follows from the action of the geochemical flows (as described in Chapter 8) during the historical evolution of a landscape (as described in Chapter 11) which together produce a geochemical landscape. The other concepts of landscape geochemistry also contribute to the classification process. For example, geochemical gradients and geochemical barriers are of considerable interest in relation to specific geochemical landscape types.

A Geochemical Classification of Landscapes

Sukachev and Dylis (1964, p. 43) stressed the difference between their concept of biogeocoenose (Chapter 4) and the concept of the geochemical landscape as described by Perel'man:

> 'A geochemical landscape' may be 'defined as a paragenetic association of combined elementary landscapes, linked together by migration of elements and adapted to a single type of mesorelief.' Therefore, watersheds, slopes, valleys, lakes are not separate sections of territory, isolated in nature, but closely linked interdependent part of one whole. . . . This quotation [from Perel'man] is adduced because it stresses the exchange of matter and energy between biogeocoenoses.

Hence the major theoretical and conceptual difference between the biogeocoenoses concept and that of a geochemical landscape is the stress placed on the migration of elements and energy between different components of the landscape by the latter.

Basic Principles of the Geochemical Classification of Landscapes

Perel'man (1966) described the principles of geochemical classification of landscapes in some detail. Because the rate of migration of materials in landscapes varies seasonally, the classification is based on the most rapid rate of migration characteristic of a particular landscape. Perel'man (1966) noted that summer migration rates are characteristic of taiga landscapes in contrast to subtropical grey earth steppes where the ephemeral plants grow during the wet spring season. He indicated that although there are not two completely identical landscapes in the world this is not a drawback—except that one has to distinguish between systematic landscape classification features and nonsystematic ones. In taiga landscapes a single taxonomic rank includes the very common type of acid landscape as well as the very rare type found in the vicinity of weathering sulfide mineral deposits. Although most fundamental landscape geochemistry research is carried out in common landscape types much useful information may also be gained from the study of rare types—for example, hot springs, mineral deposits, or isogeochemical landscapes. In general, the closer a landscape component is to the daylight surface the greater is its importance both in element migration and taxonomic significance.

Perel'man (1966) divided landscapes into two general series: abiogenic ones in which physical, mechanical, and chemical weathering are paramount and biogenetic ones in which the living biosphere plays a significant role in the migration of the elements. Four "landscape groups" are distinguished from each other by the type of circulation of the atmospheric migrant elements within them: (1) wooded landscapes (biogenetic); (2) meadow and steppe (biogenetic); (3) tundra (biogenetic); and (4) primitive deserts (abiogenetic). This classification is similar to that used by ecologists based upon the morphology of plants in ecosystems although the geochemical classification is typified by different general criteria. For example, from the geochemical viewpoint tundra is typified by abundant water and soils lacking in oxygen and carbon dioxide exchange, whereas deserts are typified by abundant oxygen, a lack of water, and a relatively slow circulation of carbon dioxide.

General Characteristics of the Four Landscape Groups

In landscape geochemistry the eluvial landscapes are always used as the basis for general classification and superaqual landscapes are considered subordinate to them. Consequently, Perel'man (1966) stressed the importance of eluvial landscapes with respect to the four landscape groups. Salient points of his classification scheme are as follows:

1. Wooded landscapes. Perel'man (1966) noted that in forests there is a great biological accumulation of organic matter which far exceeds the annual primary production. Such landscapes require decades to reach the mature state and involve two kinds of biogeochemical cycles: (a) annual cycles of leaf fall and related activity and (b) whole tree cycles which are much longer. In these land-

scapes the mineral elements (calcium, magnesium, phosphorus, etc.) may be removed from the soil in the woody parts of the trees for a relatively long time. In forests the main life forms are angiosperms and gymnosperms which produce their own microclimate with elevated levels of carbon dioxide and water vapor. In such areas most of the organic matter is above the soil surface.

2. Meadow and steppe landscapes. Meadow and steppe landscapes have no reserves of living matter and the total amount of living organic matter is usually not more than 30–40 tons/ha. The annual accumulation of organic matter in this type of landscape may equal, or exceed, that of forests although, because the majority of plants are annuals, the elements are kept out of circulation for short periods only. In these areas angiosperms predominate and have little effect on the microclimate. Organic matter tends to accumulate below the Earth's surface in these landscapes.

3. Tundra landscapes. Tundra landscapes have only a few tons of living matter per hectare and a low annual production. The biological circulation rate is low and the life forms of the plants are relatively simple (e.g., many mosses, lichens, and algae, although some higher plants do occur). Such landscapes accumulate dead organic matter below the water table as peat.

4. Primitive desert landscapes. Primitive desert landscapes are landscapes which have less biological production and circulation of elements than tundra. Included in this group are stone deserts, dried up salt lakes, and crags coated with desert varnish.

Much research is required before the general patterns for the circulation of each migrant element in each landscape group is worked out, although, eventually, such circulation patterns may be described by means of simulation models on the global, regional, or local scale of study.

Hierarchical Geochemical Classification for Landscapes

A hierarchical system for the geochemical classification of landscapes is shown on Table 12.1.

This system appears similar to the land classification systems discussed in Chapter 4 but may be directly applied to any landscape on the Earth's land surface. Series and group have already been described. Now consider the more specific levels:

1. *Landscape types* are distinguished on the basis of plant cover. The criteria used include the total mass of living matter, its chemical composition, and the characteristics of the biogeochemical cycling which occurs. From the viewpoint of geochemistry each plant species has a distinctive type of biogeochemical cycle with respect to the whole plant and with respect to the litter fall. Similarly, each plant community has a distinctive biogeochemical cycle which varies from landscape type to landscape type. Agreement between a geobotanical (plant ecological) classification of plant cover and a geochemical classification of landscapes follows because "the type of plant cover is a special form of biological circulation

Table 12.1. Basic Taxonomic Units for the Geochemical Classification of Landscapes.[a]

Hierarchical Level	Name	Criteria for Distinction
I	Series	Form of motion of matter (physical, chemical, biological) related to element migration in landscape
II	Group	Biological circulation of air migrants, relation of the total mass of living matter to annual production, organism types involved in biological circulation
III	Type	Biological circulation of air migrants, annual production of living matter, decomposition rate of remains of organisms
IV	Family	Living matter production within type
V	Class	Typical elements and ions of water migration
VI	Genus	Rates of water circulation and mechanical migration
VII	Species	Secondary aspect of migration (to be defined on a local basis)

[a] From Perel'man (1966).

of air migrants or a special type of chemical reaction (i.e. response) in the landscape'' (Perel'man, 1966, p. 155). Within a landscape group there may be a number of distinct landscape types, e.g., black earth steppes, subtropical steppes, and alpine meadows all belong to the same group.

2. *Landscape families.* A more detailed description of the plant cover types found within a landscape reveals the presence of landscape families of associated plant and soil cover. In Russia these are often associated with latitude, for example, the northern, middle, and southern taiga types. Usually not more than four of five families are defined within a given landscape type. Families, like landscape types, are distinguished not wholly on morphology but also include reference to the rate of circulation of air migrant elements within them.

3. *Landscape classes.* Within a given family the geochemical differences are due to the variation in the rate of circulation of the water migrants (Perel'man, 1966). Each landscape class has a definite kind of matter migration which is distinguished by the predominant ion species present. Perel'man gives as examples the hydrogen class, the calcium class, and the sodium class which occur when these elements become the dominant ion present which may also be typomorphic. He noted that in the southern taiga family hydrogen, calcium, and hydrogen–iron classes of landscapes exist. In the hydrogen class the soils and water are acid, the soil fertility is low, and the waters are rich in colloids, while domestic animals tend to suffer from calcium deficiency. In the calcium class the soil fertility is higher and the reaction of the waters is neutral or slightly alkaline. Where there are hydrogen–iron waters there are accumulation of peaty material and the biological productivity is low (Perel'man, 1966). In general, within any landscape family, there are usually no more than 10 landscape classes. Landscape classes are quite large taxonomic units and may be further subdivided into genera and species.

4. *Landscape genera.* Mechanical migration and water circulation rates vary within landscapes and the different genera within a landscape are based on such

differences. Perel'man (1966) recognized two extreme types of genera: those in which there is little mechanical migration where the water circulation is slow and where movement of ions in solution is important and those in which there is considerable mechanical migration (e.g., washouts) where the water circulation is rapid both above and below the surface. In general, the genera of landscapes are distinguished on the basis of topography and, consequently, relate directly to the concept of catenas in pedology.

5. *Landscape species.* More subtle changes in the migration rate of elements in landscapes are used to distinguish landscape species. For example, the landscape morphology may not be affected by the migration of elements derived from a mineral deposit although this may give rise to a different species within the landscape. Landscape species may be related directly to rock type changes which in turn affect the chemistry and rate of waters flowing through them. Physical and chemical disturbance of landscape by man may also give rise to different species of landscapes, for example, in the vicinity of a major highway or around landfills.

General Comments on the Perel'man Geochemical Classification of Landscapes

Geochemical classification of landscapes is currently in its infancy and the ideas just described are quite preliminary. The merit of the Perel'man approach, which is based directly upon the ideas of Polynov before him, is that it provides relatively simple criteria for the classification of landscapes as an alternative to those usually employed in land classification. By emphasizing the role of different chemical elements in landscape formation together with the rates at which they migrate in relation to the geochemical environments and the morphology of landscapes the Perel'man approach becomes more flexible as a descriptor of the environment than schemes based upon the static features of landscapes alone. It is the difference between basing a description of the geochemistry of a watershed on static materials (e.g., 100 soil samples taken at random according to a statistical design) and on dynamic materials (such as stream sediments) which represent mixtures of minerals, organic materials, and elements which result from dynamic processes in the landscape. One might say that the geochemical approach to landscape classification is more concerned with the migration of elements within the landscape to the exclusion of attention to details of morphology—in contrast to the normal approach to land classification which stresses morphology often to the exclusion of the dynamics of epigenetic processes.

Perel'man (1966) considered a geochemical landscape to be a paragenetic association of elementary landscape types (eluvial, superaqual, and aqual) which occur together within an area. Using the classification scheme just described it is possible to distinguish between landscapes of low contrast where all elementary landscapes belong to the same class and landscapes of high contrast in which eluvial and superaqual landscapes belong to different types or groups. For example, Perel'man (1966) noted that eluvial landscape in dry steppe belongs to Group

2 (meadow and steppe) while a subordinate landscape of dried-up salt lakes belongs to Group 4 (primitive deserts). Perel'man (1966) also discussed the problem of conjunction of landscapes; if the eluvial landscape is used as a basis of classification and the landscape has developed along with it the landscape is said to have perfect conjunction. But, a river, a flood plain, and a delta are conjugated to the slopes of the valleys through which the river runs in its lower reaches as well as the whole basin upstream. The water in such a river may then be considered as an independent unit in a geochemical classification of the catchment area as a whole because in its lower reaches the river water reflects the geochemistry of the river as a whole—not just the area of country through which it flows at the time. For example, Holmes (1971) discussed the salinity of the Murray River in Australia as measured at 24 stations along its length (Table 12.2) spanning 1385 miles. He listed data for 2 years—one of which had normal rainfall in the catchment area (1966–1967) and the other which was a drought year (1967–1968). The salinity was found to be much greater in the drought year compared with the normal year and, as a result of saline drainage from irrigation areas, the salinity

Table 12.2. The Mean Salinity of the Murray River along Its Length for the Years 1966 and 1967 and 1967 and 1968.[a]

Station	River Mileage	Total Dissolved Solids in ppm	
		1966–67	1967–68
Below Hume Dam	1385	37	37
Cobram	1197	35	39
Torrumbarry Weir	1020	65	60
Barham	945	122	75
Pental Island	903	116	161
Swan Hill	875	224	211
Tooleybuc	818	185	189
Boundary Bend	761	138	178
Euston Weir	692	139	176
Red Cliffs	564	151	225
Lock 9	479	199	328
Lake Victoria	—	254	342
Lock 6	388	260	400
Chowilla Homestead	381	299	na
Berri	327	324	498
Lock 3	268	335	560
Waikerie	238	381	690
Morgan	199	375	738
Blanchetown	171	414	744
Walker's Flat	129	397	753
Mannum	93	383	756
Murray Bridge	70	399	822
Tailem Bend	55	391	776
Goolwa	8	759	2993

[a] From Holmes (1971).

increased markedly in some parts of it, for example, between Torrumbarry Wier and Swan Hill (Table 12.2). The improvement of the water quality between Swan Hill and Boundary Road was caused by dilution flow from tributaries. From lock 9 downstream saline flow into the river resulted from a combination of escape drainage from flow irrigation waters, artificial flows of groundwaters caused by head differences of the river levels at wiers and control structures, and natural groundwater flow into the river (Holmes, 1971). This example indicated the complexity of water chemistry which may occur along a single river passing through lands used for agriculture. In this case the salinity of the river water often may not relate directly to the surrounding catchment area at all. This example focuses attention upon one of the problems of the geochemical classification of landscapes which are considerably modified as a result of man's activities.

With respect to terminology for the description of specific geochemical landscapes Perel'man (1966) suggested that local names should be used to distinguish geochemical landscape genera and species. This has the distinct advantage of short-circuiting cumbersome theoretical terminology while at the same time allowing for the description of type areas for different kinds of landscapes which occur together within a region. Such type areas may be compared later on on a local, a regional, or, in some cases, a global basis directly without reference to other levels in the classification.

Glazovskaya's Geochemical Classification of Landscapes

So far we have considered only the general principles of geochemical landscape classification as described by Perel'man (1966). A more detailed conceptual scheme for landscape field mapping was described in a classic paper by Glazovskaya (1963). Her classification (Table 12.3) is based upon the relationship between landscape types as described by Polynov (Fig. 2.6) and the kinds of geological substrate upon which they are developed.

Like the modern paleopedologists, Glazovskaya (1963) distinguished between two types of landscapes: (1) those originating with a single weathering cycle she called homogeneous landscapes, and (2) those which contain relict structures from previous weathering cycles she called heterogeneous landscapes. The main classification scheme (Table 12.3) includes a somewhat complicated, but logical, nomenclature for different kinds of landscapes as they are mapped in the field. A detailed example of how this nomenclature is applied is described in Chapter 18.

Glazovskaya (1963) divided eluvial landscapes into four kinds: (1) those which are truly eluvial (i.e., the tops of hills), (2) transeluvial landscapes on the upper parts of slopes, (3) eluvial accumulative landscapes in the bottoms of valley slopes, and (4) accumulative landscapes which occur in valley bottoms where the layer of accumulated material is deep. She divided superaqual and aqual landscapes into two groups: those which have running water and those where the water is stagnant.

A diagram showing the relationships of the daylight surface and the water

Table 12.3. Classification Pattern for Elementary Landscapes by Type of Migration of Chemical Elements.[a]

Landscape Groups, after B. B. Polynov	According to Type of Geochemical Integration	According to Migration Cycles of Elements in Original Rocks		
		Primary (ortho)	Secondary (para)	Superimposed Secondary (neo)
Eluvial	Eluvial (flat tops, well drained ancient plains)	Orth-el (ortho-eluvial on massive igneous rocks)	Par-el (para-eluvial on dense sedimentary rocks)	N-el (neo-eluvial on loose sediments)
	Trans-eluvial (upper parts of slopes)	Trans-orth-el (trans-orthoeluvial)	Trans-par-el (trans-paraeluvial)	Trans-n-el (trans-neoeluvial)
	Eluvial-accumulative (parts of slopes and dry gulleys)	Orth-el-ac (trans-orthoeluvial-accumulative	Par-el-ac (trans-paraeluvial-accumulative)	N-el-ac (neoeluvial-accumulative)
	Accumulative-eluvial (local confined lower ground with deep groundwater table)	Orth-ac-el (ortho-accumulative-eluvial)	Par-ac-el (para-accumulative-eluvial)	N-ac-el (neo-accumulative-eluvial)
Superaqual	Trans-superaqual (trans-hydromorphic)	Trans-hydr-oth (trans-ortho-hydromorphic)	Trans-hydro-p (para-trans-hydromorphic)	Trans-hydro-n (neo-trans-hydromorphic)
	Superaqual (confined lower ground with weak water exchange)	Hydr-orth (ortho-superaqual)	Hydro-p (para-superaqual)	Hydro-n (neo-superaqual)
Subaqual	Trans-aqual (streams, flowing lakes)		Trans-aqual	
	Aqual (stagnant lakes)		Aqual	

[a] From Glazovskaya (1963).

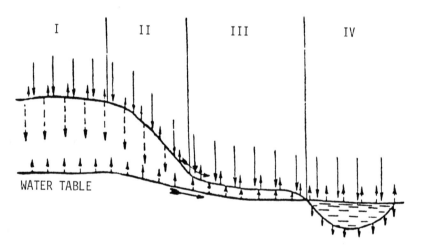

Figure 12.1. Diagram showing the flow patterns for waters in four elementary landscapes (**I**) Eluvial landscape, (**II**) transeluvial landscape, (**III**) superaqual landscape, and (**IV**) aqual landscape (see Table 12.3) (Glazovskaya et al., 1961).

table to the flow patterns of ground and surface waters in a hypothetical landscape section (Figure 12.1) provides a convenient starting point for the discussion of the application of Glazovskaya's geochemical classification of landscapes. Figure 12.2 is a conceptual model similar to Figure 12.1 except that it includes reference to both the hydrological and geological conditions and indicates general relationships between geology and Polynov's terminology for the substrate of landscapes.

Figures 12.3 and 12.4 are similar conceptual models drawn to illustrate detailed application of the terminology for geochemical landscape description included in Table 12.3. Figure 12.3 relates to an area underlaid by igneous and massive sedimentary rocks whereas Figure 12.4 relates to an area of unconsolidated sediments. Glazovskaya (1963) also described terminology for the description of landscape developed in areas where more than one weathering cycle has been active. In heterogeneous landscapes of this type the terms used become quite complicated but could readily be shortened for field use (Table 12.4).

In her 1963 paper, Glazovskaya provided several examples of the migration patterns of elements in landscapes in relation to landscape types. For example, in Figure 12.5 vertical distribution patterns for a number of elements in soils collected in two related landscapes are shown. Clearly, position in the landscape is directly related to the behavior of the elements concerned. A more complicated example was illustrated by Fortescue (1974b) (Figure 12.6). A soil profile has developed during the past 10,000 years in a layer of glacial neoeluvium 0–2 m thick lying on a glacially paved surface of Pre-Cambrian orthoeluvium. The vertical distribution patterns have been plotted for lead zinc, copper, nickel, and

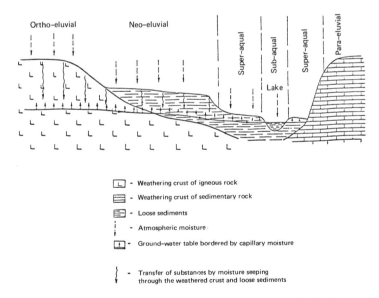

Figure 12.2. Relationships between topography, landscape type, hydrology, and geology illustrated by means of a conceptual model (Glazovskaya, 1963).

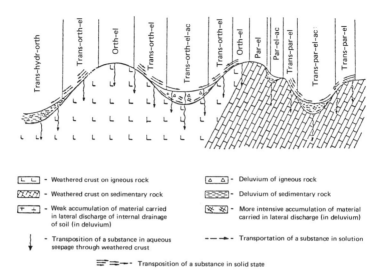

Figure 12.3. Conceptual model for the application of the terminology listed on Table 12.3 in an area of igneous and massive sedimentary rocks (Glazovskaya, 1963).

manganese extracted by a 12% hydrochloric acid solution from oven dried −80 mesh soil material taken from soil profiles located in each of 11 pits. The landscape section is located at the Montgomery Trail landscape site at the Petawawa Forest Experiment Station in Ontario, Canada, where the climate is cool and humid and the layer of glacial neoeluvium varies in texture from sand to silt in different parts of the section (Figure 12.6). In spite of this, each element was found to have a characteristic vertical distribution pattern which tended to be modified by the topographic setting of particular sample pits. In the case of nickel, an accumulation at the surface was followed by a decrease in content

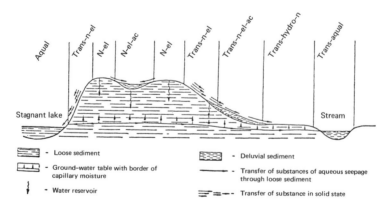

Figure 12.4. Conceptual model for the application of the terminology listed on Table 12.3 in an area of unconsolidated sediments (Glazovskaya, 1963).

Table 12.4 Genera of Heterogenous Landscapes in Relation to Geochemical History.[a]

	Current Evolution Phase: Eluvial	
	Former Evolution Phase	
Type of Eluvium	Eluvial Different from the Present	Superaqual
Orthoeluvium (on igneous rocks)	Superimposed orthoeluvial	Secondary othoeluvial
Paraeluvium (on sediments)	Superimposed paraeluvial	Secondary paraeluvial
Neoeluvium (on unconsolidated rocks)	Superimposed neoeluvial	Secondary neoeluvial

[a] From Perel'man (1966) based on Glazovskaya (1962).

below the surface humus layer and then a gradual increase in content with increasing depth. In general, the deeper the soil profile the greater the increase in nickel content (Figure 12.6). A modification of the vertical distribution related to a change in plant cover type was found in the case of manganese at Petawawa. At the top of the south facing slope (Figure 12.6) the coniferous forest cover is

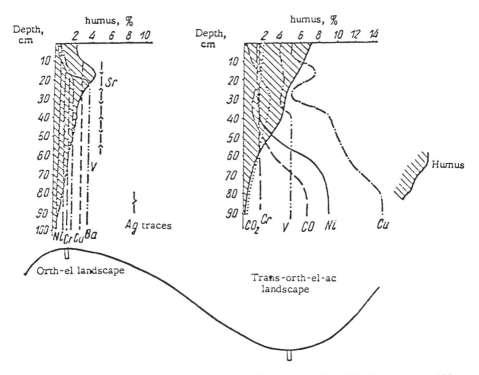

Figure 12.5. Vertical distribution patterns for elements in related landscape types (Glazovskaya, 1963).

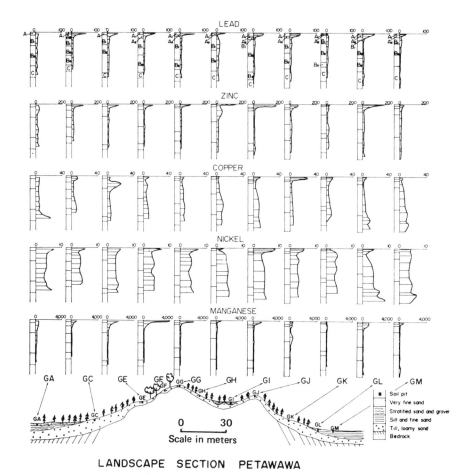

Figure 12.6. The vertical distribution of trace elements in soils taken along the Montgomery Trail, near Petawawa, Ontario, Canada (Fortescue, 1974b).

absent and in its place there is an oak *(Quercus rubra)* forest. The humus layer associated with the presence of the oak was found to be enriched in manganese—an element known to be accumulated by oak trees growing in the Petawawa area.

Glazovskaya's Conceptual Models for Landscape Features

Glazovskaya (1963) described additional concepts useful in the geochemical description of landscapes (e.g., she divided eluvial landscapes into four types):

1. Permaciedal landscapes in which the precipitation continually, or intermittantly, reaches the water table during the seasonal cycle.
2. Impermaciedal landscapes in which the precipitation goes below the zone of rooting but does not reach the water table during the seasonal cycle.

3. Illuvial landscapes in which the annual precipitation is less than the evaporation and the predominant movement of salts is upward toward the daylight surface.

4. Surface eluvial frozen landscapes which occur in cold climates where a frozen layer lies within the landscape just below the daylight surface.

These landscape types are shown diagramatically in Figure 12.7. In superaqual landscapes Glazovskaya (1963) distinguished between (1) areas with a preponderantly reducing environment, (2) areas with a preponderantly oxidizing environment, and (3) areas in which the conditions fluctuate with season.

Like Perel'man, Glazovskaya (1963) stressed that the classification of geo-

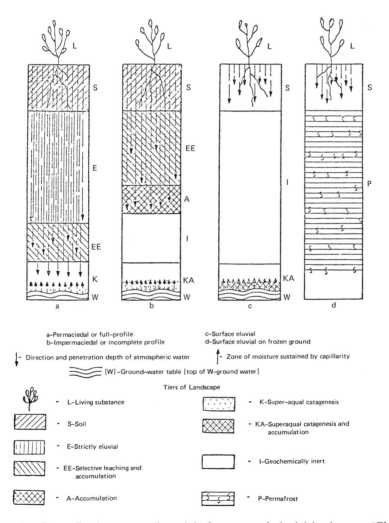

Figure 12.7. Generalized conceptual models for types of eluvial landscapes (Glazovskaya, 1963).

chemical landscapes is determined by the eluvial landscapes in a particular area of country, except for certain areas (such as extensive muskegs in organic terrain) in which eluvial landscapes may be exceedingly rare. She described the features of an ideal eluvial landscape as follows:

> Eluvial landscapes are produced on the more highly raised elements of relief, where the groundwater table lies far below the surface and exerts no influence on the vegetation or soils. Accessions of substance to an eluvial landscape occur only from the atmosphere (atmospheric precipitation or dust). There is no lateral inflow either with surface or groundwaters. The weathering crust, in an eluvial landscape has a residual character; it becomes impoverished with respect to all of the easily mobile elements in the course of its development; soils of an eluvial landscape are also leached to this or that depth, depending on the aqueousthermal environmental conditions. The passive (abiogenic) and the active (biological) accumulations of elements impede their leaching from an eluvial landscape. (p. 1404.)

Clearly this ideal situation, which is approached relatively closely in some arid, or semiarid climates, may be difficult to find in humid areas. Nevertheless, this concept is of value when one is searching for reference types of geochemical landscapes in a new area.

In another paper Glazovskaya (quoted in Perel'man, 1966) described how a series of conceptual models of geochemical landscapes may be described as an aid to mapping in complex terrain of taiga landscapes (Figure 12.8). In this case, attention is focused on changes in landscape features with increasing relief. Diagrams of this type are most useful, in combination with air photographs, for the training of personnel to carry out geochemical field mapping in a particular locality or region.

Glazovskaya (1963) also summarized the criteria for the description of the basic taxonomic units of elementary landscapes as shown on Table 12.5 and 12.6.

Table 12.5. Basic Taxonomic Units of Elementary Eluvial Landscapes.[a]

Rank	Name	Criterion[b]
I	Type	Hydrothermal conditions related to planetary migration of air migrants
II	Subtype	Degree of autonomy of landscape related to material gain and loss in solid and liquid form along relief elements
III	Class and subclass	Geochemical background and element migration conditions produced by mineralogical, chemical, and physical properties of the initial rocks
IV	Genus	Geochemical history of the landscape

[a] From Perel'man (1966) after Glazovskaya (1964).
[b] The criteria for distinguishing each rank is the relation between the biological circulation and the geological circulation due to the features stated.

Geochemical Landscape Classification

Table 12.6. Basic Taxonomic Units for Elementary Superaqual Landscapes.

Rank	Name	Criterion[a]
I	Type	Redox conditions and water salt content
II	Subtype	Degree of geochemical subordination
III	Class	Chemical composition of dissolved salts related to geological features of the salt catchment areas
IV	Genus	Geochemical history

[a] The criteria for distinguishing each rank is the relation between the biological circulation and the geological circulation due to the features stated.

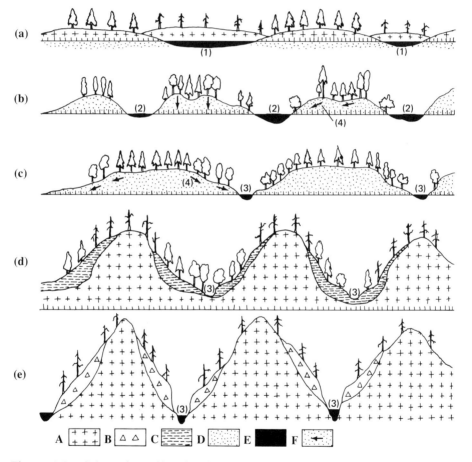

Figure 12.8. Schematic profiles of various local taiga landscapes (a) taiga+peat beds, (b) taiga+marsh and lake, (c) taiga-floodplain, (d) mountain taiga+valleys, (e) steep mountain taiga+valleys. (A) Bedrock, (B) gravitation drifts, (C) deluvial-solifluxion drifts, (D) moraines, (E) lakes, (F) groundwaters. (1) peat, (2) lake, (3) river, (4) groundwater (Perel'man, 1966).

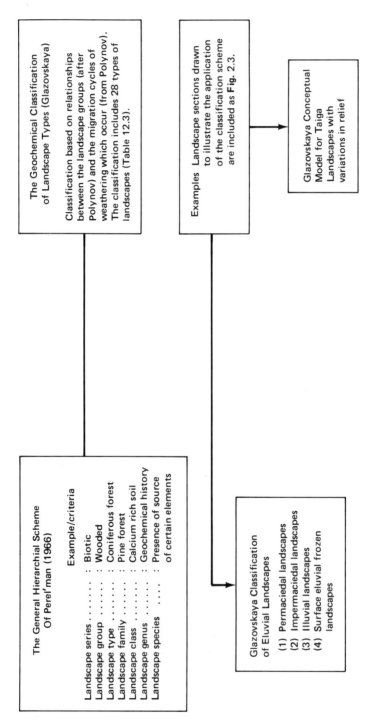

Figure 12.9. Summary of the schemes for the geochemical classification of landscapes and landscape components described in the text.

She lays greater stress than Perel'man (1966) upon relationships between the geological circulation and the biological circulation of elements in landscapes.

In the future, the concepts of geochemical mapping (combined with information from aerial photographs and remote sensing) may provide a firm basis for the preparation of maps which will be of value for the description of the circulation of elements in natural or man-modied landscapes based on simulations models.

Summary and Conclusions

This chapter has covered the most important and holistic concept in the discipline of landscape geochemistry. The geochemical classification of landscapes is still in an early, descriptive stage of development and has not yet reached the degree of sophistication of plant or soil cover type mapping. A summary of the different approaches to geochemical classification and mapping of landscapes described by Perel'man and Glazovskaya described in this chapter is included for reference (Figure 12.9). In general, the geochemical classification of landscapes still requires much research before its full scientific significance can be utilized in environmental geochemistry. The geochemical classification of landscapes provides a general conceptual foundation for the discussion of all other concepts of landscape geochemistry at the local, regional, or global scales of detail. Although the geochemical classification of landscapes has, as yet, received little attention outside Russia, it has potentially great value in environmental geochemistry. An indication of the potential of geochemical mapping in exploration geochemistry is provided by the worked example described in Chapter 18.

Discussion Topics

1) Is the Perel'man geochemical classification for landscapes, summarized in Table 12.1 suitable for all landscapes including those seriously disturbed by man's activities?

2) The establishment of a simple terminology for the Glazovskaya classification listed on Table 12.3.

3) The apparent importance of schematic profiles such as that in Figure 12.8 in the interpretation of geochemical anomalies in exploration geochemistry in humid areas.

13. Chemical Complexity and Landscape Geochemistry

"For the human race, classification is a natural and inherent, intuitive process; to create some semblance of order from an otherwise disorderly matrix by the pigeon-holing and catagorization of the matrix entities. But when it comes to vegetation and natural communities, can this process work in a similar manner to the classification of colors or species? Webb (1954), the eminent taxonomist, views the scene with a certain amount of exasperation and contends that until problems of unit diagnosis and delimitation have been solved all attempts at precise classificatory procedures are useless. Like the ordinary taxonomist the plant sociologist must select his characters for description and delimitation of units."

David W. Shimwell, *The Description and Classification of Vegetation* (Seattle, Wash.: University of Washington Press, 1972), p. 42.

Introduction

The purpose of this chapter and the three which follow it is to examine more closely each of the four aspects of the philosophy for environmental geochemistry (outlined in Chap. 3) and to discuss each aspect in relation to landscape geochemistry. In this chapter the role of chemical complexity in environmental geochemistry is related to three aspects of chemical complexity: (1) the hierarchy of chemical complexity with respect to chemical entities studies in the environment (see Table 3.2); (2) the accuracy and precision of chemical data which was listed under the heading of scientific effort on Table 3.2; and (3) the type of data base upon which chemical information is obtained. These three will be discussed in relation to eight examples, each of which is concerned with a particular level in this hierarchy of environmental goechemistry.

Example I. The Isotope Level of Complexity

The use of isotopes, particularly radioactive isotopes, for the study of the details of the migration patterns of nutrients, and other elements, in landscapes is well established. A simple experiment of this type was described by Kolehmainen *et al.* (1969) who described a tracer experiment with ^{131}I which was added to a small lake in Finland at a particular place and time. Briefly, 17 mCi of carrier-free ^{131}I was pumped from a boat into Lake Pitkannokanlampi (N 60 40', E 24 40') during a period of 1 hr and 40 min and then the content of the radioactive iodine was measured in the plants and fish of the lake and its waters during the following 41-day period. A map of the lake showing the depth curves and points where the samples were taken is shown in Figure 13.1. The study showed that peak radioactivity in all organisms was reached within 5 to 7 days. The effective half-life of the isotope in the lake waters was 6.5 days. Data for the decrease in radioactivity (due to ^{131}I) in plants and fish are shown in Figure 13.2a and b. Concentration factors calculated by comparing the peak radioactivity of the organism to the initial radioactivity of the water were sponge 200, green algae 200, moss 90, yellow water lily 60, crucian carp 25 (Kolehmainen et al., 1969).

In the terminology of landscape geochemistry, the study is descriptive, carried out at the isotope level of complexity at the local scale of intensity during a short period (i.e., 41 days) of ecological time. From the viewpoint of the discipline, the study was carried out in an aqual landscape and concerned the short-term BAC of the plants and fish. The data plotted on Figure 13.2 are an example of a geochemical gradient of ^{131}I in the organisms over time.

The chemical data would be classed as test data because no limits for decision or accuracy of the determination of radioactivity of the ^{131}I were given in the paper. This experiment is a simple example of a landscape study with data bases connected both in time and space. The data bases for the water plants and fish relate directly to the ^{131}I added to the lake as a whole, and the variation in content of ^{131}I measured in the organisms relates directly to the instant time of 1 hr and 40 min during which the ^{131}I was added to the lake.

This experiment is most effective from the viewpoint of environmental geochemistry and chemical complexity because: (1) it is concerned with a chemical entity (^{131}I) which was not present in significant amounts in the lake prior to the

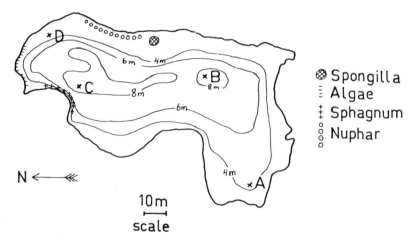

Figure 13.1. Map of Lake Pitkannokanlampi (N 60 40′, E 24 40′) in Finland used for the experiment showing depth curves and points where the water samples were collected (Kolehmainen et al., 1969).

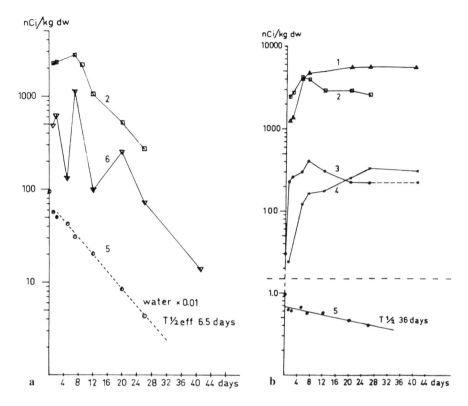

Figure 13.2. (a) The ^{131}I content in a green algae (Oedogonium sp.) (2) in a moss (*Sphagnum recurvium* coll.), in nCi/kg (6), dry weight and in the water 10^{-2} nCi/liters. (b) The ^{131}I content in a sponge (*Spongilla lacustris*) (1), in a green algae (*Oedogonium* sp.) (2) (3), in the yellow water lily (*Nuphar luteum*) (4), in the crucian carp (*Byprinus carassium*) in nCi/kg dry weight (5), and in the water in nCi/liter all corrected to the moment of labeling (Kolehmainen et al., 1969).

commencement of the experiment; (2) as a result of the design of the experiment the data were obtained from data bases connected in both space and time; and (3) the precision and accuracy of the analytical method used were likely to be high because to the type of instrumentation used to measure the radioactivity of the samples even though no precise information on the performace of the methods was given in the paper.

Example II. The Element Level of Complexity

There are two types of study which concern the level of abundance of elements in landscapes. The first (and more common) type is geochemical census which involves the collection of numerous samples of the same kind of material which are subsampled and analyzed chemically for the absolute abundance of particular elements on a weight percent basis. The data obtained are then either plotted on a map (or chart) and/or examined statistically for patterns within it. Such a study may be termed passive because it does not relate to any change in the chemistry of the environment studied. An example of an active geochemical census study is one in which a series of observations are taken (as in a passive study) and then sampling is repeated later after the chemistry of the environment has been changed (due to natural causes, or as a result of man's activity). If the repeats are made on a regular basis the procedure is called environmental monitoring.

Tidball and Sauer (1975) provided a good example of a regional level of geochemical census with respect to the concentration of total phosphorus in the surface horizon (0 to 15 cm depth) of agricultural soils of Missouri. A total of 1140 samples of soil were collected at the rate of 10 samples per county. The data were then analyzed statistically and plotted using a three-symbol system on a base map showing the outlines of the counties (Figure 13.3). This map is one of a series of 32, each of which was plotted for a different element using the same general format and sample material. Geochemical maps of this type have numerous general applications in environmental geochemistry.

In relation to landscape geochemistry the approach described by Tidball and Sauer (1975) must be considered a first approximation even though it was planned using a statistical model and involves a study of 32 elements in each of the 1140 soil samples. It is a first approximation because no information was given on the partial abundance (e.g., "available" P), or speciation of the elements involved which is needed for studies of element migration. Also no attention was given to the details of the vertical distribution of the elements in the soils studied. But in spite of these limitations geochemical census studies of this type are of considerable interest from the viewpoint of regional environmental geochemistry particularly when the absolute abundance of elements in the soil cover is related to the abundance of elements in the underlying rocks.

Example III. The Partial Element Level of Complexity

The partial abundance of an element in a natural material is often of considerable practical significance, provided that the method of chemical analysis used to collect the information is suitable for the local conditions involved. Some idea of the care required to select such an analytical method may be obtained from the following example. Daughtry et al. (1973, p. 438) described the research involved in setting up a partial soil test method for plant available phosphorus in acid organic soils of North Carolina. The abstract of their interesting paper reads as follows:

> Studies with 16 soils ranging from poorly drained mineral soils high in organic matter to colloidal mucks were conducted to evaluate soil-testing methods for available P. It was found that P extracted in a single extract of water or dilute calcium chloride gave no indication of whether crops grown on these soils would respond to P fertilization. Phosphorus content of the twelfth extract of successive extractions with water was much better related to percent yield of crops grown on control plots of these soils.

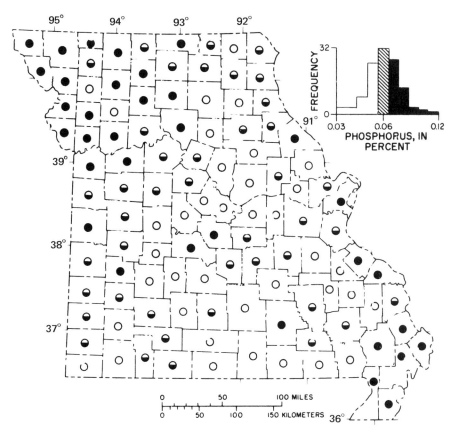

Figure 13.3. An example of a geochemical census element abundance survey the phosphorus content of the surface horizon of agricultural soils of Missouri. Map symbols express concentrations as county averages. Solid dots represent the highest of three concentration classes on the histogram, open circles the lowest. Values on the abscissa are the minimum, the geometric mean, and the maximum, respectively (Tidball and Sauer, 1975).

Dilute acid (0.05 N HCl + 0.025 N H_2SO_4) extractable P was the best indicator of potential P supply, accounting for 87% of the variation in percent yield for three greenhouse crops grown in these soils. Air-dried samples were superior to field moist samples because the drying process led to mineralization of organic P and this mineralization was reflected by increase in dilute-acid-extractable P. The increases in dilute-acid-extractable P upon drying were highly correlated with organic P mineralized during an incubation period ($r = 0.91$). Thus, extraction of air dried organic soils with dilute acid indirectly made a measure also of potential contribution of organic P to plant-available P. Correlations between plant growth and P extracted with dilute acid were much better when volume measurements of soil were used rather than when weighed measurements of soil were used.

Table 13.1 shows data for total P, organic P, inorganic P, dilute extractable P, P content at the twelfth extraction, soil pH, and organic matter. It was concluded that North Carolina soils with low (i.e., less than 15 lb/acre) phosphorus soluble

Table 13.1. Total P, Organic P, Dilute Acid-Extractable P, P Content of the Twelfth Extraction, Soil pH, and Organic Matter of Some North Carolina Soils.[a]

Soils	Total P (lb/acre)	Field-Moist		Soil Test (lb/acre)	Twelfth Equilibration (lb/acre)	pH	Loss on Ignition (%)	Apparent Density of Dried Samples (g/cm^3)
		Organic P (lb/acre)	Inorganic P (lb/acre)					
Inorganic								
1	259	163	57	7.2	0.8	4.1	11.1	0.83
2	305	176	62	4.8	0.5	4.0	12.7	0.82
3	504	377	107	8.4	0.9	4.0	18.2	0.75
Muck								
6	223	157	52	9.6	1.2	3.8	85.8	0.29
9	322	200	51	14.4	1.1	4.0	82.7	0.42
10	250	189	48	26.4	1.0	4.3	89.5	0.36
13	322	227	69	7.2	1.4	4.0	57.1	0.47
Reed bed								
5	640	500	138	42.0	7.6	3.5	57.2	0.39
11	516	448	98	15.6	2.3	4.6	41.7	0.57
12	678	500	182	30.0	3.0	4.6	45.3	0.54
Intergrade								
4	302	194	63	19.2	3.2	3.8	23.1	0.71
7	830	550	244	62.4	3.0	4.3	68.9	0.48
8	1040	557	400	201.0	7.8	5.4	22.8	0.73
14	647	380	258	45.2	1.8	4.8	42.2	0.59
15	458	385	87	22.4	1.9	4.1	63.7	0.44
16	682	338	418	57.6	3.1	4.7	27.1	0.69

[a] From Daughtry et al. (1973).

Table 13.2. Values of Dilute Extractable P under Varying Extraction Conditions.[a]

Soils	Dilute Acid-Extractable P (lb/acre)			
	P_{1w}[b]	P_{1m}[c]	P_1[d]	P_2[e]
Inorganic				
1	5.3	1.8	7.2	9.6
2	2.6	2.4	4.8	7.2
3	4.8	6.0	8.4	18.0
Muck				
6	5.6	14.4	9.6	15.6
9	4.7	9.6	14.4	25.2
10	11.0	9.6	26.1	28.8
13	3.0	9.6	7.2	24.0
Reed bed				
5	17.5	36.0	42.0	74.4
11	3.6	9.6	15.6	26.4
12	4.3	21.6	30.0	60.0
Intergrade				
4	17.1	16.8	19.2	26.4
7	18.5	40.8	62.4	88.7
8	65.5	145.0	201.0	199.0
14	16.0	36.0	45.2	74.4
15	8.4	12.0	22.4	26.4
16	20.9	40.8	57.6	98.4

[a] From Daughtry et al. (1973).
[b] P_{1w} = P extracted from a weighed, air-dry 5.0g sample using a 1:4 soil:extracting solution ratio for 5 min.
[c] P_{1m} = P extracted from a moist scooped 4-ml sample using a 1:4 soil:extracting solution ratio for 5 min.
[d] P_1 = P extracted from an air-dry scooped 4-ml sample using a 1:4 soil:extracting solution ratio for 5 min.
[e] P_2 = P extracted from an air-dry scooped 4-ml sample using a 1:8 soil:extracting solution ratio for 10 min.

in dilute acid contained relatively small amounts of both organic and inorganic phosphorus. Soil #3 (Table 13.1) was an exception because it contained high total phosphorus (540 lb/acre) and only 0.9 lb/acre of dilute extractable phosphorus. The mineral soils (#1, #2, and #3) and the muck soils (#6, #9, and #13) contained low dilute extractable phosphorus and the reedbed and intergrade soils varied from 15.6 to 62.4 lb/acre phosphorus (Tables 13.2 and 13.3). Soil #8 was not included in this grouping because of its extremely high level of labile inorganic phosphorus. In general, Daughtry et al. (1973) found that acid-extractable phosphorus values reflected those for total and inorganic phosphorus. Soil #4 was an exception with a relatively low total amount of total phosphorus (302 lb/acre)—probably associated with the low clay content in a soil composed of sand and organic matter.

Even after dilute acid-extractable phosphorus had been chosen as the most suitable technique, further information was required to determine the optimum conditions for the field test. Data from these experiments are provided on Table

Table 13.3. Amounts of Organic P Mineralized during Incubation and Increases in Organic P, Inorganic P, and Dilute Acid-Extractable P after 7 Days of Air Drying.[a]

Group	Soil Number	Organic P Mineralized (lb/acre)	Changes with Air-Drying		Increase in Dilute Acid-Extractable P (lb/acre)
			Organic P (lb/acre)	Inorganic P (lb/acre)	
Inorganic	1	11.5	−3	+8	6.0
	2	19.0	+1	+6	4.8
	3	23.0	−19	+15	9.6
Muck	6	11.5	−12	+6	4.8
	9	20.0	−24	+13	12.0
	10	5.5	−12	+8	7.2
	13	24.5	−24	+10	9.6
Reed bed	5	61.0	−44	+22	26.4
	11	39.0	−28	+12	13.2
	12	67.0	−38	+13	24.0
Intergrade	4	35.5	−40	+27	14.4
	7	40.0	−43	+16	20.4
	8	22.5	−40	+41	67.0
	14	100.0	−14	+24	26.4
	15	47.5	−10	+24	12.0
	16	77.0	−10	+4	24.0

[a] From Daughtry et al. (1973).

13.2. Briefly, it was found that changing the soil/solution ratio from 1:4 to 1:8 and the extraction time from 5 to 10 min gradually increased the amount of dilute acid-soluble phosphorus. However, this did not improve the correlation obtained between phosphorus extracted and either plant growth or phosphorus uptake.

This example was described in some detail because it demonstrates the numerous parameters which may require study in the choice of extraction methods for a particular nutrient element from a soil, particularly so that firm relationships between partial soil element abundance and plant availability can be established.

Example IV. The Persistent Chemical Level

Pesticide residues and PCBs are examples of persistent chemical substances which occur in the environment. One material in which such substances tend to concentrate is sewage sludge. If the sludges are then spread as a fertilizer on farmers fields these chemicals are a potential hazard to health of plants, animals, and man. A good example of a baseline study of the level of PCBs in sewage sludge was a study by Liu and Chawla (1976) who determined the PCB content of samples of this material from five sewage plants in Southern Ontario, Canada, between June 1973 and March 1974 (Figure 13.4). It was found that the sewage

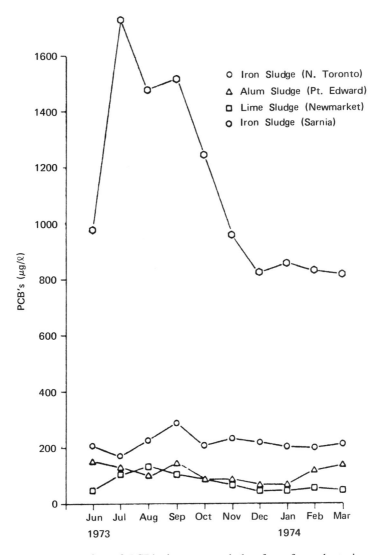

Figure 13.4. Concentration of PCB's in sewage sludge from four plants in southern Ontario during a 10-month period (Liu and Chawla, 1976).

sludge from three of the areas (North Toronto, Pt. Edward, and Newmarket) had below 300 μg/liter of PCBs during the 10-month period and each plant tended to have a characteristic mean content of PCBs regardless of season. The mean for North Toronto was 214 μg/liter, for Point Edward it was 108 μg/liter, and for Newmarket it was 74 μg/liter. In the sewage from the Sarnia plant (which is situated in an industrial area) a relatively high amount of PCBs (average 1122 μg/liter) were found which showed a marked higher seasonal trend occurring in July,

August, September, and October of 1973 compared with the other months studied. As Liu and Chawla (1976, pp. 248 and 259) stated:

> The resistance of PCB's to biodegradation in the aquatic environment is well known but their relative inertness toward sewage treatment processes is astonishing where bacterial population and activity are extremely high. The ability of PCB's to accumulate and concentrate in the digested sludge is probably due to their olephilic character which enables them to be absorbed or trapped by the hydrophobic groups of lignins or humic substances in the digested sludges. Most lignins have been shown to have ability to combine with the oleophilic groups of the lipid material.... This would probably explain why sludges have hundreds to thousands times higher levels of PCBs than the incoming raw sewage. The levels of PCB's and chlorinated pesticides in one Swedish sewage treatment plant have been studied and it was found that the sludge had 2700 times more PCB's and 330 times more DDT than the incoming wastewater.

Thus, the PCBs act as chemical entities in the environment which are stable to processes of chemical and biochemical degradation and which may be accumulated within landscape components without losing their chemical identity.

Persistent chemicals in the environment may pose a threat to man's well being particularly if they pass in significant amounts from soils to plants and then into foods. It is one role of landscape geochemistry to predict the behavior of such chemicals in particular landscapes and another role to provide information upon remedial measures required based on a knowledge of the environmental behavior of the chemicals concerned and a knowledge of geochemical barriers which may contain them naturally.

Example V. The Nutrient Level

Many chemical substances are used by man as foods for plants, animals, or himself. Such substances often contain trace impurities which may, or may not, be harmful to the organisms concerned; in some cases the food is refined which results in an increase in purity. A good example here is domestic sugar which is used as a food by man in either the raw or refined state. Hamilton and Minski (1972, 1973) provided information on the content of 34 elements in four samples of cane sugar (Table 13.4). Their data suggested that granulated sugar is a relatively pure substance in comparison with Barbadoes brown sugar which contains relatively high levels of a number of elements including potassium, sulfur, magnesium, calcium, silicon, phosphorus, and iron. This is an example of another level of chemical complexity in the environment—that of trace elements within chemical substances used as sources of foods.

Food analyses usually constitute disconnected data bases from the viewpoint of their origin (e.g., as in the case of the cane sugar) although some foods—such as vegetables—may reflect the chemistry of the environment in which man who eats them lives. However, once a food has been eaten elements in it either become a part of the organism which eats it, or occur in the wastes from that organism. In either case, the trace elements which are toxic will circulate in the environment. The classic case of such circulation is the element mercury, which,

Table 13.4. Semiquantitative Estimates for the Content of 34 Elements in Different Types of Cane Sugars[a] (Hamilton and Minski, 1972, 1973).

Element	Barbados Brown Sugar (μg/g dw)	Demerara Sugar (μg/g dw)	Refined Sugar (μg/g dw)	Granulated Sugar (μg/g dw)
Pb	0.2	0.02	<0.001	0.002
Ba	0.8	0.3	0.01	0.01
Cs	0.04	0.02		
I	0.03	0.003	<0.001	<0.001
Sb	0.08	0.006	<0.002	<0.002
Sn	0.1	0.03	<0.004	0.01
Cd	0.2	0.06	<0.007	<0.007
Ag	0.03	0.004	<0.001	0.002
Nb	0.01	0.003	<0.007	<0.007
Zr	0.03	0.004	<0.001	0.008
Y	0.07	0.006	<0.005	0.001
Sr	9	2	0.03	0.1
Se	0.2	0.03	0.005	0.007
Ge	0.1	0.03		
Ga	0.2	0.04		
Zn	3	0.5	<0.02	<0.02
Cu	3	0.3	0.4	0.08
Fe	49	8	11	0.1
Mn	15	2	0.02	0.01
Cr	3	0.4	0.02	0.08
V	0.4	0.02	0.002	<0.001
Ti	7	1	0.5	<0.007
Sc	2	0.1	0.001	0.003
Ca	1650	81	12	26
K	15000	1900	18	123
S	2040	500	5	3
P	250	25	3	0.2
Si	735	60	2	4
Al	0.7	0.90	0.02	0.007
Mg	1760	144	2	0.5
F	0.05	0.06	0.02	0.003
B	5	2	0.4	0.3
Be	0.03	0.006	0.002	0.0002
Li	0.1	0.03	0.001	0.0002

[a] From Hamilton and Minski (1972, 1973).

when present in trace amounts in fish or bread, may seriously affect the health of man.

Example VI. The Tissue Level

If an organism eats food grown locally and breathes the air from one place and drinks liquids obtained from one environment for a length of time some part of the organism may concentrate particular elements from the environment. Much scientific interest today focuses upon relationships between the abundance of

Table 13.5. Comparison between the Strontium Content of Human Bone and Water Quality in the United Kingdom.[a,b]

Area	ppm Sr (μg Sr/g ash) 2 + − 20y	Total Water Hardness[c]	Category of Hardness[c]
Invernesshire	229	3–160	S–MH
Wigton-Aspatria	166 ± 34 (A)[d]	12–60	MS
London-S.E. England	146 ± 6	294	SH–VH
Thames Valley	139 ± 6	150–350	MH–VH
Carlisle S. Fringe	124 ± 9 (A)[d]	12–60	MS
Carlisle	119 ± 8	65	MS
Stirlingshire	133 ± 8	30–55	S–MS
Lanarkshire	110 ± 7	25–50	S–MS
Glamorgan	110 ± 8	110–180	S–H
Devonshire	108 ± 9	60–250	S–H
Lancashire	107 ± 10	25–108	S–MH
S. Wales (Cardiff)	106 ± 6	110	S–H
S.E. Lancashire	105 ± 10	19–72	S
Blackpool, Fylde	104 ± 9	24–90	S–MS
Birmingham, Warwickshire	104 ± 5	100	S–H
Worcestershire	104 ± 28 (A)[d]	80–305	S–VH
Westmorland	99 ± 3	18–84	S–MS
Shropshire	94 ± 8 (A)[d]	240	SH–VH
N. Wales	93 ± 4	50	S–SH
Glasgow, Renfrewshire	89 ± 8	9–58	S–MS

[a] From Hamilton and Minski (1972, 1973).
[b] Values for total hardness refer to data reported for particular major water companies. Categories of hardness provide a general guide for the quality of water in various counties of the United Kingdom.
[c] S = Soft, 0–50 ppm; MS = moderately soft, 50–100 ppm; SH = slightly hard, 110–150 ppm; MH = moderately hard, 150–200 ppm; H = hard, 200–300 ppm; VH = very hard, >300 ppm.
[d] A = Adult bone.

elements in organs (or tissues) of man and the geochemistry of the environment in which he lives. For example, Hamilton and Minski (1972, 1973) made a comparison between the content of strontium of human bone and the hardness of waters in the areas from which the samples of bone were collected (Table 13.5). Hard waters, containing relatively high contents of calcium, strontium, and magnesium are characteristic of limestone/chalk terrains, while soft waters are relatively low in these elements. Hamilton and Minski (1972, 1973) concluded that high strontium and hard water are correlated in southeast England but not in northwest England where domestic waters are piped in from soft water sources. These writers also quoted earlier work by Mole (1965) who studied relationships between calcium and strontium in human bone. Hamilton and Minski (1972, 1973, p. 388) plotted data for the strontium content in human bone on a map of England, Scotland, and Wales showing outcrops of limestones and chalk (Figure 13.5) and concluded that:

> (1) High strontium in bone in southeast and northeast England is related to the distribution of strontium in calcareous rocks, i.e. chalk and limestones;
> (2) Two areas of non calcareous sandstone, one in central England, and the other

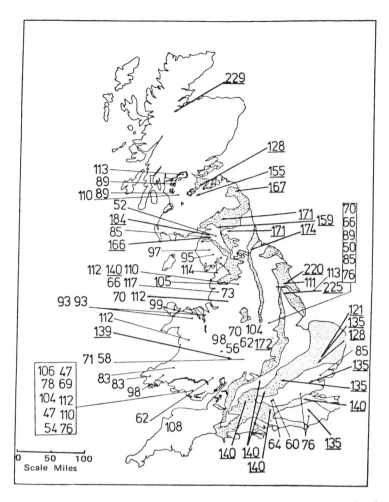

Figure 13.5. Possible relations between the concentration of strontium (μg/g of ash) in samples of ashed human bone and the distribution of calcareous rocks (shaded areas) in the United Kingdom (Hamilton and Minski, 1972, 1973).

in northwestern England, tend to be associated with high levels of strontium in bone. Both areas are near outcrops of calcareous rocks; and

(3) The level of strontium in bone from individuals residing in nonlimestone/chalk areas is low relative to that from calcareous regions.

These examples are included here because they illustrate a first approximation of relationships which may exist between environmental components. Pattern comparisons of this type at the regional scale of study may often suggest directions for more detailed environmental research which could be initiated in no other way. The links between calcareous rocks and waters and man indicated by Hamilton and Minski (1972, 1973), may be valid—although other alternative solutions to the problem of strontium-rich bones might be found (e.g., the breathing of dust direct from the atmosphere rather than the intake of strontium ions in waters).

Table 13.6. Trace Elements in Soils from Sites from Which the Fungi Were Collected.[a]

Element	Average, σ	n[b]	Range	After Bowen
As	22.4 ± 56	7	13.7–27.6	6
Br	15.2	2	13.8–16.6	5
Cd	1.37 ± 1.18	7	0.32–3.8	0.06
Cu	25.4 ± 8.3	7	15.5–40.9	20
Hg	0.20 ± 0.10	5[c]	0.075–0.33	0.03–0.8
Mn	813 ± 354	7	565–1425	850
Se	0.60 ± 0.16	7	0.43–0.83	0.2
Zn	126 ± 46	7	91–214	50
V	102 ± 50	7	38–176	100

[a] From Bryne and Ravnik (1976).
[b] n = the number of sites included in the average.
[c] Excluding Bela and Idrija sites.

Regional study involving the determination of specific elements in human tissues in relation to the distribution of the same elements in other materials is an aspect of landscape geochemistry which requires further study involving a definition of a series of successive approximations which may be used on a global scale.

Example VI. The Organism Level

In order to complete the discussion of the chemical complexity of landscapes let us consider the chemical composition of whole organisms. Bryne and Ravnik (1976) provided data for the distribution of ten trace elements in 27 species of higher fungi collected from sites in Slovenia, Yugoslavia. The data for the fungi were compared with similar data for the soils upon which the fungi were growing

Table 13.7 Data for Trace Elements in Fungi and Soil (ppm Dry Weight).[a]

Sample	As	Br	Cd	Cu	Hg	I	Mn	Se	Zn	V
Cemšenik										
1 *Macrolepiota p.* (cap.)	1.52	1.13	11.0	18.9	2.0	1.55	13.0	1.82	86.9	0.14
2 *Hygrocybe punicea*	0.29	2.40	39.9	9.37	0.48	1.71	11.2	0.62	79	0.33
3 *Coprinus commatus*	0.75	1.46	14.0	76.2	2.07	4.7	21.7	0.34	75.2	0.64
5 *Lycoperdon perl.* (spores)	3.80			160	1.43	3.88	39.0	3.09	192	0.21
Soil	27.0	13.8	3.84	40.9	0.22		488	0.56	160	58.5
Smlednik										
8 *Amanita phall.* (cap)	0.39	34.2	1.55	27.4	0.18	0.64	31.3	0.33	133	0.42
	0.25[b]			18.1	0.13		31.8	0.26	106	0.58
7 *Amanita phall.* (stalk)	0.10			8.80	0.08		32.3	0.19	79	0.74
9 *Amanita musc.* (cap)	1.83	70.6	12.6	28.5	2.07	1.56	19.7	3.24	132	138
	1.18	124	9.78	18.7	1.45	1.42	16.55	2.15	95.5	115.5
10 *Amanita musc.* (stalk)	0.53	178	6.95	8.92	0.82	1.27	13.4	1.06	59	93
Soil	26.6	16.6	0.32	26.8	0.25		690	0.55	106	85.4
Ljubljana, city center										
33 *Agaricus camp.*	0.35	1.86	1.38	221	14.1		9.8	4.16	75.0	0.09
Soil	13.7		0.72	30.5	0.33		365	0.60	214	38.4

Table 13.7. (Cont.)

Sample	As	Br	Cd	Cu	Hg	I	Mn	Se	Zn	V
Dvor. Dolenjska										
55 *Macrolepiota p.* (cap)	3.9		5.72	225	6.0		7.77	2.9	91.4	0.097
39 *Boletus edulis*	1.04	19.6	3.5	12.4	0.89		77.0	19.8	67.4	1.94
40 *Lactarius piperatus*	0.21		10.3	59.6	1.62		45.2	0.50	75.9	1.17
41 *Lactarius volenus*	0.14		6.3	516	0.079		30.7	0.69	52.7	0.69
61 *Cantharellus cibarius*	0.16	2.5	0.62	35.5	0.038		25.7	0.04	65.0	0.32
Soil	15.7		0.92	22.1	0.071		1425	0.43	95	99.4
Kureščck										
18 *Macrolepiota p.* (cap)	1.17		7.24	127	3.52		11.5	2.39	381	0.19
19 *Lactarius deliciosus*	0.37	2.78	4.6	15.6	0.75		7.5	0.71	93.6	0.18
24 *Amanita musc.* (cap)	1.0	88.0	14.1	27.7	1.07		17.3	4.21	147	199
	0.73	*142*	*10.8*	*20.9*	*0.81*		*13.6*	*2.85*	*106*	*201*
25 *Amanita musc.* (stalk)	0.46	196.5	7.4	14.1	0.55		9.8	1.49	65.8	202
26 *Suillus bovinus*	0.20		0.53	6.18	0.14		12.5	0.49	66.3	0.22
27 *Lactarius torminosus*	0.45		1.30	13.6	0.54		6.67	1.28	120	0.07
28 *Ramaria pallida*	3.66		9.70	24.2	0.81		32.3	1.88	50.8	0.11
29 *Lycoperdon perlatum*	4.33	4.48	4.61	94.7	2.32		20.7	3.56	227	0.27
31 *Melanoleuca evenosa*	2.08		10.8	80.5	5.63		32.5	1.62	102	0.13
32 *Calvatia utriformis*	1.01		3.55	154	2.94		166	1.76	128	0.15
Soil around No. 18	24.3		2.3	24.0	0.18		1515	0.39	157	155
Soil around Nos. 19, 27	22.3			20.2	0.13		820	0.49	120	
Soil around Nos. 24, 25	17.6		0.76	11.2	0.13		890	0.45	86.6	167
Soil around Nos. 26, 29	22.6			14.7	0.11		810	0.38	133.5	
Soil around No. 28	34.4			32.4	0.16		917	0.46	123	
Idrija										
30 *Lactarius deliciosus*	1.07		1.58	32.2	37.6		9.16	0.18	84	0.33
43 *Russula cyanoxantha*	0.085		3.30	32.8	0.31		3.75	0.88	132	0.38
44 *Lactarius del.*	0.19		0.88	9.0	2.95		17.3	1.28	104	0.077
Soil Idrija 4	27.6		0.47	21.3	6.6		765	0.78	91	97
Bela										
11 *Hypholoma fasc.*	0.29	4.07	0.76	35.3	7.9	1.49	12.8	0.30	92.6	0.51
12 *Lycoperdon perl* (spores)	6.83	1.40	2.75	127	40.3	3.2	49.7	7.02	262	0.50
13 *Lactarius del.*	1.63	1.27	2.63	13.9	4.0	4.3	6.41	0.88	160	0.30
14 *Amanita pantherina*	0.42	78.4	5.15	22.9	12.7	1.67	6.73	2.0	110	0.17
15 *Sarcodon imbricatum*	0.35		3.1	35.5	18.7	11.0	5.49	1.69	168	0.13
16 *Cortinarius praestans*	0.81		4.16	45.5	35.0	2.0	7.59	1.03	123	0.19
17 *Cortinarius saturatus*	2.16		11.8	40.2	9.5		8.08	0.63	74.9	0.21
35 *Russula cyanoxantha*	0.065		2.15	47.8	8.6		11.9	0.43	43.8	0.054
36 *Collybia dryophila*	0.96		4.8	53	45.1		15.2	0.37	147	2.03
46 *Hydrium repandum*	0.85		0.53	17.0	2.83		30.8	0.07	57.4	0.31
47 *Cortinarius multiformis*	0.86		8.05	30.3	10.6		12.6	0.88	103.7	0.077
48 *Lactarius scrobiculatus*	2.6		0.59	8.4	2.84		25.3	0.58	59	0.24
20 Soil around No. 13	24.1			10.9	2.81		877	1.0	71.8	136
21 Soil around Nos. 12, 15	16.2		0.92	11.1	4.2		1185	0.90	81.9	
23 Soil around No. 16	17.3			12.8	1.50		924	0.81	93.9	
22 Soil around No. 17	30.2			23.6	0.55		883	0.60	113	217
38 Soil	20.8		1.29	14.0	4.03		1160	2.3	103	
Kolinska[c]										
66 *Boletus ed.* (NE Slovenia)	0.49	30.0	1.55	20.4	2.27		24.0	18.7	72	0.37
67 *Boletus ed.* (Serbia)	0.42	25.8	1.72	16.6	2.47		29.7	21.9	62	1.11

[a] From Bryne and Rovnik (1976)
[b] Values in italics represent the mean of stalk and cap values, and were used in averages.
[c] Commercially collected for food industry.

Table 13.8. Average Element Content of up to 27 Fungi Species from 7 Sites in Slovenia (ppm Oven Dry Weight).[a,b]

Element	Average, σ	n[c]	Range	Median
As	1.25 ± 1.5	38	0.065–6.8	0.80
Br	2.3 ± 1.2[1]	10	1.13–142	2.1
Cd	4.8 ± 3.9[2]	37	0.53–39.9	3.6
Cu	53.4 ± 57	38	6.2–225	33
Hg	2.2 ± 3.0[3]	24	0.04–45	1.4
I	3.1 ± 2.8	12	0.64–11	1.8
Mn	20.0 ± 15.2	37	3.8–166	13.5
Se	1.5 ± 1.4[4]	35	0.04–22	1.0
Zn	110 ± 66	38	44–381	93
V	0.40 ± 0.46[5]	36	0.05–201	0.23

[a] From Bryne and Ravnik (1975).
[b] The following accumulators are omitted from the averages: for Br the genus *Amanita*, for Cd *Hygrocybe punicea*, for Hg results from Bela and Idrija, for Se *Boletus edulis*, and for V *Amanita*.
[c] n = the number of samples of all species used for average (some duplication).

(Tables 13.6, 13.7, and 13.8). It is interesting that within this collection of fungi several element accumulator species were found. The generalized data shown in Table 13.8 takes this into account because species which were found to be accumulators of one of the following elements—bromine, cadmium, mercury, selenium, or vanadium—were not included in the table of average concentration levels. This difficulty is not uncommon when the level of many elements in a number of species of similar organisms is determined.

Another difficulty with the preparation of biogeochemical baselines of this type is contamination during the sampling process. For example, because of relatively high amounts of the elements of interest in fungi, Bryne and Ravnik (1976) decided that contamination from the environment is not likely to effect the results—except perhaps in the case of vanadium and manganese (Tables 13.6 and 13.7). It seems clear that fungi are of considerable interest to the study of heavy metals in the environment and the use of fungi as extractors of such elements form soils in polluted areas might be feasible.

The Relationships between Landscape Geochemistry and the Hierarchy of Chemical Complexity

In this chapter seven examples were described of different levels of chemical complexity at which environmental geochemical investigations may occur. These seven are, of course, a very small sample of different ways in which chemical entities in landscapes are studied, although they do represent common types of investigation.

In order to focus attention on relationships between the discipline of landscape geochemistry and the seven examples a table has been drawn up relating

Table 13.9. Relationships between the Basics of the Discipline of Landscape Geochemistry and Information Described for Seven Examples of Studies in Environmental Geochemistry Included in Chapter 13

Example	1 Element Abundance	2 Element Migration	3 Geochemical Flows	4 Geochemical Gradients	5 Geochemical Barriers	6 Historical Geochemistry	7 Geochemical Classification of Landscapes
I ^{131}I in the lake waters	1	4	3	3	1	3	1
II Total P in soil	4	3	2	2	2	2	1
III Partial abundance of P in soil	4	3	2	2	2	1	1
IV Persistent biochemical species study	4	4	2	4	4	1	1
V Trace elements in food	4	1	1	1	1	1	1
VI Strontium in human bone	4	3	2	3	3	4	4
VII Trace elements in fungi	4	1	1	1	3	1	1

each of the seven basics of landscape geochemistry to each of the seven examples (Table 13.9). Value judgments for the importance of each basic concept or principle to each example are indicated on a scale of 1 to 4 where 1 denotes little (or no) importance and 4 is representative of maximum importance.

Several conclusions can be drawn from Table 13.9. In the first place the concept of "Element Abundance" was, as might be expected, the most important in six out of the seven examples. Similarly, the geochemical classification of landscapes which is a concept unique to landscape geochemistry was of least importance in the examples studied. Otherwise the importance of the concepts vary with the subjects studied; some, such as the trace elements in fungi (Example VII), were of relatively little importance in landscape geochemistry in contrast to the study of strontium in bone (Example VI) which involved all seven concepts.

It is concluded that, although environmental geochemical investigations may be selected because they are concerned with one or more of levels of the hierarchy of chemical complexity which relates to the philosophy of environmental geochemistry, the information involved is also of importance in relation to the basics of the discipline of landscape geochemistry as well.

Discussion Topics

1) Discuss the likely reasons for the value judgments listed on Table 13.9 and suggest which decisions should be altered to make the table conform with your experience in environmental geochemistry.

2) Rank the investigations in order of importance in (a) fundamental landscape geochemistry, and (b) applied landscape geochemistry and provide reasons for your decisions.

3) Suppose an example had been cited for metal ions migrating in soils from a mineral deposit in an arid landscape. Which of the basics would be of greatest importance and which of least importance to the example? If possible list ionic species for cations and anions and indicate if your scale of importance varies from ion to ion.

14. Scientific Effort and Landscape Geochemistry

"In most farming systems two different types of management policies can be recognized. First, there are short term or tactical policies which relate to the action(s) to be taken when confronted with a specific situation. Secondly, there are long term or strategic policies which relate more to the overall planning horizon of the farmer."

A. Wright, "Farming Systems Models and Simulation," in *Systems Analysis and Agricultural Management* J. B. Dent and J. R. Anderson, eds. John Wiley & Sons, Sydney, Australia, p. 17

Introduction

Scientific effort within a discipline, or disciplines, relates to one or more of a series of approximations each of which provides a more complete world picture than that which preceded it. In Chap. 3, four levels of scientific effort in environmental geochemistry were identified.

1. Descriptive and empirical
2. Descriptive and statistical
3. Descriptive by means of sytems analysis
4. Predictive by means of systems simulation models

In general, these four levels reflect the evolution of environmental geochemistry from a purely empirical and descriptive science to one which is increasingly predictive. This scheme is clearly a compromise, designed to stimulate thinking rather than a universal and comprehensive classification of all aspects of geochemistry. Readers who grasp the nature of each level and some of the relationships between them will be able to relate this simple classification to the methodology used in their specialties.

This chapter illustrates the methodology involved at each level. The reader is expected to consider the methodology employed to solve the problems as being more important than the subject matter. However, the examples have been chosen to be typical studies which frequently occur in both environmental geochemistry and ecology.

The Descriptive/Empirical Level of Scientific Effort

A distinction can be drawn between isolated and connected descriptive studies of the geochemistry of the environment. Isolated studies provide information on the abundance of elements (or some other geochemical parameters) in samples of a single environmental material collected at a single instant in time. This often involves methods of geochemical surveying or geochemical census. Connected studies result in the collection of samples of more than one material once, or many times, from the same locations. Chemical data obtained from such studies may be used to establish direct relationships between the geochemistry of components of the same or similar landscapes.

A simple example of an isolated study was described by Volk et al. (1975) who described the partial abundance of 16 elements in a series of soils from the Everglades Wildlife Management Conservation Area No. 3 in Florida. Briefly, surface soil samples (depth 0–15 cm) were collected with a soil core sampler and transported to the laboratory in plastic bags. Nine individual samples from each plot (each 91.5 m² in area) were air dried and mixed, and subsamples were taken for chemical analysis. The soil subsample mixtures were dry ashed at 435°C and the residue was dissolved in 5 N HCl and then evaporated to dryness prior to being redissolved in 1 N HCl. This solution was then used to determine element content by atomic absorption (Table 14.1). The data in Table 14.1 show the abundance of elements in soils underlying plots supporting different vegetation cover types in the Florida Everglades.

Table 14.1. Concentration of Elements Extractable by Hydrochloric Acid from Soils Collected in the Everglades Wildlife Management Area in Florida.[a]

	Location				
	Burned Saw-Grass Plot		Unburned Saw-Grass Plot		Muck-Burn Plot
Element	A	B	C	D	E
N (%)	3.29 ± 0.05	2.70 ± nd[b]	3.02 ± 0.01	1.16 ± 0.14	2.96 ± 0.08
Ca (%)	30310 ± 598	31030 ± 394	34180 ± 1270	16800 ± 1370	47240 ± 1250
Mg (%)	1400 ± 24	1080 ± 18	1420 ± 44	624 ± 46	1850 ± 36
K (%)	368 ± 13	344 ± 18	327 ± 8	242 ± 16	349 ± 14
P (%)	585 ± 8.4	463 ± 13	548 ± 15	374 ± 24	752 ± 25
Cu (%)	10.1 ± 1.9	6.3 ± 0.4	5.5 ± 0.2	3.5 ± 0.3	7.5 ± 0.3
Fe (%)	6780 ± 90	9940 ± 975	5500 ± 276	7520 ± 589	7790 ± 263
Mn (%)	297 ± 21	147 ± 25	218 ± 5	97 ± 12	241 ± 10
Zn (%)	12.4 ± 0.6	9.3 ± 0.9	12.4 ± 0.5	10.2 ± 2.3	12.7 ± 0.5
Na (%)	428 ± 14	442 ± 29	322 ± 16	209 ± 19	584 ± 18
Al (ppm)	2220 ± 169	6150 ± 204	2720 ± 270	4870 ± 410	2530 ± 105
Co (ppm)	4.8 ± 0.1	5.5 ± 0.3	4.3 ± 0.1	2.9 ± 0.2	5.5 ± 0.2
Cr (ppm)	4.6 ± 0.3	8.8 ± 0.3	5.0 ± 0.3	8.7 ± 0.6	5.0 ± 0.2
Ni (ppm)	6.8 ± 0.1	9.1 ± 0.3	7.1 ± 0.3	6.3 ± 0.5	9.1 ± 0.2
Pb (ppm)	27.5 ± 0.6	17.9 ± 0.7	21.1 ± 0.4	15.6 ± 1.2	25.3 ± 0.5
Sr (ppm)	210 ± 4	114 ± 0.3	219 ± 8	55 ± 4.5	300 ± 1
Ash (%)	11.9 ± 0.4	17.2 ± 0.6	13.2 ± 0.7	69.2 ± 3.3	14.0 ± 0.4
pH, moles/liter (1:2 soil–water)	6.7	6.2	6.2	6.2	7.0

[a] From Volk et al. (1975).
[b] nd, not determined.

These studies are important in themselves for the solution of the problem studied but, from the viewpoint of landscape geochemistry, tney are isolated incidents in both space and time. They cannot be related directly to similar studies carried out in other landscapes, largely owing to a lack of standardization in reporting of morphological data on the conditions of the landscapes from which the samples were collected.

An example of an isolated biogeochemical investigation was described by Weiner et al. (1975, p. 536). They reported upon the wet and dry weight concentrations of macronutrients and micronutrients in whole-body samples from 27 specimens of white-tailed deer (*Odocoileus virginianus*) shot at the Savanna River Plant site in Georgia. The methods and materials used were as follows:

> Wet- and dry-weight whole-body concentrations of six macronutrients (Ca, K, Mg, N, Na, and P) and five micronutrients (Cu, Fe, Mn, Mo, and Zn) were determined for 27 white-tailed deer shot at the Savanna River Plant. Carcasses (stomach and intestinal contents included) were frozen whole, sawed into smaller units, and ground in a meat grinder. Samples were then freeze dried and further homogenized in a Wiley mill, and three to six subsamples of the homogenate, each weighing approximately 250g, were analyzed for each deer. Chemical analysis was performed by arc spectrography . . . for all elements except nitrogen, which was measured by the Kjeidahl method (Jackson, 1956). Chemical analyses were conducted at the Georgia Soil Testing and Plant Analysis Laboratory in Athens, Georgia.

The data obtained from the investigation (Table 14.2) give information for the chemical composition of the deer and for the exportation of nutrients from the area by the sampling. This type of descriptive information is of interest particularly when it is related to other components of the landscape from which it was obtained. It may then be used as a basis for comparative studies or incorporated in sytems analysis (or systems simulation) models.

These examples belong to a large set of similar studies which are of general interest in landscape geochemistry. Because most of the data obtained from such investigations do not relate to a uniform set of environmental descriptors, it is often of little value in detailed comparative studies.

A simple example of environmental data which were collected with detailed comparisons in mind was described by Richardson and Lund (1975) for soil leachates from six plots located in the woodland areas of the Michigan Biological Station, Pellston, Michigan (Table 14.4). These writers described general features of the landscapes and soils studied as shown in Table 14.3 and the project as follows:

> The effects of clear-cutting of aspen were studied at three sites on the forest land of the University of Michigan Biological Station near Pellston, Michigan (latitude 45°34′N). The three well-drained upland sandy soil sites chosen were considered representative of the moraine and glacial outwash typical of millions of acres of aspen forests in the northern Great Lakes region. The soils of the research site were labeled good, intermediate, and poor on the basis of aspen production and are typical of areas where mature aspen stands (50 to 80 years old) have developed after logging and

Table 14.2. Average Whole-Body Nutrient Composition and Standing Crops for White-Tailed Deer and Exportation of Nutrients by Harvest of Deer at the Savannah River Plant.[a]

	Wet-Weight Concentration[b] (mean ± 2 SE)	CV[c]	Dry-Weight Concentration[b] (mean ± 2 SE)	CV[c]	Standing Crop[d]	Nutrient Loss[e]
Macronutrient	mg/g	%	mg/g	%	g/ha	g/ha/year
Nitrogen	31.2 ± 0.8	6.6	104.0 ± 2.3	5.7	222.5	16.2
Calcium	9.28 ± 0.76	21.3	30.9 ± 2.6	21.5	66.2	4.8
Phosphorus	6.78 ± 0.47	18.1	22.6 ± 1.6	18.3	48.3	3.5
Potassium	2.86 ± 0.15	13.8	9.53 ± 0.50	13.6	20.4	1.5
Sodium	1.16 ± 0.05	11.5	3.88 ± 0.19	12.9	8.3	0.60
Magnesium	0.27 ± 0.03	29.4	0.91 ± 0.10	28.9	1.9	0.14
Micronutrient	µg/g	%	µg/g	%	mg/ha	mg/ha/year
Iron	49.5 ± 3.8	19.9	164.5 ± 10.8	17.1	353	26
Zinc	20.6 ± 1.5	19.5	68.4 ± 4.7	17.8	147	11
Manganese	8.56 ± 1.23	37.3	28.5 ± 4.0	36.2	61	4.5
Copper	7.83 ± 0.53	17.5	26.1 ± 1.8	17.9	56	4.1
Molybdenum	0.91 ± 0.11	30.8	3.0 ± 0.3	29.9	6	0.5

[a] From Weiner et al. (1975).
[b] Concentration data are means for 27 deer.
[c] Coefficient of variation.
[d] Based on an estimated density of 0.16 deer/ha with a mean live weight of 44.54 kg/deer.
[e] Based on an estimated annual removal of 900 deer from the 780-km² Savannah River Plant.

repeated fires. Two homogenous study areas of approximately 1 ha each were located on each of the three soils. Homogeneity in terms of forest-species composition was determined by sampling all trees 5.00 cm (2.0 in) and greater in diameter at breast height (dbh). Understory and ground cover were analyzed in 20 randomly placed 9- and 0.25-m² plots, respectively. Slopes on all sites range from 0 to 6%.

In the fall of 1972, paired 0.5 ha subplots established inside each 1-ha study area were utilized for treatment comparisons. One of the plots on each site was commercially clear-cut, and the slash material was left. The other site remained undisturbed as a control. A 25-m buffer strip separated the control from the clear-cut areas. The nutrient inputs and outputs of the control and clear-cut subplots were monitored in terms of bulk precipitation and soil leachate.

Soil leachate from each of the sites was recovered with a series of pressure-vacuum lysimeters. The samplers consist of a one-bar entry value, porous, ceramic cup with a 1-µ size pore. The design and development of the lysimeter is described by Wagner (1962) and Parizck and Lane (1970). The control and clear-cut plots contain 20 and 30 lysimeters, respectively. Samplers were placed in pairs along five equidistant grid lines to uniformly divide the plots. Each lysimeter was placed in the upper portion of the C horizon (~90 cm deep). The lysimeters were installed at a 60° angle so that the porous ceramic tip would be positioned under an undisturbed profile.

Leachate was collected as soon as winter snow melt allowed. Samples were taken at monthly intervals or when sufficient leachate was available. All samples were frozen within 2 hr after collection, stored, and analyzed within 2 months. Cation analyses (Ca^{2+}, Mg^{2+}, K^+, and Na^+) were completed following standard procedures for atomic absorption spectrophotometry (Perkin-Elmer Corp., 1973).

Table 14.3. Physical Soil Analyses of Good, Intermediate, and Poor Aspen Sites in Northern Lower Michigan.[a]

Horizon	Sand (%)	Silt and Clay (%)	Depth (cm)	Bulk Density	Soil and Series Type	Physiography
			Good Soil			
A1	99.1	1.1	0–7	0.88	Menominee	Well-drained
A2	95.8	4.1	7–20	1.37	loamy sand	moraine
B	98.0	2.0	20–78	1.34		
C	85.0	15.0	78+	1.77		
			Intermediate Soil			
A1	96.9	3.1	0–5	0.62	Emmet	Well-drained
A2	94.8	5.2	5–20	1.47	loamy sand	outwash over
B	98.4	1.6	20–88	1.44		till
C	84.0	16.0	88+	1.83		
			Poor Soil			
A1	98.4	1.6	0–3	0.80	Rubicon	Well-drained
A2	98.2	1.8	3–10	1.51	sand	outwash
B	99.7	0.3	10–80	1.58		
C	99.8	0.2	80+	1.71		

[a] From Richardson and Lund (1975).

A three-way layout was used to analyze the effects of time, site, and treatment on nutrient concentrations of soil leachate. All leachate values given are means of at least 15 replicates per treatment plot. Statistical analyses follow Sokal and Rohlf (1969).

The mean concentrations of Ca^{2+}, Mg^{2+}, K^+, and Fe^{2+} in soil leachate for each site (control and clear-cut) during the 1973 growing season are given in Table 14.4. Considerable variation in leachate values among both treatment and control samplers at each site was found. A minimum of 15 and 20 individual replicate samplers were used per site during each month on the control and clear-cut plots, respectively. An analysis of variance showed no significant difference ($P<0.05$) between the control and clear-cut plots for all nutrient leachate values. A statistically significant ($P<0.05$) difference for month and site was found for all cations except sodium.

On the average, for the good and intermediate soils, the concentration of calcium leachate was 3 and 1.5 times as great, respectively, as that for the poor soil (Table 14.4). Iron leachate concentration was approximately 30% higher in the good soil than in the intermediate and poor soils (Table 14.4). The poor and intermediate soil leachate concentrations were only 25% of the Mg^{2+} leachate from the good soil (Table 14.4).

Seasonal trends for leachate vary with each nutrient. Two general peaks in leachate concentration were noted in the April and July-to-August sampling periods. The first reflected spring thaw, and the second may be accounted for by the above average rainfall in July (Table 14.4). (p. 538)

This example includes reference to many environmental parameters which were relevant to the study and which were considered when the geochemical aspect

Table 14.4. Mean[a] Monthly Concentrations of Four Cation in Soil Leachate from Control and Clear-cut Aspen Plots on Good, Intermediate, and Poor Soils in Northern Lower Michigan.[b]

Site and Treatment	March $\bar{X}$	SE[c]	April $\bar{X}$	SE[c]	May $\bar{X}$	SE[c]	June $\bar{X}$	SE[c]	July–August $\bar{X}$	SE[c]
Calcium (ppm)										
Good										
Control	6.62	0.50	10.42	9.23	9.17	1.76	9.93	1.97	13.56	2.72
Clear-cut	13.70	2.58	13.02	1.75	12.49	2.35	13.63	2.37	13.41	2.34
Intermediate										
Control	6.35	0.44	10.69	1.44	6.21	1.13	6.67	1.06	9.19	0.77
Clear-cut	5.99	0.53	9.27	0.77	6.37	1.13	5.62	0.97	6.94	0.80
Poor										
Control	6.06	0.59	6.65	0.53	3.45	0.41	3.35	0.47	5.21	0.50
Clear-cut	5.86	0.30	6.17	0.39	3.39	0.36	2.04	0.32	3.45	0.34
Magnesium (ppm)										
Good										
Control	2.26	0.25	3.54	0.70	3.25	0.76	4.66	1.24	7.84	2.98
Clear-cut	5.65	1.64	3.83	0.58	4.72	1.11	4.23	1.03	7.44	2.21
Intermediate										
Control	1.31	0.09	2.43	0.41	1.58	0.31	2.03	0.26	2.09	0.30
Clear-cut	1.10	0.13	1.80	0.11	1.16	0.16	1.36	0.14	1.09	0.16
Poor										
Control	1.15	0.07	1.38	0.08	0.89	0.08	0.95	0.08	1.01	0.09
Clear-cut	1.14	0.07	1.42	0.06	0.85	0.06	1.05	0.06	0.61	0.07
Potassium (ppm)										
Good										
Control	1.32	0.18	1.44	0.24	1.24	0.16	1.48	0.19	1.74	0.26
Clear-cut	1.78	0.21	3.83	0.45	1.57	0.19	1.85	0.31	2.09	0.28
Intermediate										
Control	1.57	0.11	2.44	0.28	1.72	0.20	1.91	0.17	2.23	0.14
Clear-cut	1.79	0.22	2.89	0.14	2.00	0.24	2.24	0.24	2.26	0.23
Poor										
Control	1.67	0.26	1.67	0.15	1.35	0.11	1.19	0.09	1.35	0.11
Clear-cut	1.40	0.10	2.00	0.10	1.20	0.13	1.09	0.10	0.94	0.08
Iron (ppb)										
Good										
Control	59.68	4.70	63.41	6.12	96.84	15.23	47.09	9.86	69.76	8.61
Clear-cut	63.73	7.20	63.32	8.33	83.17	10.37	59.73	5.05	63.38	4.97
Intermediate										
Control	60.14	9.18	56.63	4.97	86.14	17.19	27.48	2.80	41.80	4.33
Clear-cut	56.41	5.10	57.73	6.44	67.62	9.49	37.48	2.91	51.96	5.70
Poor										
Control	53.69	8.52	50.38	4.54	93.89	8.78	29.38	3.45	39.09	2.78
Clear-cut	55.07	4.60	47.15	3.83	64.94	6.21	26.30	3.14	44.47	3.32

[a] A minimum of 15 replicates were used in each calculation.
[b] From Richardson and Lund (1975).
[c] One standard error of the mean.

aspect of the work was planned and carried out. Experimental designs of this type could be described for landscape geochemistry which would be general enough to provide wide application and yet be relatively simple to complete in the field.

The Statistical Level of Scientific Effort

Environmental geochemistry, particularly in agriculture and forestry, has benefited greatly from the statistical (biometric) approach to the study of the behavior of elements and chemical substances in the environment. This aspect of the scientific effort hierarchy stems from the pioneer work of Fisher and others after 1910 [e.g., Fisher (1958)] who originally described factorial experiments to test the effects of chemical treatments upon the growth rate and yield of crops in agriculture. In these experiments fertilizer elements (singly or in various combinations) were added in treatments to plots containing plants of one or more plant species and then the yield from all plots was measured at harvest time. In such experiments the plots and treatments are allocated according to a statistical design and the experimental data are analyzed by descriptive statistical methods and a regression formula is derived. This type of experiment is essentially empirical because no detailed hypothesis is required to take into account underlying causes of variations before the commencement of the experiment. Constraints of this approach include the need for a large number of treatment conditions and often the difficulty of relating the results of the statistical analysis to environmental conditions of the experiments. A good example of this approach was described by Fuehring (1960) who carried out a factorial experiment involving the growth and mineral content of corn plants under 27 treatments. Five micronutrient elements (Zn, Mn, B, Cu, and Fe) were combined in each treatment which was applied to the plants for a 100-day period. A statistical analysis of the data obtained from this experiment revealed a second-order multiple regression equation which fitted the data fairly well and enabled Fuehring to show that as the amount of some elements increased over the deficiency range, other elements held constant in the environment and still others showed a decreasing concentration in the plant tissues, apparently as a result of dilution by growth (Fortescue and Martin, 1970).

A modern, more detailed example of the successful application of statistical analysis to a difficult environmental geological problem was described by Fu-Su Yen and Goodwin (1976). They described the use of discriminant statistical analysis to distinguish between very similar tuff beds in the Green River Formation in Utah. This approach was adopted only after the common methods for the identification of pyroclastic deposits, such as (1) ranges and model values of the refractive index of volcanic glasses; (2) the bulk chemical composition of ash or pumice including major and minor elements; (3) the characteristics of the phenocryst assemblage; and (4) the stratigraphic position, degree of weathering, thickness, and grain size of the rocks, all failed to distinguish between the different tuff beds in the Green River Area. The discriminant statistical analysis was based on the trace element content of biotities separated from samples of tuff. Only those trace elements known to occupy the octahedral and tetrahedral biotite lattice sites were chosen for study. Other trace elements, which occur as interlayer cations in the biotites, were not included because their content can change as a result of postdepositional alteration of the rocks.

The field area from which the samples were collected is indicated in Figure 14.1 and details of the chemical anlaysis of the biotites are given in Table 14.3. The difficulty of separating out the different biotites by inspection is apparent from Table 14.5. Table 14.6 lists the slope and correlation coefficients of several paired elements from the biotites.

Fu-Su Yen and Goodwin (1976, pp. 348–349) discussed the statistical interpretation of their data as follows:

> To be useful for correlation, the element ratios determined for a tuff layer must be constant throughout the entire lateral extent of the layer, and show sufficient differences from those of other tuff layers to allow distinction. The selection of element ratios with these characteristics was carried out using fundamental statistical methods.
>
> The greatest differentiation between tuff layers is obtained from pairs of elements whose concentrations in biotites vary inversely. Furthermore, if the paired elements

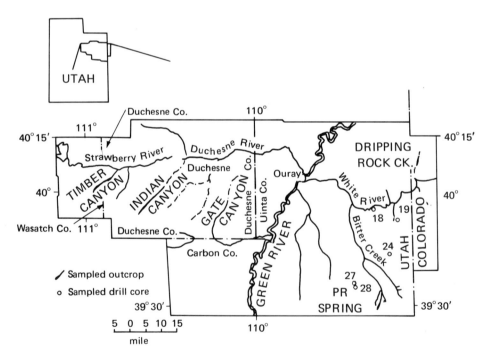

Figure 14.1. Index map of study area showing locations of sampled outcrops and drill holes (Fu-Su Yen and Goodwin, 1976).

Table 14.5. Minor and Trace Element Composition of Biotites from Tuffs of the Green River Formation.[a,b]

	Fe	Mg	Ti	Al	K	Na	Ca	FeO	MgO	TiO$_2$	Mn¹	Ni¹	Rb¹	Cr¹	Sr	Mn*	Ni*	Cr*
GE$_{11}$	13.2	7.53	3.06	5.55	5.40	0.10	2.32	17.2	12.5	5.1	1285	36	300	90	200	23.4	0.61	1.73
GE$_{12}$	13.0	7.73	2.80	5.51	5.10	0.04	2.20	16.7	12.8	4.7	1265	30	350	94	240	22.9	0.51	1.81
IE$_{21}$	15.7	5.48	2.48	5.34	8.40	0.32	0.65	20.2	9.2	4.3	1380	25	584	85	60	25.1	0.43	1.63
IE$_{22}$	15.6	5.38	2.52	5.26	8.32	0.32	0.66	20.0	9.0	4.2	1400	26	480	86	42	25.5	0.44	1.65
GE$_{23}$	15.1	5.48	2.58	5.14	7.25	0.38	1.12	19.4	9.2	4.3	1390	32	540	90	60	25.3	0.55	1.73
GE$_{21}$	15.2	5.38	2.62	5.64	7.92	0.83	0.83	19.5	9.0	4.2	1375	25	544	91	50	25.3	0.43	1.75
19E$_{25}$	15.0	5.40	2.62	6.46	7.61	0.65	1.10	19.3	9.1	4.2	1410	32	315	103	30	25.6	0.55	1.98
DE$_{26}$	15.4	5.63	2.60	4.80	6.04	0.55	1.04	19.8	9.5	4.4	1380	30	505	92	160	25.1	0.51	160
TE$_{31}$	14.2	6.27	2.24	5.70	6.80	0.30	0.98	18.3	10.4	3.8	1470	36	554	46	70	26.8	0.61	0.88
TE$_{32}$	15.2	5.73	2.26	5.92	6.10	0.10	0.56	19.5	9.5	3.8	1400	35	550	43	70	25.5	0.60	0.83
IE$_{33}$	14.1	5.27	2.24	5.55	6.84	0.84	0.84	18.2	8.7	3.8	1410	36	184	44	15	25.7	0.61	0.84
DE$_{31}$	14.3	5.27	2.18	5.60	6.65	0.64	0.50	18.4	8.7	3.7	1490	29	546	46	65	27.1	0.49	0.88
TP$_{11}$	15.4	5.65	2.65	5.70	7.70	0.77	0.76	19.9	9.4	4.5	1360	30	540	88	40	24.8	0.51	1.68
TP$_{12}$	15.0	5.73	2.71	5.50	7.40	0.64	0.66	19.3	9.5	4.5	1350	36	550	90	40	24.6	0.61	1.73
IP$_{13}$	15.1	5.63	2.60	5.12	7.63	0.15	0.78	19.4	9.3	4.3	1370	31	355	85	50	24.9	0.52	1.63
IP$_{14}$	14.5	5.57	2.70	5.38	7.32	0.44	1.36	18.6	9.2	4.5	1340	30	510	90	55	24.4	0.51	1.73
GP$_{15}$	15.2	5.94	2.62	5.30	7.11	0.44	1.12	19.5	9.8	4.3	1380	32	520	88	80	25.1	0.55	1.69
19P$_{16}$	14.6	5.76	2.74	6.41	6.79	0.72	0.59	18.8	9.6	4.5	1330	28	410	92	50	24.2	0.47	1.77
TP$_{21}$	16.7	6.20	2.48	5.04	7.10	0.56	1.26	21.5	10.3	4.1	1315	28	420	92	40	24.0	0.47	1.76
TP$_{22}$	16.8	6.10	2.54	5.10	7.18	0.84	1.28	21.6	10.1	4.2	1405	28	600	90	60	25.6	0.47	1.73
19P$_{23}$	16.8	5.73	2.68	5.72	7.82	1.04	0.50	21.6	9.5	4.5	1290	28	510	98	40	23.5	0.48	1.88
IP$_{31}$	16.6	5.81	2.46	5.36	6.06	0.70	1.46	21.3	9.6	4.1	1550	40	325	88	50	28.2	0.68	1.71
IP$_{32}$	16.6	5.69	2.54	5.38	6.04	0.60	1.64	21.3	9.5	4.2	1510	42	400	90	50	27.5	0.71	1.73
27D$_{11}$	13.1	7.01	2.82	6.26	8.00	0.82	0.84	16.8	11.6	4.7	938	40	330	112	80	17.1	0.68	2.15
24D$_{12}$	12.8	6.89	2.80	6.42	7.78	0.22	0.30	16.5	11.4	4.6	928	46	292	114	70	16.9	0.78	2.78
PD$_{13}$	13.4	7.45	3.00	5.70	7.00	0.65	0.93	17.4	12.4	5.0	900	54	404	110	70	16.4	0.91	2.12
27D$_{21}$	13.7	6.35	1.16	6.48	4.52	0.70	2.08	17.6	10.7	1.9	1780	64	610	126	130	32.4	1.10	2.42
28D$_{22}$	15.0	6.66	0.95	6.40	3.20	0.40	2.00	19.3	11.2	1.5	1900	58	400	100	90	34.5	0.98	1.92
181	13.2	5.00	1.96	5.56	4.94	0.64	2.06	17.0	8.3	3.3	1010	42	150	94	60	18.8	0.71	1.81
191	15.9	5.12	2.50	6.66	6.54	1.08	0.48	20.5	8.5	4.0	1000	38	304	124	50	18.7	0.64	2.39
272	16.6	5.14	1.04	6.44	4.26	0.32	0.92	21.4	8.5	1.7	1560	60	284	140	50	28.4	1.02	2.70
274	15.7	4.53	0.98	6.76	4.34	0.58	37.6	20.2	7.5	1.7	1880	94	620	106	170	34.2	1.57	2.04
275	18.5	4.28	1.06	6.50	4.86	0.32	1.18	23.6	7.1	1.8	1660	60	650	122	100	30.2	1.02	2.35
284	19.1	5.56	1.28	6.24	3.54	0.82	1.00	29.6	9.2	2.1	2020	70	530	150	90	31.8	1.20	2.89
286	17.2	5.16	1.96	6.52	2.86	1.38	0.75	22.2	8.6	3.3	3000	56	530	130	210	44.6	0.94	2.58

[a] From Fu-Su Yen and Goodwin (1976).
[b] Sample numbers consist of sample location abbreviations: I, Indian Canyon; T, Timber Canyon; G, Gate Canyon; D, Dripping Rock Creek; P. P. R. Spring; 19, a drill hole number (see Fig 14.2). Member abbreviations: E, Evacuation Creek; P, Parachute Creek; D, Douglas Creek. *, denoted values given in ppm equivalents; 1, denoted values in ppm. All other values as weight percent.

Table 14.6. Slopes and Correlation Coefficients of Several Paired Elements, from Biotites in Green River Tuffs.[a]

Elements	Elements	Slope	Correlation Coefficient
Fe + Mg + Ni	Cr + Mn + Ni	-3.26×10^{-3}	$-.3364$
Fe + Mg	Cr + Mn + Ni	-2.21×10^{-3}	$-.2484$
Mg	Cr + Mn + Ni	-7.14×10^{-3}	$-.6208$
Ti	Cr + Mn + Ni	-1.04×10^{-3}	$-.6282$
Mn	Cr ± Ni	4.92×10^{6}	$-.6852$
Mn	Mg + Fe	3.64×10^{4}	$-.0356$
Mn	Mg	-6.80×10^{4}	$-.6797$
Mn	Ti	-1.45×10^{1}	$-.5914$
Cr	Mn	-7.42×10^{2}	$-.5914$
Ni	Mn	-2.10×10^{2}	$-.5474$
Fe	Mg	-2.62×10^{4}	$-.3374$
Mn[b]	Cr + Mn + Ni	9.03×10^{4}	$-.9937$

[a] From Fu-Su Yen and Goodwin (1976).
[b] The highly correlated Cr = .9937 and positive slope relationship indicates a relatively low discriminant ability.

show comparatively low correlation and have a small variance within the biotite group from a single tuff layer, then the ratio among those elements will have even greater discriminant utility. Table 14.6 lists calculated slopes and correlation coefficients of several paired elements. Most element ratios of high correlation and positive slope are not listed.

Fu-Su Yen and Goodwin also provided a series of diagrams which include mean values for the various element ratios for each correlated tuff layer (Figure 14.2). The writers interpret the diagram as follows (p. 352):

> Mean values are at the centres of the crosses and the length of the cross arms show the range of one standard deviation about the mean. In general, the scatter diagrams show that ratios involving minor and trace elements in biotite have smaller ranges of standard deviation and greater separation of biotite groups. The poorest discrimination of all seems to be shown by the diagram FeO versus MgO (Figure 14.2f). The best discrimination is shown by the diagrams for Mn/Fe + Mg (Figure 14.2c) and Cr + Mn + Ni/Fe + Mg + Ti (Figure 14.2a). It should be noted, however, that the positions of the points in these diagrams are in approximately the same positions relative to each other and the consideration of Cr, Ni, and Ti in (Figure 14.2a) does little to improve the discriminant ability which already exists in Figure 14.2c. These diagrams suggest that statistical methods of discrimination using combinations of trace element ratios are necessary for complete distinction of one tuff unit from another.

Although somewhat complicated, this example describes a clear and successful solution to an extremely difficult problem in geology, and indicates the power of statistical methods to provide valid answers where purely descriptive geology, petrology, and geochemistry failed.

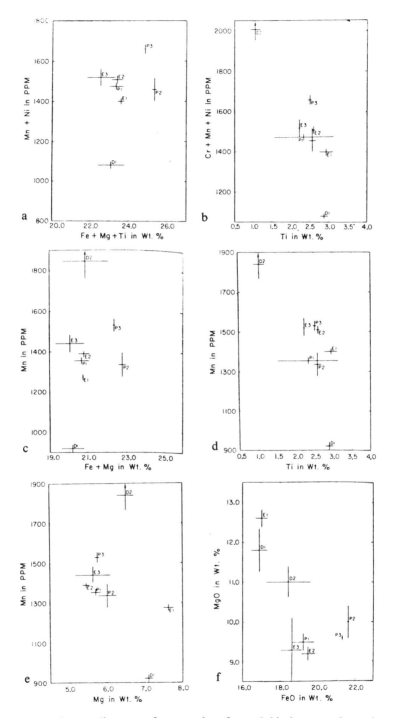

Figure 14.2. Scatter diagrams of mean values for each biotite group for various element pairs and groups. (Fu-Su Yen and Goodwin, 1976).

The Systems Analysis Level of Scientific Effort

Unlike the statistical approach, the mathematics of the systems approach is always based on a hypothesis derived from real-world processes. In general, the procedure first describes the system by means of a flow diagram (such as that shown in Fig. 14.4) which includes reference to all the different components of the system and the pathways which are possible for flows between the components. A mathematical model is then set up to mimic the system based on transfer equations devised to describe the flow of energy or materials between the components. The effectiveness of the model as a whole cannot surpass either the validity of its component transfer equations or the realism of the interactions between components upon which the model is based (Fortescue and Martin, 1970).

Let us consider a general ecological example of the application of the systems analysis approach. Webster et al. (1975) discussed in theory the concepts of "resistance" and "resilience" in relation to the cycling of nutrients in ecosystems and described a systems model to provide a hypothetical example of the relative importance of the two concepts in each of eight common types of ecosystems (p.4).

> The first aspect concerns the resistance of an ecosystem to displacement. An ecosystem that is easily displaced has low resistance, whereas one that is difficult to displace is highly resistant and is, in this sense, very stable. The second aspect of relative stability concerns return to the reference state, or resilience. An ecosystem that returns to its original condition rapidly and directly following displacement is more resilient, more stable in this sense, than one that responds slowly or with oscillation. . . . This view of ecosystems identifies two alternatives for persistence. Resistance to displacement results from the formation and maintenance of large biotic and abiotic structures. Resilience following displacement reflects inherent tendencies for the dissipation of such structure, but, because it is related to ecosystem metabolism, it also reflects rates with which structure is reformed following its destruction. In the closed biogeochemical cycles of the biosphere, the observable structural and functional attributes of ecosystems are determined by the realized balance between factors favoring resistance and resilience. Nutrient cycling, a fundamental process inherent in ecosystems, thereby becomes a central issue in the consideration of mechanisms of macroscopic relative stability.

They developed their argument with reference to eight different kinds of ecosystems:

> We suggest that three factors are involved in nutrient utilization in ecosystems: 1) the presence or absence of large abiotic nutrient reserves, 2) the degree of localization of nutrients within the biota, and 3) the turnover rate of the actively recycling pool of nutrients. Figure 14.3 schematically depicts these factors. In this figure a specific ecosystem type is associated with a given combination of factors. This conceptual scheme is clearly idealized since there exists a great range of each of these distinct types of ecosystems. However, this scheme is consistent with current ecological theory and represents a convenient method of examining relationships between nutrient cycling and stability.

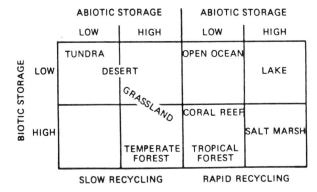

Figure 14.3. Alternative properties of nutrient cycles. Shown in each box is an idealized ecosystem type that seems to exhibit the indicated combination of properties (Webster et al., 1975).

Methods

To facilitate quantitative comparisons among these various idealized ecosystems, we constructed a general model of nutrient cycling (Figure 14.4). In this diagram the food base (x_1) may be either primary producers or detritus. Consumers (x_2) are organisms that feed directly on the food source. The $F_{3,1}$ is either death or mechanical breakdown of the food base to detritus (x_3). In an ecosystem with internal primary production, detritus is essentially dead primary producers (litter). In detritus-based systems this component is fine particulate organic matter. Decomposers (x_4) are those organisms which feed directly or indirectly on detritus. Available nutrients (x_5) are directly available for use in primary production. Nutrients in reserve (x_6) are not available but are tied up in sediments, primary minerals, clay complexes, or other refractory materials (e.g., humics). However, they may become available through transfer to x_5. Inflows and outflows occur primarily through the available nutrient pool.

We have quantified this general model for seven of the ecosystem types shown in Figure 14.7 (Table 14.7). We also applied this model to an idealized stream, which typifies an ecosystem without large abiotic reserves, with low biotic localization of nutrients, with little or no recycling, and with large nutrient throughflows. Standing-crop values were normalized to an available nutrient pool of 100 units. All transfers were per year. The values given in Table 14.7 are relative estimates that reflect differences among the idealized ecosystems, rather than exact, absolute estimates of nutrient transfers and standing crops. A variety of sources was consulted for each ecosystem type (Table 14.7). However, gaps and inconsistencies existed which were filled from general references or qualitative considerations. Each system was assumed to be at steady state.

From these numbers we derived several indexes which reflect structural characteristics of the eight ecosystems and which quantify the concepts of abiotic storage, biotic storage, and recycling (Table 14.8). Both the turnover time of the reserve ($T_6 = 1|a_{6,6}|$) and the proportion of nutrients localized in the two abiotic pools [$(x_5 + x_6)/\Sigma x$] are indexes of abiotic storage. Reserve turnover varies from slow in forests to fast in oceans and streams. The proportion of total nutrients in abiotic compartments is highest in temperate forests and lakes and lowest in tundra.

Biotic storage, given by the turnover time of biotic compartments [$(x_1 + x_2 + x_3 + x_4)/F_{1,5}$], is higher in terrestrial ecosystems and lower in aquatic ecosystems.

We calculated two indexes of recycling. The turnover rate of the detritus pool ($F_{1,5}/x_3$) is higher in aquatic systems and generally lower in terrestrial ecosystems, except for tropical forests where there is a rapid turnover of detritus. The ratio of

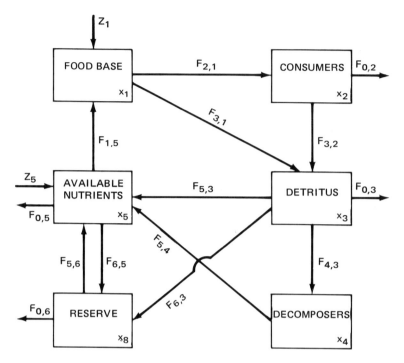

Figure 14.4. General nutrient-flow model of an ecosystem. x_i is the size of the ith compartment; z_1 is inflow to compartment x_1; $F_{1,5}$ is the flow from x_5 to x_1; and $F_{2,1}$ and $F_{3,1}$ are the outflow to the environment from x_1.

recycling to input ($F_{1,5}/\Sigma z$), as used in stochastic analyses, is approximately the inverse of the other recycling index. However, since systems with larger biotic pools typically recycle more nutrients than do systems with smaller biotic standing crops, this index partially confounds storage and recycling. This index ranges from 500 for grasslands to 0 for streams.

Two other useful indexes are the ratios of total standing crop to recycling material ($\Sigma x/F_{1,5}$) and total standing crop to total inflow ($\Sigma x/\Sigma z$). Both indexes estimate system turnover time. Longest turnover times occur in temperate forests and grasslands, whereas there is rapid turnover in stream and ocean ecosystems.

The specific values given in Table 14.7 have obvious deficiencies. Each idealized ecosystem represents a wide spectrum of actual ecosystems differing in many important characteristics. Similarly the kinetics of specific nutrients within a given ecosystem differ, quantitatively and qualitatively. In quantifying the general model shown in Figure 14.4, we have attempted to suppress such specific details and to focus instead on the alternative properties of nutrient cycles depicted in Figure 14.3. Our emphasis is thus on macroscopic properties of ecosystems rather than on specific differences between systems or nutrients. Comparison of the structural indexes (Table 14.8) with Figure 14.3 reveals that the chosen values agree well with the idealized conceptualization.

Table 14.9 shows that the eight hypothetical ecosystems, ordered from least to most resistant (largest to smallest ω_n), were stream, ocean, lake, tundra, salt marsh, tropical forest, grassland, and temperate forest. The four terrestrial ecosystem models were, on the whole, much more resistant than the four aquatic models. Analyses did not reveal such a clear separation of ecosystems with high and low resilience, nor did the eight systems differ as much with respect to the resilience aspect of relative stability as they did in relation to resistance. From least to most resilient (largest

Table 14.7. Summary of Relative Values Used in Quantifying the General Nutrient Cycle Model in the Eight Idealized Ecosystems Investigated.[a]

Parameter[b]	Tundra	Grassland	Temperate Forest	Tropical Forest	Ocean	Lake	Salt Marsh	Stream
x_1	200	500	100	500	10	10	1,000	500
x_2	15	50	0.5	2.5	10	1	25	50
x_3	200	1,000	25	5	10	25	1,000	10
x_4	20	100	1	1	0.5	25	100	20
x_5	100	100	100	100	100	100	100	100
x_6	100	1,000	5,000	1,500	50	2,000	50,000	1,000
z_1	0	0	0	0	0	0	0	1,000
z_5	1	1	1	1	110	100	75	100,000
$F_{2,1}$	20	100	1	5	500	20	100	200
$F_{3,1}$	30	400	5	46	545	180	900	800
$F_{3,2}$	20	100	1	5	500	20	100	190
$F_{4,3}$	50	480	5.5	49.9	50	180	500	300
$F_{5,3}$	0	10	0.4	1	900	10	400	600
$F_{5,4}$	50	480	5.5	49.9	50	180	500	300
$F_{5,6}$	1	10	0.6	1.1	10	20	1,000	100
$F_{6,3}$	0	10	0.1	0.1	50	10	50	0
$F_{6,5}$	1	0	0.5	1	20	10	950	100
$F_{1,5}$	50	500	6	51	1,045	200	1,000	0
$F_{0,2}$	0	0	0	0	0	0	0	10
$F_{0,3}$	0	0	0	0	45	0	50	90
$F_{0,5}$	1	1	1	1	5	100	25	100,900
$F_{0,6}$	0	0	0	0	60	0	0	0

[a] From Webster et al. (1975).
[b] x_i represents the size of the ith compartment; z_i is the input to x_i; $F_{i,j}$ is the flow of nutrients from x_j to x_i; and $F_{o,j}$ represents nutrient loss to the environment from x_j. All values are normalized against x_5, which was set to 100 units/unit area for each system.

Table 14.8. Index Summarizing Various Structural Characteristics of the Eight Hypothetical Ecosystems and Differentiating among the Properties of the Nutrient Cycles Shown in **Fig. 14.5**.[a,b]

System	Abiotic Storage		Biotic Storage	Recycling		System Turnover	
	T_6	$\frac{x_5 + x_6}{\Sigma x}$	$\frac{x_1 + x_2 + x_3 + x_4}{F_{1,5}}$	$\frac{F_{1,5}}{x_3}$	$\frac{F_{1,5}}{\Sigma z}$	$\frac{\Sigma x}{F_{1,5}}$	$\frac{\Sigma x}{\Sigma z}$
Tundra	100	0.31	8.7	0.25	50	12.7	635
Grassland	1,000	0.86	3.3	0.5	500	23.5	11,750
Temperate forest	8,333	0.97	21.1	0.24	6	870.93	5,226
Tropical forest	1,364	0.76	9.97	10.2	51	41.34	2,108
Ocean	0.714	0.83	0.029	104.5	9.5	0.173	1.64
Lake	100	0.97	0.305	8	2	10.80	21.6
Salt marsh	50	0.96	2.12	1	13.3	52.22	696
Stream	10	0.65	∞ (0.58)[c]	0 (99)[c]	0 (0.0098)[c]	∞ (1.70)[c]	0.017

[a] From Webster et al. (1975).
[b] x_i is the size of the ith compartment; z_i is the imput to x_i; $F_{i,j}$ is the flow of nutrients from x_j to x_i; $F_{o,j}$ is the nutrient loss to the environment from x_j; and 1_6 is the time constant of x_6.
[c] Since $F_{1,5} = 0$ for the stream, the indicated index was recalculated using the total loss from x_3 instead of $F_{1,5}$.

Table 14.9. Results of Relative Stability Analysis of Nutrient-Cycling Models for Eight Hypothetical Ecosystems.[a]

System	Critical Root	Mean Root	Natural Frequency	Damping Ratio
Temperate forest	−0.0001	−1.312	0.000227	1.2174
Grassland	−0.0001	−2.218	0.000228	1.1794
Tropical forest	−0.0003	−10.456	0.001039	1.2585
Salt marsh	−0.0013	−5.128	0.003898	1.1852
Tundra	−0.0015	−0.810	0.004413	1.1840
Lake	−0.0083	−9.718	0.02924	1.2954
Stream	−0.0999	−188.350	6.2947	1.4700
Ocean	−0.7678	−61.85	1.8478	1.1404

[a] From Webster et al. (1975).

to smallest ζ), the ecosystems were stream, lake, tropical forest, temperate forest, salt marsh, tundra, grassland, and ocean. This factor is tied to system characteristics (such as recycling) which do not differ strictly between aquatic and terrestrial ecosystems. Although several of the aquatic models were more resilient than most terrestrial ones, the lake model showed on the the smallest resilience values, probably related to slow turnover of the large abiotic storage pool. These results should be interpreted cautiously, in light of the data used in this analysis. Certainly the order-of-magnitude differences in the natural frequencies would seem to reflect real differences in the idealized ecosystems.

This example illustrates how the system approach can be used in modern ecology to discuss comparatively global problems which relate to ecosystem dynamics. The general ecosystem model (Figure 14.4) is a typical example of the box and arrow diagram which is the conceptual model of the system studied, and around which the mathematical model is structured. This level of study of the geochemistry of landscapes, particularly relating to the flows described by Kozlovskiy (1972) in Chapter 8, currently presents a challenge to students of landscape geochemistry.

The Systems Simulation Level of Scientific Effort

The starting point for systems simulation studies is a box and arrow diagram similar to that used in systems analysis. A relatively simple and clear example of the application of systems analysis and systems simulation to a problem in ecology was described by Dahlman and Sollins (1971) who studied the problem of nitrogen cycling in grassland ecosystems:

> Many factors influence the processes of nitrogen accumulation, component nutrition, seasonal dynamics, microbial transformation, and conservation in grassland soils. Because comprehensive empirical data on nitrogen cycling are unavailable for single ecosystems, results from different grassland studies were synthesized to formulate a generalized model of nitrogen dynamics for a native bluegrass ecosystem. A compartmental modeling technique was used to illustrate important transfers of nitro-

gen and also to identify substantial voids in literature data, for example, seasonal variations of microbial biomass and transformation rates. Nine state variables, twenty-three flows, and seven input-output transfers were modeled (Figure 14.5) from literature data; references from which parameters were calculated are listed in Table 14.10. All flows were described with linear functions, except for several constant inputs and several internal transfers (e.g., litter fall) approximated with annual pulse functions. In the absence of conclusive data on certain transformations (e.g., nitrification rate), linear fractional transfers were estimated on the basis of textbook descriptions of the respective processes. (p.71)

After describing the technical details of the computer programs and hardware used for the modeling procedure, the writers continued their account of the model:

The initial objective is to simulate accumulation of nitrogen as residual humus, a phenomenon which is characteristic of grassland ecosystems. That this was accomplished is illustrated in Figure 14.6, where nitrogen of the humus compartment (Q_{13}) increased even under circumstances of moderate humus mineralization (0.01 to 0.5%/year), denitrification, leaching and erosion loss, and small predator harvest. This simulation indicates that a moderate precipitation input and free-living biological fixation will account for nitrogen accumulation in a carbon-rich system. In a further test of the model's verity, the mineralization step ($Q_{13,5}$) was blocked, thus increasing immobilization, which is the more common situation in grassland systems. This made residual humus nitrogen unavailable to the system, and biological nutrition, which diminished by 75%, was maintained at this level by natural inputs and rapid turnover of

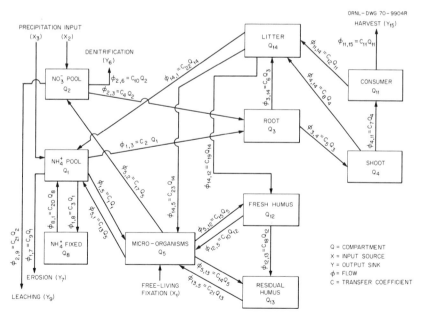

Figure 14.5. Compartments and flows of a generalized nitrogen cycle in grasslands (Dahlman and Sollins, 1971).

Table 14.10 Initial and Final Compartment Conditions for Simulation Model for the Nitrogen Cycle in Grasslands[a]

Donor Compartment	Fractional Transfer Coefficient
NH_4^+ pool (Q_1)	$C_{1,2,3,9} = 1.4, 0.32, 0.36, 0.1$
Nitrate pool (Q_2)	$C_{4,10,24} = 0.7, 0.2, 0.2$
Root (Q_3)	$C_{5,6} = 0.5, 0.4$
Shoot (Q_4)	$C_{7,8} = 0.2, 0.8$
Microorganism (Q_5)	$C_{13,14,15,17} = 0.12, 0.15, 0.6, 0.35$
NH_4^+ fixed (Q_8)	$C_{20} = 0.05$
Micro consumer (Q_{11})	$C_{11,12} = 0.1, 0.9$
Fresh humus (Q_{12})	$C_{16,18} = 0.05, 0.1$
Residual humus (Q_{13})	$C_{21} = 0.1$ or 0.001^b
Litter (Q_{14})	$C_{19,22,23} = 0.1, 0.1, 0.8$

[a] From Dahlman and Sollins (1971).
[b] No transfer in the event of substantially reduced mineralization.

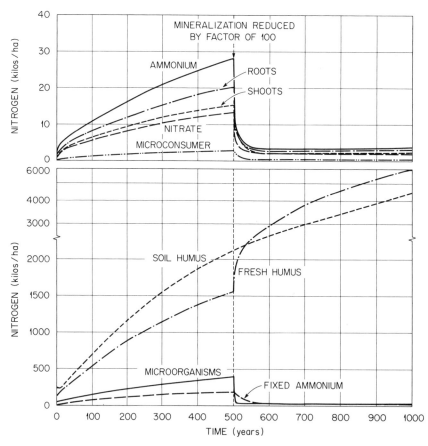

Figure 14.6. Simulated nitrogen accumulation by selected components of a bluegrass ecosystem based on designated transformation rates and system response to reduction of mineralization by a factor of 100 (Dahlman and Sollins, 1971).

more dynamic compartments. New equilibration occurred within 75 years. Although simulation oversimplifies transformations and environmental movements of nitrogen, this technique is useful for quantifying the principal flows of the nitrogen cycle. The generalized model also serves as a useful reference for evaluating subsequent impacts of other management practices, for example, increased harvest, grazing stress, fertilization, and potential detriments of biocidal treatments such as herbicidally induced reduction of nitrification.

Although the model in this case was put together from existing sources of information without the inclusion of field study data, it does suggest the potential of the systems simulation level of study of the biogeochemistry of ecosystem types. From the viewpoint of landscape geochemistry, the other uses for the model described at the end of the account by Dahlman and Sollins (1971) are of particular interest.

A second relatively simple example of a simulation study was described by Bartos (1973) for the forest succession which occurs in aspen forests of the central Rocky Mountains area of the United States. In this case, it is the biomass which is the subject of simulation without specific reference to nitrogen or any other nutrient. Nevertheless, this example provides some idea of the potential of systems simulation in forestry in general and in the area of about six million acres of aspen forests which occur in the western Rocky Mountains region in particular.

As in the previous case, the analysis begins with a box and arrow diagram (Figure 14.7) which indicates flows of matter which occur within the total system. In this case there are five major compartments: (1) aspen, (2) conifer, (3) shrubs, (4) annual herbage, and (5) perennial herbage. These components are all considered above ground biomass and are expressed as weight (or volume) per unit area. The three dominant processes in the model are growth, mortality, and regeneration and they are influenced by the various compartments involved. The flows between the components of the model were developed by expressing the informational flows via closed equations. This procedure is adopted during the initial stage of model building prior to the replacement of the fixed equations by actual data. Examples of the functions used to determine the scaling factors utilized in the development of the process equations are given in Figure 14.7b and the results for a 400-year simulation of aspen succession appear in Figure 14.7c. Both examples of systems simulation models are hypothetical but they do serve to describe in relatively simple terms the essence of the systems simulation approach to the description and prediction of landscape evolution. Although both relate to particular types of ecosystems, it is not difficult to see how models of this type could be linked together to describe and predict the circulation of chemical entities in areas of country where eluvial, superaqual, and aqual landscape type occur together.

Summary and Conclusions

This chapter includes examples chosen to illustrate the four levels of the hierarchy of scientific effort described in Chapter 3. In general, the sequence from

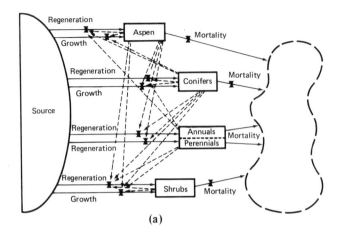
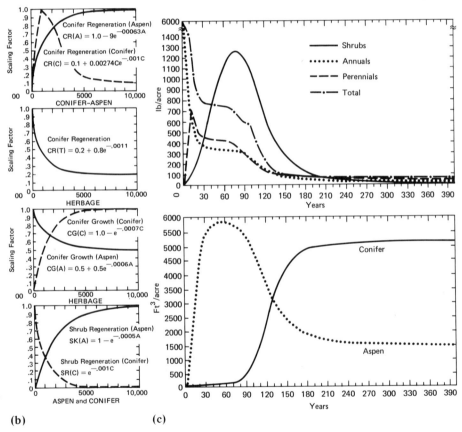

Figure 14.7. An example of the use of a systems simulation model. (a) Box and arrow diagram of aspen succession. (b) Examples of functions used to obtain scaling factors utilized in the development of process equations. (c) Results from a 400-year simulation of aspen succession (Bartos, 1973).

descriptive investigations of particular landscape components to simulation modeling of ecosystems (or landscapes) can be viewed as a series of successive approximations each requiring a different kind of scientific thinking. All have their place in landscape geochemistry, but the need for standardization of approach is greatest at the descriptive level so that information obtained from landscapes of the same general type can be related to similar information collected elsewhere. In order to do this it is not always practical to employ the detailed uniformity of constraints used by Richardson and Lund (1975). However, some uniformity of sampling of landscapes can be facilitated by the preparation of conceptual models prior to the commencement of field work, as are described in Chapter 18.

As in the case of Chapter 13, a table is included to emphasize relationships between the basics of landscape geochemistry and the subject matter of the examples cited in this chapter (Table 14.11). As in the previous two examples, a system of value judgments from 1 to 4 is used to indicate the relationships. In general, the analysis in Table 14.11 is similar to that in Table 13.7, indicating close relationships between the subject matter of the examples and the basics of landscape geochemistry.

I conclude that one way in which landscape geochemistry might develop in relation to the hierarchy of scientific effort is by the description of a series of type field studies each of which relates to some aspect of the hierarchy of scien-

Table 14.11. Relationships between the Basics of the Discipline of Landscape Geochemistry and Information Described for Five Examples of Studies in Environmental Geochemistry Included in Chapter 14.

Example	1 Element Abundance	2 Element Migration	3 Geochemical Flows	4 Geochemical Gradients	5 Geochemical Barriers	6 Historical Geochemistry	7 Geochemical Classification of Landscapes
I Element in Everglades soils	4	3	1	2	2	1	1
II Elements in whole deer	4	1	1	1	3	1	1
III Elements in soil leachates	4	3	2	1	1	1	1
IV Trace elements in tuff beds	4	4	1	1	1	3	1
V Nutrient cycling in ecosystems	4	4	4	3	3	1	2
VI Simulation model of tree growth	3	3	4	2	2	4	2

tific effort. These types could be cited as starting points or guides to later studies so that eventually a relatively small number of such experimental designs would evolve which would be used commonly by all students of the discipline of landscape geochemistry. As Perel'man (1966) suggested, these type studies could be identified by the localities within which they were first described so that the number of type studies may grow without the need for a normal and complex terminology and classification system to develop with them.

Discussion Topics

1) Formalization of the hierarchy of scientific effort for the systematic study of (a) eluvial landscapes, and (b) aqual landscapes.

2) The study of element migration rates in landscape geochemistry.

3) The establishment of valid and complex relationships between components of the same landscape.

15. Space and Landscape Geochemistry

"What matters about measurement is the ability of a metric to generate comparative data, given that certain conventions apply. The real difficulty in comprehending and accepting the approach to large scale decision making. . . . is not that the measures do not measure the quantities with which people are familiar, nor are they more or less exact. The problem is that the conventions under which the measures are taken and quoted are not the tacit assumptions of us all."

Stafford Beer, *Decision and Control* (New York: John Wiley & Sons, 1966) p. 495

Introduction

Chapter 3 noted that field investigations may occur at the local, regional, or global scale of detail; this chapter describes common ways in which studies of environmental geochemistry are carried out in space. Although the examples selected for inclusion are not formally involved with landscape geochemistry, they include approaches which can be applied directly to our discipline.

Units

The hierarchies of chemical complexity and scientific effort are not defined using a formal scheme. In contrast, the hierarchies of space and time are based on units which have been adopted almost universally throughout science. In the cases of space, the metric system is usually employed, although the older system is still used for certain investigations in both Europe and North America. For this reason a general conversion table is included (Table 15.1).

Examples of Approaches to the Study of Environmental Geochemistry in Space

Observations at a point

Certain investigations made at a single point within a landscape may provide significant and pertinent geochemical information which can be related to the surrounding area. For example, Cannon (1970), in a discussion of trace element excesses and deficiencies in different geochemical provinces in the United States, described interesting geochemical relationships in the rocks, soils, and

Table 15.1. Conversion Factors Used in Environmental Investigations.[a]

To Convert Column 1 into Column 2, Multiply by	Column 1	Column 2	To Convert Column 2 into Column 1, Multiply by
		Length	
0.621	kilometer, km	mile, mi	1.609
1.094	meter, m	yard, yd	0.914
0.394	centimeter, cm	inch, in	2.54
		Area	
0.386	kilometer^2, km^2	mile^2, mi^2	2.590
247.1	kilometer^2, km^2	acre, acre	0.00405
2.471	hectare, ha	acre, acre	0.405
		Volume	
0.00973	meter^3, m^3	acre-inch	102.8
3.532	hectoliter, hl	cubic foot, ft^3	0.2832
2.838	hectoliter, hl	bushel, bu	0.352
0.0284	liter	bushel, bu	35.24
1.057	liter	quart (liquid), qt	0.946
		Mass	
1.102	ton (metric)	ton (U.S.)	0.9072
2.205	quintal, q	hundredweight, cwt (short)	0.454
2.205	kilogram, kg	pound, lb	0.454
0.035	gram, g	ounce (avdp), oz	28.35
		Pressure	
14.50	bar	lb/inch^2, psi	0.06895
0.9869	bar	atmosphere, atm	1.013
0.9678	kg (weight)/cm^2	atmosphere, atm	1.033
14.22	kg (weight)/cm^2	lb/inch^2, psi	0.07031
14.70	atmosphere, atm	lb/inch^2, psi	0.06805
		Yield or Rate	
0.446	ton (metric)/hectare	ton (U.S.)/acre	2.24
0.892	kg/ha	lb/acre	1.12
0.892	quintal/hectare	hundredweight/acre	1.12
		Temperature	
	Celsius	Fahrenheit	
$\left(\frac{9}{5}°C + 32\right)$	−17.8°C	0°F	$\frac{5}{9}(°F − 32)$
	0°C	32°F	
	100°C	212°F	
		Water Measurement	
8.108	hectare-meters, ha-m	acre-feet	0.1233
97.29	hectare-meters, ha-m	acre-inches	0.01028
0.08108	hectare-centimeters, ha-cm	acre-feet	12.33
0.973	hectare-centimeters, ha-cm	acre-inches	1.028
0.00973	meters^3, m^3	acre-inches	102.8
0.981	hectare-centimeters/hour, ha-cm/hour	feet^3/sec	1.0194
440.3	hectare-centimeters/hour, ha-cm/hour	U.S. gallons/min	0.00227
0.00981	meters^3/hour, m^3/hour	feet^3/sec	101.94
4.403	meters^3/hour, m^3/hour	U.S. gallons/min	0.227

Plant Nutrition Conversion—P and K

P (phosphorus) × 2.29 = P$_2$O$_5$

K (potassium) × 1.20 = K$_2$O

[a] From Thomas and Law (1977).

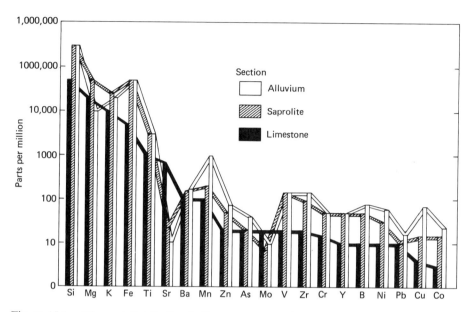

Figure 15.1. Element distribution in limestone, saprolite, and alluvium in Hangerstown Valley, Maryland (Cannon, 1970).

waters at a locality in Maryland where residual soils occur upon limestone. These soils are rich in phosphorus and potassium and require the addition of lime to be productive for agriculture. Absolute abundance of two elements in samples of limestone, saprolite, and alluvium taken from a single outcrop (i.e., a point in space) is plotted in Figure 15.1 to indicate changes of metal percentages which occur in the soil material as a result of the weathering process. Other interesting geochemical observations made at this site involved calcium (not listed in Figure 15.1), strontium, and molybdenum. These elements were dissolved out of the limestone during the formation of the saprolite leaving the clay with an increased metal content. When this material was reworked by water and redeposited to form the alluvial soil, the percentage of metal (e.g., iron and manganese) was again increased (Figure 15.1). This was accompanied by a further decrease in levels of calcium, strontium, and molybdenum. As a result, the groundwaters in the area were relatively high in calcium, strontium, and molybdenum (Cannon, 1970). This shows how related geological, pedological, and geochemical information on rocks, soils, and waters collected from a point in space may be interpreted in relation to details of weathering soil genesis and groundwater chemistry.

Observations along a Traverse Line

In exploration geochemistry, it is common practice to collect samples of similar material at regular intervals along one or more traverse lines located parallel to

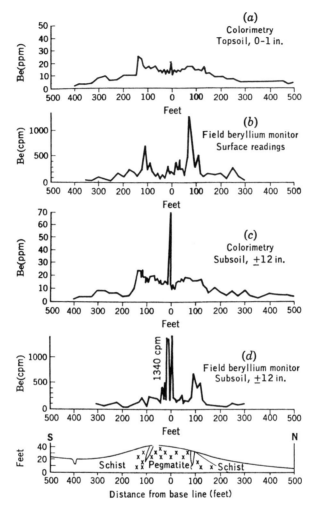

Figure 15.2. The content of beryllium in samples of soil (−80 mesh) taken at two depths in the vicinity of a vein as determined by two analytical methods (Hawkes and Webb, 1962, p. 207).

one another. Hawkes and Webb (1962) described an interesting example of this approach in which soil material was collected from two depths along a traverse line located at right angles to the strike of a pegmatite which had weathered in a tropical climate (Figure 15.2). Samples of soil from each of the two depths were analyzed by two quite different techniques (one colorimetric and the other radiometric) for content of beryllium. The most definitive pattern was obtained in samples collected 12 inches below the surface from the "field beryllium monitor." The data show that there is a vertical gradient in the beryllium content of the soil above and to the side of the vein. More generally, this example illustrates how a first approximation survey of this type will indicate the areal extent of geochemical anomalies.

Observations along a Landscape Section

When information collected along a traverse line includes detailed reference to the vertical as well as the horizontal distribution patterns for chemical elements, the base for the study is referred to as a landscape section. Fortescue and Usik (1969) described a landscape section in a superaqual landscape at the Mer Bleu peat bog near Ottawa, Ontario, Canada (Figure 15.3 and 15.4). Figure 15.3 details the surface topography of the bog, the bog depth, and the kinds of woody material and peat found there together with the macrostructures of the bog section. The location of sample points from which core material was collected are included in Figure 15.3.

Figure 15.4 illustrates by means of "saw tooth" diagrams vertical distribution patterns for each of nine trace elements in the organic samples. An interesting finding was that similar vertical distribution patterns occurred for the same element in adjacent boreholes and that these patterns were characteristic for each element. As in the previous example, this appraisal level study would provide a valuable guide to the planning of more detailed studies which could be designed to explain the reason for the distribution patterns of the elements studied.

Observations at Numerous Points within an Area of Country

Not all environmental geochemical investigations are carried out on a local scale. For example, Shacklette (1971) described the distribution and absolute abundance of a number of elements in samples of soils collected from the continental United States. One of his maps—for copper—is reproduced as Figure 15.5. On this map the lowest 20% of the values (i.e., below 12 ppm) appear as open triangles, the middle 60% are shown as open circles (i.e., from 12 to 37 ppm), and the top 20% (i.e., above 37 ppm) are shown as black squares. It is interesting that, even at this general and continental level of study, it is possible to identify regions of relatively high (or low) total copper in the soil. Broadscale geochemical surveys of this type may be interpreted in relation to data from remote sensing or in relation to the incidence of certain diseases in plants, animals, or man. As in the other examples in this series, the data presented by Shacklette (1971) are considered at the appraisal level of study only.

Information Obtained from Regular Sampling on a Grid Pattern Laid out in an Area

In exploration geochemistry, it is often desirable to provide contour maps for the distribution of chemical elements, minerals, or other chemical entities within an area of country. For example, in Figure 15.6 the nickel content of soils of an area is contoured based on samples collected in a regular grid pattern. Test geochemical surveys of this type are important to landscape geochemistry because they indicate the homogeneity of geological or pedological conditions. The presence of geochemical gradients or geochemical barriers may occasionally be discovered by a visual inspection of maps of this type. Geochemical maps may also be used to indicate anomalies resulting from a point source of pollution, such as from a smelter, or for effects of the disposal of wastes in a landfill. (Figure 16.1.)

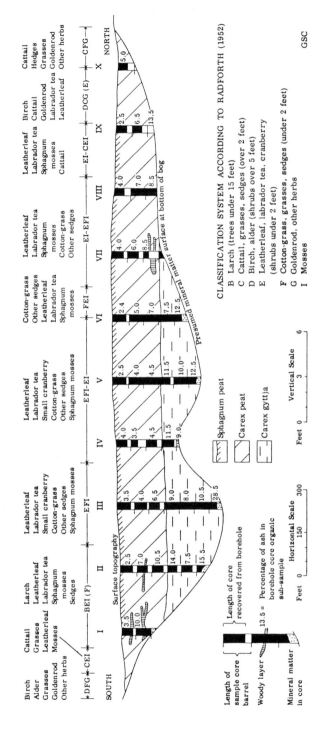

Figure 15.3. Section across the Mer Bleu peat bog showing vegetation cover types and stratification of the peat material (Fortescue and Usik, 1969).

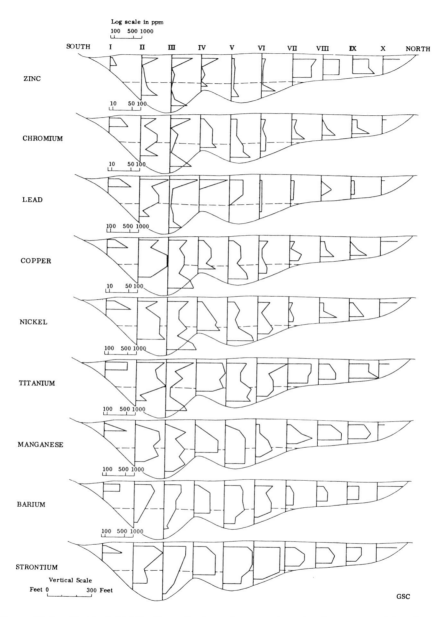

Figure 15.4. Vertical distribution of minor elements in ash of peat material collected from boreholes located across to South Arm of the Mer Bleu peat bog near Ottawa, Ontario, Canada (Fortescue and Usik, 1969).

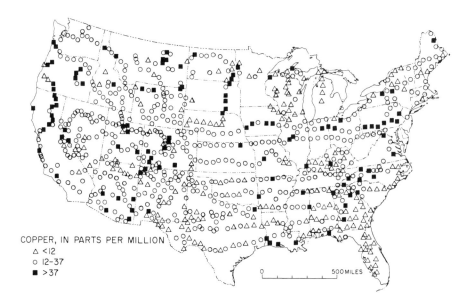

Figure 15.5. Abundance of copper in soils and other surficial materials of the conterminious United States (Shacklette, 1971).

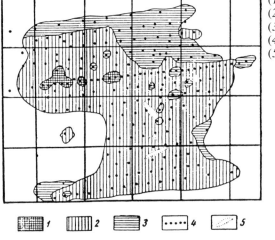

(1) Nickel content 0.5% or above;
(2) Nickel content of 0.1% to 0.5%;
(3) Nickel content less than 0.1%;
(4) Sampling points;
(5) Dykes.

Figure 15.6. The content of nickel in soil indicated as isoconcentration contours. (Malyuga, 1964).

Continuous Geochemical Sampling along Traverse Lines

In all the previous examples samples of material for chemical analysis were collected at one or more points in the landscape. Recent advances in instrumentation have allowed for schemes which involve continuous observations along traverse lines. Sampling occurs continuously and in some cases as immediate instrumental response is plotted as the traverse proceeds. For example, Barringer (1976) described the use of optical correlation remote sensors for the detec-

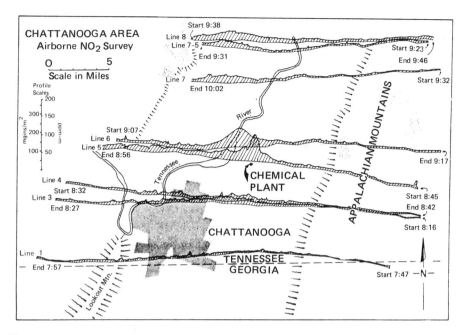

Figure 15.7. Airborne profiles across Chattanooga, Tennessee showing remotely sensed burden of nitrogen dioxide over an industrial area (Barringer, 1976, p. 321).

tion of nitrogen dioxide in the atmosphere in the vicinity of the city of Chattanooga, Tennessee (Figure 15.7). In this case, at the time of sampling relatively large concentrations of the gas were found at points along traverses over the city, particularly near a chemical plant situated to the northwest of it (Figure 15.7). At present, other systems of geochemical surveying of the atmosphere or the surface microlayer are being developed by Barringer Research of Toronto and other organizations. In time these techniques may replace the appraisal level sampling methods described in this chapter because they are quicker and cheaper to conduct. They may also produce more reliable appraisal level information regarding natural or man-made geochemical anomalies.

Aerial Survey of Landscapes

Aerial photographs have been used for over 50 years as an aid to mapping of geological, soil, and plant covers in landscapes. More recently, the orthodox methods of aerial photography have been supplemented by numerous techniques of remote sensing, some of which are based on quite novel principles. Consequently, today environmentalists often use aerial photographs and data from remote sensing as an aid to all kinds of land management. An elementary summary of the scope of such studies was provided by Colwell (1968) who discussed the use of particular photographic emulsions for the identification of specific landscape resource features (Table 15.2):

> ... The photo-interpretation approach to tree species identification is much the same as that already discussed for crop type identification. However, in two notable

Table 15.2. The Feasibility of Identifying Various Natural Resource Features by Aerial and Space Photography Using Different Photographic Emulsion[a].

	Photographic Scale and Film-Filter Combination											
	Scale 1:10,000				Scale 1:30,000				Scale 1:1,000,000			
Type of Resource Feature: Vegetation Resources	Pan-12	IR-89B	Ekta-HF-1	IR Ekta 12	Pan-12	IR-89B	Ekta-HF-1	IR Ekta 12	Pan-12	IR-89B	Ekta-HF-1	IR Ekta 12
Vegetated or not	++[b]	++	++	++	++	++	++	++	++	++	+	++
Wild or cultivated	++	++	++	++	++	++	++	++	++	++	++	++
Fields with crops	++	++	++	++	++	++	++	++	++	++	+	++
Fallow fields	++	++	++	++	++	++	++	++	+	+	−	+
Mature conifers	++	++	++	++	− +	+	++	++	− +	+	−	++
Mature hardwoods	++	+	++	++	++	++	++	++	++	++	+	++
Open brushfields	+	++	++	++	− +	++	++	++	− +	++	+	++
Riparian vegetation	+	++	++	++	+	+	++	++	+	+	+	+
Aquatic vegetation	++	++	++	++	− +	++	+	++	+	+		
Meadow or grassland												
Sparse or drying vegetation less than 1 m high	−	−	+	++	− −	− −	−	+	− −	− −	− −	+ −
Vegetation not yet in leaf	−	−	+	++	− −	− −	+	++				
Herbaceous vegetation in standing water	−	++	+	++	−	+	++	++	−	−	−	+ −
Sprayed brushfields	+	+	++	++	+	−	+	++	−	−	−	
Dead or dying vegetation more than 1 m high	−	+	+	++	−	−	+	++	−	−	+	+
Snags or other downed timber	+	− −	++	++	− +	− +	+	++				
Burned areas	++	++		++	− +	++	++	++	+ −	+ −	+ −	++
Windrowed brush	++	++	++	++	++	++	++	++	+ −	+ −	−	+ −

[a] From Colwell (1968a,b).
[b] Legend: ++ generally and easily identifiable; + generally and easily identifiable, but often requiring very close study; − inconsistently identifiable; − − unidentifiable.

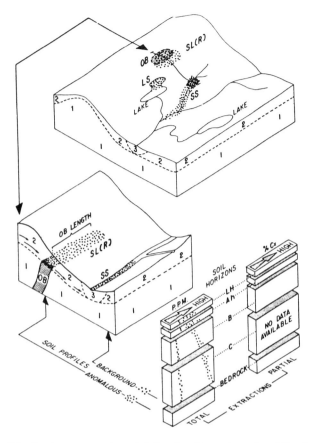

Figure 15.8. Idealized conceptual model for the dispersion of mobile elements in well-drained residual soils of the Canadian Cordillera. (Bradshaw, 1975, p. 26).

respects the problem is more difficult for the forester. . . . A given piece of farmland is usually occupied by a single type of crop, and its photo identification is facilitated by explaining the uniform mass effect that this single type of crop produces on aerial photographs. In contrast a given piece of forest land is often occupied by a random mixture of many tree species, making both the photo identification of the component species and a classification of the timber into meaningful categories most difficult. Most crop plants on a farm have been purposefully spaced so as to receive full sunlight. Consequently, these plants are uniformly exposed to the aerial view and exhibit a remarkably uniform appearance on aerial photographs. In contrast any plants the forester wishes to identify as to species or type from study of aerial photos are in the "forest understory" since they are observed to varying degrees by overtopping trees, their appearances also are quite variable. This makes species identification difficult. . . . Topography, stand density, elevation range, and the size and configuration of the dominant trees in the stand are more easily interpreted from aerial photographs than are species. A forester can establish a high correlation between these factors and species composition. Nevertheless, a great deal of confirmatory field checking is required.

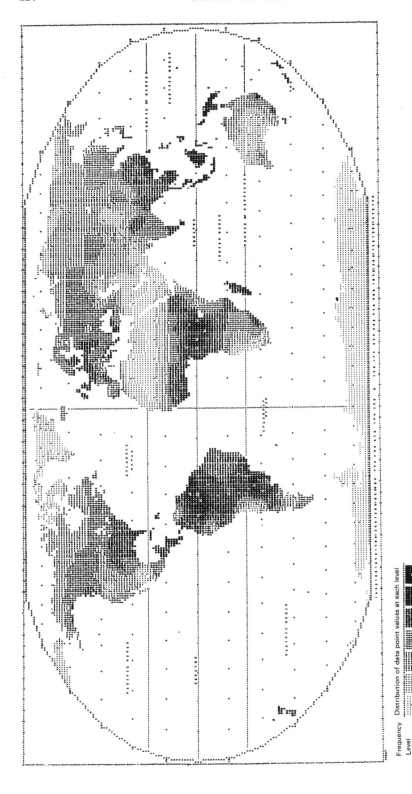

Figure 15.9. A simulation of the global productivity of the biosphere based on weather data obtained from some 600 stations around the world (Lieth, 1975).

Table 15.3. Relationships between the Basics of the Discipline of Landscape Geochemistry and Information Described for Nine Examples of Studies in Environmental Geochemistry and Related Subjects Included in Chapter 15.

Example	1 Element Abundance	2 Element Migration	3 Geochemical Flows	4 Geochemical Gradients	5 Geochemical Barriers	6 Historical Geochemistry	7 Geochemical Classification of Landscape
I The study of saprolite	4	4	2	2	2	3	1
II The study of Be near a pegmatite	4	3	1	4	2	4	1
III The vertical distribution patterns of elements in peat	4	4	2	3	3	4	2
IV Copper distribution in the U.S.	4	2	2	4	3	3	1
V Nickel in soils	4	4	2	4	4	1	1
VI Nitrogen dioxide in the atmosphere	4	4	2	4	1	1	1
VII Air photos for identification of landscapes	1	2	1	2	1	1	4
VIII Maps for biosphere productivity	1	1	4	2	1	1	4
IX Conceptual model for the behavior of elements in landscapes	4	4	3	4	4	4	1

From the viewpoint of environmental geochemistry, remote sensing is valuable largely in relation to the description of natural landscape features or visible effects of pollution of the landscape by man. Thus, the relationship between geochemistry and the data from remote sensing, although often indirect, is of considerable potential importance in landscape geochemistry.

Observations from Volumes of Landscapes

The use of landscape prisms and landscape sections to describe relationships between the morphology and geochemistry of particular landscapes has already been described (Figure 2.5). More extensive use of block diagrams to illustrate important aspects of the geochemistry of landscapes have also been attempted [e.g., Bradshaw (1975) provided an idealized conceptual model for the dispersion pattern of mobile elements around mineral deposits situated below well-drained soil in the Canadian Cordillera (Figure 15.8)]. It should be noted that this is a generalized (or strategic) conceptual model which does not describe the conditions around any particular mineral deposit, but summarizes conditions found around several mineral deposits located within the Cordillera. Such diagrams will become increasingly important in landscape geochemistry as the need to summarize descriptions of the behavior of elements migrating in the environment becomes more pressing.

Synthesis of Information on a Global Scale

Recently computerized methods of mapping have been used to describe distribution patterns on a global scale. Although such approaches have not been commonly used it is likely that such developments will occur in the future. An example of the potential of such data processing techniques was described by Leith (1975) who studied global plant productivity patterns on the basis of some 600 stations located in different parts of the world (Figure 15.9).

This study is of general interest in landscape geochemistry for three reasons: (1) it is computerized on a global scale; (2) it describes an aspect of the living biosphere (land only) on a global scale; and (3) it is a simulation based on data and information collected for other purposes which was reprocessed to produce the pattern of global productivity. In the future this approach may be used to describe productivity and the distribution of nutrients (and other chemical entities) at the regional or local levels of intensity also.

Summary and Conclusions

In this chapter nine specially selected examples of investigations in environmental geochemistry have been described which together provide an overview of current methodology involved in geochemical sampling of landscapes. A flow diagram summarizing relationships between the nine approaches (Figure 15.10) relates the examples to the classes of investigations each represents.

These examples may also be used for describing the relationship between the hierarchy of space and the seven basics of landscape geochemistry (Table 15.3). As in similar tables in Chap. 13 and 14, the concept of element abundance is seen to be the most important of the seven basics in Chapter 15.

Thus in environmental geochemistry there are many different methods of collecting environmental and geochemical information in relation to space, all of which are relevant to some aspect of landscape geochemistry.

Discussion Topics

1) What is the importance of standardization of map scale fractions, sheet sizes, and legends for the display of the data of landscape geochemistry for purposes of comparison?

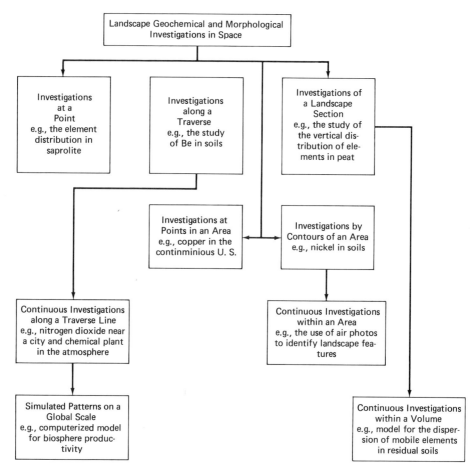

Figure 15.10. Nine approaches used by environmental geochemists with respect to space.

2) Should time intervals between observations during environmental monitoring be standardized?

3) Can one make a general plan for landscape geochemical investigations around a point source in open country where the geological conditions are uniform but the wind direction is very variable?

16. Time and Landscape Geochemistry

"In many geological applications, geological observations in a stratigraphic succession correspond to changes taking place through time. Where deposition was continuous and the rate of the deposition was relatively constant, stratigraphic thickness can be considered directly proportional to time. Therefore, time trend curves can be plotted with stratigraphic thickness corresponding to time as the independent variable and the parameter being studied as the dependent variable. The dependent variables can include such things as grain size, carbonate content, color, or fossil distribution. In modern sedimentary studies where processes are being studied through time, absolute time can be plotted as the independent variable and the dependent variables can include such things as barometric pressure, wind velocity, wave height, or current velocity."

William T. Fox, "Some Practical Aspects of Time Series," in *Concepts in Geostatics*, R. B. McCammon, ed. (Berlin: Springer Verlag, 1975), p. 70.

Introduction

It is noted in Chap. 3 that landscape processes occur in technological, ecological, pedological, or geological time—concurrently in most cases. Together, these processes result in the formation and evolution of natural landscapes and the gradual modification of man-made landscapes. This chapter considers some examples of the role of time taken fron environmental geology and geochemistry.

Geochemical investigations can be divided into three kinds: (1) those which relate to a single instant in time; (2) those which relate to several instants in time; and (3) those which involve environmental monitoring. Investigations which relate to a single instant in time produce data which when plotted on a map or graph indicate the state of a particular part of the environment at a given time. Environmental monitoring results in a series of observations of an environment which are taken continuously or at regular or irregular intervals depending upon the type of study.

With respect to the historical development of landscapes during ecological, pedological, and geological time, it is often desirable to have some indicator of the sequence of climatic, geological, ecological, or other events which have resulted in the formation of a landscape and an indication of the time period involved. Here we are concerned with a relative (or absolute) time scale relevant to a particular problem. Relationships between environmental geochemistry and time are many and varied; in this chapter some common types are considered because they are particularly important to landscape geochemistry.

Isolated Incidents in Time

Man may add chemical substances to the landscape over a period of years and the effects of such additions may persist in the landscape for a considerable length of time. For example, Le'Grand (1970) described the cumulative effect of the addition of chemical substances to a landscape near Denver, Colorado (Figure 16.1). In this case the study (which occurred at an instant in time) indicated that two chemical entities "chlorate toxicity" and "2-4-D type toxicity" had migrated in groundwater a certain distance down the slope. In addition, Le'Grand (1970) provided an indication of the anticipated area of pollution at some time in the future, although he did not indicate a time required for this to come about. This example was included here because it shows both cause and effect relationships at an instant in time as well as future projection of a distribution pattern of pollution in the environment.

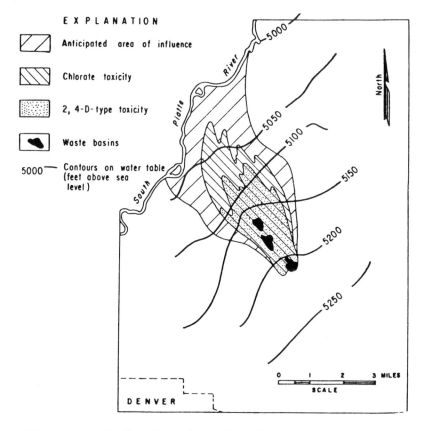

Figure 16.1. Patterns of polluted groundwater-formed seepage from waste basins near Denver, Colorado (Le'Grand, 1970, p. 311; after Walker, 1961).

Environmental Monitoring

A living plant consists largely of elements of the atmosphere which have been fixed by photosynthesis plus water and mineral elements derived from the soil in which it grows. Foresters have often studied the dry matter production of forest trees during their life cycle; studies of this type can be used to illustrate changes in the growth rate of trees. For example, Switzer et al. (1968) described the dry matter production of a stand of loblolly pine during a 60-year life cycle from observations made at intervals during this time period (Figure 16.2). They compared the current annual increment and the mean annual increment of the foliage, branches, and stems of the trees during the 60-year period and concluded that the maximum growth rate for all parts occurred before 30 years of age. Maximum rates of growth began in the foliage components of the crown which are nutrient

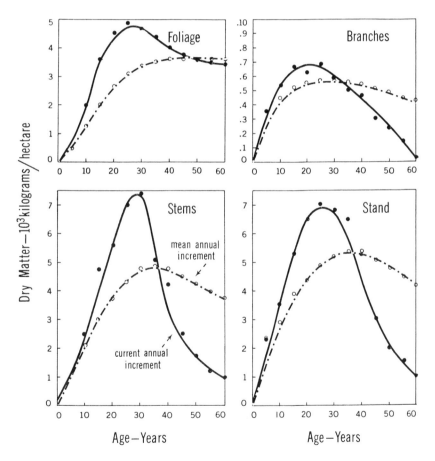

Figure 16.2. Current and mean annual increments for stand fractions, Loblolly Pine on good sites (Switzer et al., 1968).

sinks during the initial growth period. Current annual increments declined after this formative period, especially in the stems and branches, in contrast to foliage growth which tended to remain constant. This suggests that the most pressing need for chemical fertilization of a stand is during its initial stage of growth.

Retrospective Monitoring in Geological Time

Geologists, confronted with a fragmentary record of changes which occurred in a landscape in geological time, may wish to use such a record as a basis for the simulation of the events which led to its formation. The record is essentially "retrospective monitoring." A classic example, described by Briggs and Pollack (1967), was the formation of Permian salt deposits of the Cayugan series in the Michigan Basin in late Silurian time. Briggs and Pollack first described a general conceptual model for the basin connected to the sea by a single inlet. Isosalinity lines were then drawn on the model to indicate the distribution patterns for the evaporite minerals which would form during geological time including carbonate, sulfate, and chloride (Figure 16.3). The circulation of water in the basin was simulated by solving the pertinent differential equations for the motion of a fluid. In the model the precipitation of salts is assumed to occur whenever a prescribed saturation concentration is reached. When this occurs, the mass of "dissolved" material is converted into a thickness of the "deposited" precipitate. Evaporation rates from modern evaporite basins were used as a basis for calculation of the rate of deposition in the ancient evaporite basins. Although Briggs and Pollack (1967) described certain problems encountered in setting up their model, it was eventually perfected to a point at which the simulation of the deposition of salts in the Michigan basin could be carried out. The geography of the basin is shown on Figure 16.6 and the observed thicknesses for salt in the basin (from cores and other sources) is summarized on Figure 16.4. As a result of the use of the computer model, Figure 16.5 indicates thicknesses for the salt deposits obtained for the basin. The fit between the observed data and those from simulated models is relatively good. It provides a good example of the closeness with which a model based on sound chemical and physical principles (plus less rigorous geological assumptions, such as limits of the basin, location, and magnitude

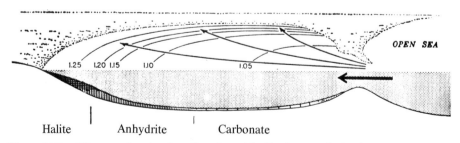

Figure 16.3. Diagram showing flow lines in an idealized evaporite basin connected to the open sea by a single outlet (Briggs and Pollack, 1967).

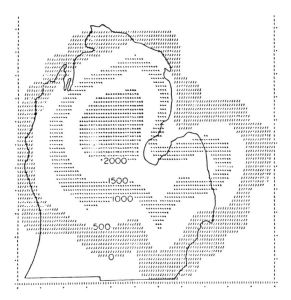

Figure 16.4. Thickness contour map based on borehold data showing distribution of halite in Cayugan series (Upper Silurian) in the Michigan Basin. Bands of symbols denote contour lines on computerprinted maps. Values in feet (Briggs and Pollack, 1967).

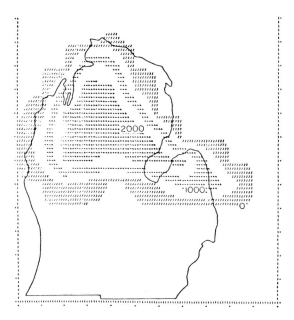

Figure 16.5. Simliated salt-thickness patterns employing "leaky reef" hypothesis. Thickness values are in feet. Compare this isopack map with Fig. 16.4 which is drawn from observed data (Briggs and Pollack, 1967).

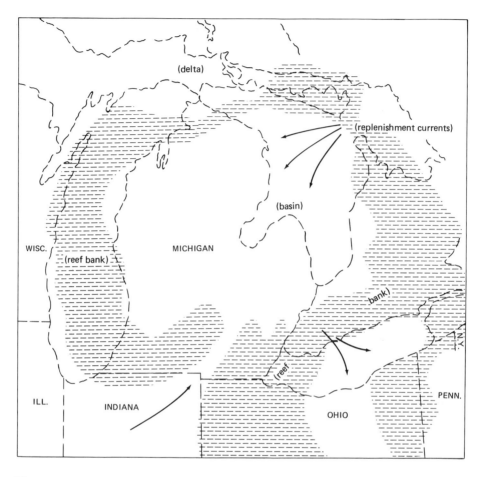

Figure 16.6. Map showing inferred geography in and adjacent to the Michigan Basin during the late Silurian (Briggs and Pollack, 1967).

of the channels to the sea, etc.) may be combined to provide a reasonable explanation of the genesis of the salt deposits observed in the Michigan Basin today. Although this is a relatively simple geological model, it indicates the potential of simulation models to describe the evolution of geological component landscapes. Similar simulation research during geological time is commonly undertaken by sedimentologists and geomorphologists (see Leopold et al., 1964).

Retrospective Monitoring in Pedological Time

Studies of long-term geochemical effects relating to pedological processes are relatively rare. An interesting example of this kind of study was described by Brass (1975) who investigated in detail the distribution of strontium isotopes and rubidium in soil profiles developed under different climatic conditions from the same geological formation in New Zealand. Seven soil profiles were studied—all developed on Mesozoic arkoses which outcrop in a band running north to south of New Zealand (Figure 16.7). The soil profiles belonged to four groups: (1) The Whangarei Group at a low altitude and highly weathered; (2) The Wellington Group located further south and less deeply weathered; (3) The Canterbury Group from the South Island which was formed at a relatively high altitude; and (4) The Southland Group collected at the southern end of the South Island. Brass (1975) found that the strontium and rubidium trends were regular in any one profile, but as might be expected differed from profile to profile; in some soil profiles the strontium increased with depth and in others it decreased. Relatively large contrasts were found in the behavior of strontium in soil profiles located only a few kilometers apart (e.g., the two Wellington profiles). Strontium was always found to be depleted and rubidium increased in the fine fraction relative to the coarse fraction of the soils studied. In general, the soil material tended to be depleted in strontium but enriched in rubidium. The $^{87}Sr/^{86}Sr$ ratios in the separated fractions followed the Rb/Sr pattern because the highest ratios were almost always found in the fine fraction. It was concluded that the removal of common (and radiogenic) strontium during weathering does not proceed at the same rate (common strontium probably weathering more rapidly) and the residual $^{87}Sr/^{86}Sr$ and Rb/Sr ratios in the weathering profile form an ''isochron'' with an ''age'' different from the age of the original parent rock (Figure 16.8). Table 16.1 lists data for the stratigraphic age and the ''isochron age'' which is less than the stratigraphic age of the parent rock of the soil. Brass (1975) observed that the stontium in solid weathering products may become several percent more radiogenic than the strontium in solution which is only slightly less radiogenic than in the source rock. He concluded that the variation in marine $^{87}Sr/^{86}Sr$ ratio over geological time is due to changes in rock isotope ratios and abundances of the source rocks and not to changes in the weathering regime (Brass, 1975).

Figure 16.7. Map of New Zealand showing location of sampling stations (Brass, 1975).

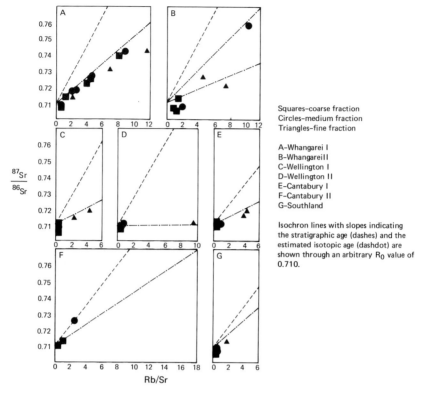

Figure 16.8. Isochron diagrams of the weathering profile data. Squares, coarse fraction; circles, medium fraction; triangles, fine fraction. (A) Whangarei I; (B) Whangarei II, (C) Wellington I; (D) Wellington II; (E) Canterbury I; (F) Canterbury II; (G) Southland. Isochron lines with slopes indicating the stratigraphic age (dashes) and the estimated isotopic age (dash–dot) are shown through an arbitrary R_0 value of 0.710 (Brass, 1975).

Table 16.1. Stratigraphic Age, Isochron Age, and n for New Zealand Weathering Profiles.[a]

Profile	Stratigraphic Age (t)	Isochron Age (t')	n
Whargarei I	Permian, 220 m.y.	100 m.y.	1.8
Whargarei II	Permo-Triassic, 220 m.y.	upper, 50 m.y.	1.3
		lower, 120 m.y.	2.5
Wellington I	Triassic, 200 m.y.	60 m.y.	1.4
Wellington II	Triassic, 200 m.y.	0 m.y.	1.0
Canterbury I	Jurassic, 150 m.y.	60 m.y.	1.7
Canterbury II	Jurassic, 150 m.y.	75 m.y.	2.0
Southland	Jurassic, 150 m.y.	100 m.y.	3.3

[a] From Brass (1975).

Although this paper was mainly concerned with the relationship between the stratigraphic age and the "isochron" age of the rock and soil materials taken from the seven profiles, it provided details of the behavior of rubidium and the two strontium isotopes in old soils and related materials. These are interpreted as a general indication of the behavior of these elements during long-term weathering and in general the value of n (Table 16.1) may relate directly to the cumulative effect of soil-forming processes at the different sample sites.

The Uniformity of Environmental Conditions during Ecological Time

The uniformity of climate will influence to a significant extent the behavior of elements and chemical substances as they migrate through landscapes. Consequently, landscape geochemistry is concerned with the generalities and details of environmental change during the formation of a landscape. Perhaps the most effective tools for the description of environmental change during pedological and ecological time are: (1) radiocarbon dating, (2) palynology, and (3) dendrochronology.

Radiocarbon Dating

Some of the advantages and pitfalls of radiocarbon dating of environmental samples of organic matter are illustrated by the following example. Runge *et al.* (1973) studied the organic clay complexes which occur in loess deposits on the Timaru Downs in New Zealand. The Darling No. 1 section was sampled at (1) the lower increment of the surficial peat, (2) the interface of the surficial peat and the first loess (sampled from an adjacent bog), (3) the interface of the first loess and the buried peat, (4) the middle of the buried peat, (5) the interface of the buried peat and the second loess, (6) 0.60–0.75 m below the bottom of the buried peat, and (7) 1.00–1.10 m below the bottom of the buried peat (Runge *et al.*, 1973) (Figure 16.9).

Whole peat samples, as well as humic and fulvic acid fractions and the residue (humin) from extractions, were dated by the ^{14}C method (Runge *et al.*, 1973) and the dates obtained are listed in Fig. 16.9. The writers summarized their conclusions as follows (p. 745):

> The clay-organic complex can be used to approximate the age of buried paleosols *if contamination can be accounted for* (for example, dry climate or presence of soluble salts including $CaCO_3$. The corrected MRT (mean residence time) of 16,400 years for the clay-organic complex of the first paleosol developed in loess two is reasonable considering the soil formed during the interval 11,800 to 31,000 radiocarbon years ago. However, the correct MRT of 18,500 for the second paleosol developed in loess three is not reasonable since it must be older than 31,000 radiocarbon years when all other dates and stratigraphic relationships are considered. The technical problems of fractionating clay-organic complexes (approximately 1% C) into

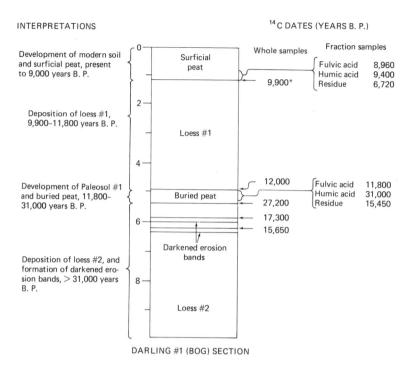

Figure 16.9. Mean residence times, position of samples, and interpretation of dates measured on the Darling No. 1 (bog) section from Timaru Downes, New Zealand (Runge et al., 1973).

fulvic acid, humic acid, and residue (humin) are great but their effect on MRT's may be considerable.

Conclusions and interpretations based on MRT of dispersed organic matter from the soil horizons must be reached with caution. For example, the buried peat in our study and MRT's of 11,800, 15,450, and 31,000 years BP based on fulvic acid, residue, and humic acid, respectively. The MRT of the humic acid fraction is 94% greater than the MRT of the residue fraction. The results of this study suggest that some of the conclusions in the literature based on residue dates may be in error.

Clearly radiocarbon dating is very important in dating paleosols (and similar materials), although, for practical reasons, such dates are often somewhat difficult to obtain and cannot be made in large numbers.

Palynology

An alternative method of dating ecological events in a landscape is on the basis of the fossil pollen record which may be studied in samples of peat or lake sediment cores collected under favorable conditions from superaqual or aqual landscapes. An interesting case of dating by pollen was described by Anderson (1974) who studied the chestnut pollen decline as a time horizon in lake sediments in eastern North America. Briefly, chestnut [*Castanea dentata* (Marsh.) Borkh.]

was a dominant forest tree of eastern North America until the early part of this century when it was eliminated by the blight fungus *Endothia parasitica* Anders. The blight was first observed in New York City in 1904 and within 50 years had spread throughout the entire range of the chestnut species. Anderson (1974) documented the decline of *C. dentata* on the basis of a study of pollen records in Lakes Ontario, Erie, and Woodcliff Lake, New Jersey. According to the data from Woodcliff Lake, the decline in chestnut pollen occurs at an age equivalent to the die-out of the chestnut in an area. Anderson (1974) noted that the decline in chestnut pollen in Lake Ontario occurred in 1930 and in Lake Erie in 1935. Thus, the die-out of chestnut provides a marker in the pollen record similar to that associated with the increase in *Ambrosia* sp. which marked the change from forest cover to agriculture in southern Ontario around 1850. Anderson (1974) provided a map (Figure 16.10) showing the location of studied cores from Lake Erie and Lake Ontario and he listed depths relating the pollen horizon levels, average sedimentation rates, and annual thickness of the accumulating sediment (Table 16.2) in them.

The information in Figure 16.11 provides an insight into the procedure used to interpret pollen profiles from lake sediments under near ideal conditions. Once time marker horizons (such as that for *Ambrosia* sp. or *Castanea*) are observed within a particular region this facilitates the interpretation of information relating to changes in plant cover species during ecological time and changes in the change rate of other environmental conditions (such as sedimentation, Table 16.2) during pedological and/or technological time.

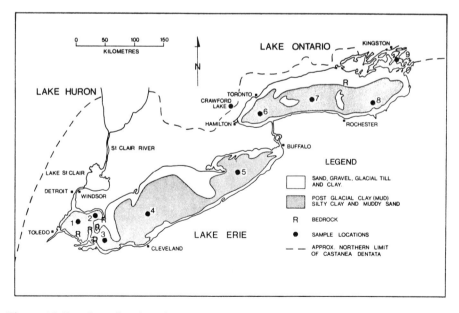

Figure 16.10. Sampling locations and general surficial sediment distribution in lakes Ontario and Erie (Anderson, 1974).

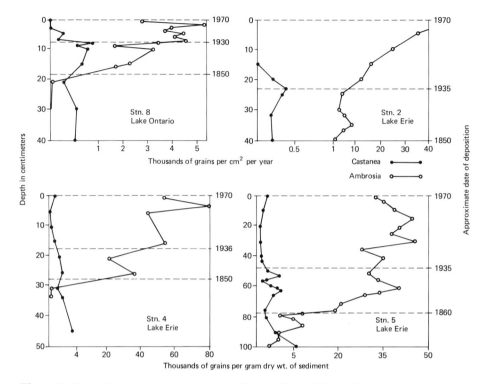

Figure 16.11. *Castanea* and *Ambrosia* pollen profiles of the surficial sediments at stations 2, 4, 5, and 8. Dates approximated on the basis of the *Ambrosia/Castanea* curves (Anderson, 1974).

Table 16.2. Pollen Horizon Levels, Average Sedimentation Rates, and Annual Thickness of Sediment Accumulation at the Nine Sampling Stations.[a]

Castanea location	Level of Castanea Location (cm)	Level of Ambrosia Horizon (cm)	Estimated Recent Sedimentation Rate Based on Castanea Horizon (g m^{-1} year^{-1})	Mean Annual Thickness of Sediment Accumulating (mm)	Absolute (maximum) Annual Thickness of Sediment Accumulating (mm)
			Lake Erie		
1	26.5	40.0	3580	7.6	16.6
2	23.5	40.0	3465	6.7	12.8
3	9.5	21.0	1109	2.7	5.8
4	18.0	28.0	1190	5.1	12.4
5	47.0	77.5	5049	13.4	24.0
			Lake Ontario		
6	7.0	15.5	420	1.7	3.5
7	7.5	22.0	423	1.8	6.2
8	7.0	19.0	366	1.7	3.3
9	19.0	27.5	1156	4.7	9.6

[a] From Anderson (1974) after Kemp et al. (1974).

Dendrochronology

Under favorable conditions dendrochronology (tree ring dating) is by far the most detailed method of documenting climatic conditions and ecological change within a particular locality, region, or on a worldwide basis (Fritts, 1976). The procedure adopted in back dating tree ring records from the present to the past is indicated diagramatically on Figure 16.12. The procedure in relating the climate in a particular year to the growth rate of a particular tree ring is not always simple and some of the complexities of the process are indicated in Figure 16.13. These conceptual models provide a general introduction to principles of this interesting subject which is described fully in the book by Fritts (1976).

Two practical examples of the application of the tree ring dating technique to environmental problems are given on Figures 16.14 and 16.15. In Figure 16.14 the tree ring record at 17 sites was used to describe the annual virgin streamflow of the Colorado River at Lee Ferry as reconstructed by using ring width index variation on trees collected from the 17 sites in the Colorado River Basin. The calibration period for this study was between 1896 and 1961 and was based on the growth for each year and the three years following. The smoothed curve below (Figure 16.14) represents a running mean for 10 years. Figure 16.15 shows an example of tree ring width chronologies used to infer statial variations in climate. To begin with, ring width chronologies which respond to the same climatic variables are chosen and growth values are scaled for purposes of comparison

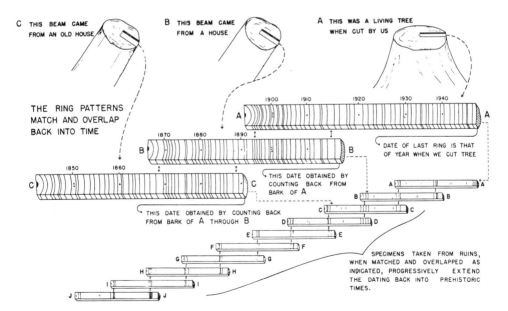

Figure 16.12. Diagram illustrating cross-dating and extension of a dated tree ring width chronology backward in time (Fritts, 1976; after Stallings, 1949).

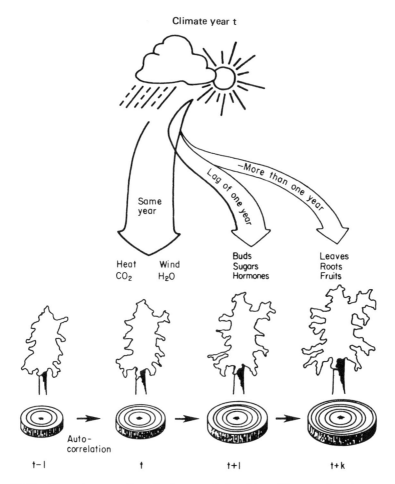

Figure 16.13. Tree ring growth and climate relationships. Climate of a given year t is modelled to have a large effect on ring width for the same year through the heat, wind, carbon dioxide, and water that impinges on the tree. However, it can also affect ring width in the following years, $t + 1$ up to $t + k$ through effects on buds, sugar hormones, and the growth of leaves, roots, and fruits. Because of these linkages with the width of the ring in year $t - 1$ is statistically related to the ring width in year t. This effect is modeled as autocorrelation in ring width (Fritts, 1976).

later. The data are then averaged by decade and these values plotted on maps of the area using isograms to facilitate analysis (Figure 16.15). The areas of higher (or lower) than average growth are inferred to be areas dominated by appropriate anomalies in climate (Fritts, 1976).

Summary and Conclusions

In this chapter eight examples of environmental investigations which involve the variable time have been described. Some of these refer to short-term variations

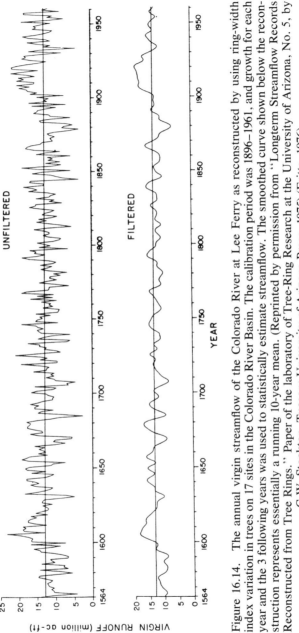

Figure 16.14. The annual virgin streamflow of the Colorado River at Lee Ferry as reconstructed by using ring-width index variation in trees on 17 sites in the Colorado River Basin. The calibration period was 1896–1961, and growth for each year and the 3 following years was used to statistically estimate streamflow. The smoothed curve shown below the reconstruction represents essentially a running 10-year mean. (Reprinted by permission from "Longterm Streamflow Records Reconstructed from Tree Rings." Paper of the laboratory of Tree-Ring Research at the University of Arizona, No. 5, by C.W. Stockton, Tucson: University of Arizona Press, 1975) (Fritts, 1976).

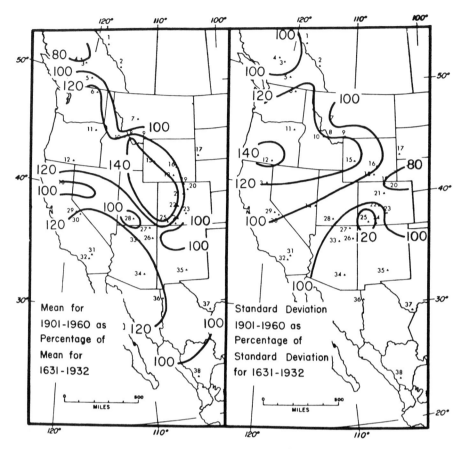

Figure 16.15. Ring-width data for different time periods can be compared to reveal particular anomalies in climate. The means and standard deviations of 38 tree-ring chronologies for AD 1901–1960 are expressed as a percentage of the means and standard deviations for the longer period, AD 1631–1932. The mean growth and, by inference, the moisture supply have been anomalously high in the extreme southwest, along the western slope of the central Rocky Mountains, and in the Pacific Northwest during recent times. The variability of moisture as revealed by the standard deviation of ring width has been high in the northwest and locally in the southwest (Fritts, 1976).

which relate to events in technological time and others involve ecological, pedological, and geological time. Under favorable conditions it is now possible to trace the history of the development of geological, pedological, ecological, and technological events in landscapes with considerable accuracy. Such studies are of fundamental importance in both landscape geochemistry and general environmental investigations.

As in the previous three chapters, a table has been drawn up to indicate relationships between the hierarchy of time and the seven basics of landscape geochemistry (Table 16.3). In this case the history of landscapes was found to be of greater importance than abundance, which is to be expected under the circum-

Table 16.3. Relationships between the Basics of the Discipline of Landscape Geochemistry and Information Described for Eight Examples of Studies of Environmental Geochemistry (and Related Topics) Included in Chapter 16.

Example	Element Abundance	Element Migration	Geochemical Flows	Geochemical Gradients	Geochemical Barriers	Historical Geochemistry	Geochemical Classification of Landscapes
I Seepage of polluted groundwaters	4	4	2	4	4	3	1
II The growth rate of pine trees	3	3	4	2	2	3	1
III The development of an evaporite basin	4	4	4	4	4	4	1
IV The weathering rate of soils developed from the same substrate	4	4	2	4	4	4	1
V Radiocarbon dating of organic soils	4	4	1	3	3	4	1
VI The demise rate of chestnut trees	1	1	1	1	1	4	1
VII Streamflow of the Colorado River	1	1	2	1	1	4	2
VIII Details of climatic change	1	1	2	1	1	4	2

stances. Otherwise, the other five basics were judged to be of variable importance according to the examples cited.

It is concluded that many landscape components may be used to describe the variable time passage and history of the development of landscapes.

Discussion Topics

1) What are the most effective ways of relating the variable time to the solution of problems of man's addition of chemical to elluvial landscapes in a humid as opposed to an arid climate?

2) What is the importance of the geological history of a landscape in relation to its soil cover?

3) What is the geochemical history of landscapes in relation to the study of migration rates for chemical entities within them?

17. Landscape Geochemistry as a Totality

"When situations arise such that a non-surficial geochemist finds it difficult to understand why one technique in exploration geochemistry is successful while another is not, the cause is often due to the fact that some of the actors involved have not developed a sufficient expertise in landscape geochemistry as a whole."

R. J. Allan, *In* book review of *Conceptual Models in Exploration Geochemistry: The Canadian Cordillera and Canadian Shield* (compiled and edited by P. M. D. Bradshaw) *Chemical Geology*, vol. 18, 1976, pp. 245–246.

Introduction

The purpose of this chapter is to provide an overview of landscape geochemistry as it has been outlined in Part II as a preliminary to the description of applications of the discipline given in Part III.

A Sequential Description of Landscape Geochemistry

Figure 17.1 shows landscape geochemistry linked to general geochemistry, environmetal geochemistry, ecology, and other disciplines by the concept of abundance. Without identification of the element or chemical entity studied and the component of the landscape, it is usually not possible to consider relationships between an investigation and landscape geochemistry. In this book, the different aspects of abundance (e.g., absolute abundance, relative abundance, partial abundance, and speciation) have been discussed and related to each other and to the other basics of the discipline. As the discipline of landscape geochemistry evolves, it is to be expected that other aspects of the concept of abundance will be added from time to time. For example, units similar to the Clarke and the Clarke of Concentration for the lithosphere might be defined for the freshwater hydrosphere, biosphere, and atmosphere; other units which relate the content of elements in natural material on a volume percent—rather than a weight percent—basis may be defined and introduced. These new units may be required when large amounts of quantitative, multielement data suitable for processing in computers in large quantities become available from environmental geochemistry, ecology, and landscape geochemistry.

The concept of abundance is strongly linked to that of element migration, and

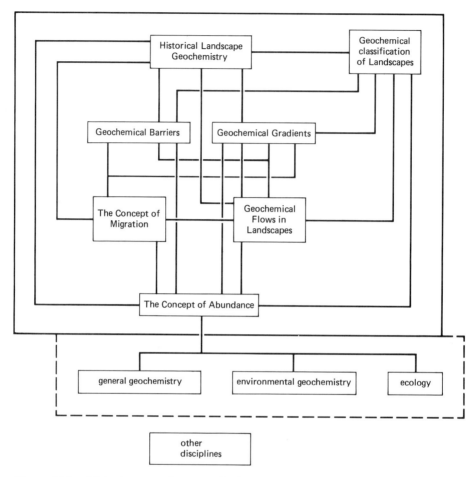

Figure 17.1. Links between the seven basics of landscape geochemistry and between landscape geochemistry and other kinds of environmental science.

usually relates to the chemical composition of a landscape component at a particular point or points in space at a single time. The concept of element migration involves the problem of the rate of movement of chemical entities within landscapes. The concept of the migration rate of chemicals in landscapes can be used in a number of ways. For example, Perel'man (1966), and Polynov before him, stressed the importance of first approximation data for the migration rate of elements and ions in all landscapes. More recently, Likens and Bormann (1972) actually measured and monitored the migration rate of elements and ions in the landscapes at the Hubbard Brook site. In the future, data on the abundance of chemicals in landscapes may be fed into generalized sumulation models which will accurately predict their effects on landscapes at various times. Such simulations will be very valuable to environmental impact statement preparation and decision making on land use.

When viewed holistically, the idea of the migration rate of chemicals in landscapes leads directly to a consideration of the geochemical flows of substances in them. Geochemical flows in landscapes, as described conceptually by Kozlovskiy (1972), provide landscape geochemistry with useful starting points for the systematic study of the migration of substances relative to the details of landscape development and evolution. In studies of this type it is the basic concepts which matter because upon them depends the utility of the data in information produced in relation to local, regional, or global considerations. Kozlovskiy's concepts combined with the landscape prism concept provide a conceptual model which can be used in relation to natural or man-made landscapes within which flows are studied holistically.

The flows of chemical substances in landscapes are not uniform with respect to space or time. Two important concepts which contribute to this lack of uniformity are geochemical gradients and geochemical barriers. Geochemical gradients are important in common landscapes which are relatively uniform geochemically, and in landscapes which include localized areas in which relatively high concentrations of certain chemicals occur. These geochemical anomalies may be due to natural causes, as in the case of mineral deposits, or due to man's activities, as in the case of areas polluted by smelter operations. The study of geochemical gradients related to such anomalies occupies many environmental geochemists and studies of this type may contribute much to the understanding of the migration patterns of common elements in normal landscapes. Similarly, the systematic study of geochemical barriers near geochemical anomalies may also contribute to the general knowledge regarding the behavior of chemicals in landscapes. Natural or man-made geochemical barriers are important in environmental geochemistry, particularly with respect to the solution of problems of disposal of solid or liquid wastes. In general, a knowledge of abundance, migration rates, geochemical flow patterns, geochemical gradients, and geochemical barriers in a landscape provides a sound conceptual basis for the study of the geochemistry of a landscape now and in the future. It is here that the holism of landscape geochemistry is of major importance because the approach involves all kinds of ecosystems which occur together within an area and the study of the flows of chemical substances between them. Indeed, as Perel'man (1966) pointed out, the study of the geochemistry of the links between landscape types is often more rewarding than the study of relatively uniform landscapes.

But a detailed consideration of the features of a landscape today should also relate to the past as well. Consequently, studies of geomorphology and geology are basic to a deep understanding of the geochemistry of landscapes, especially when they are combined with a knowledge of groundwater hydrology and soil genesis. The historical development of landscapes involves parallel consideration of eluvial, superaqual, and aqual types of landscapes. Information on the development of one part of the landscape often contributes to the solution of problems in another adjacent part. For example, information obtained from lake sediment cores may be used to explain the details of soil-forming processes which occurred in the lake area. Such information often cannot be obtained directly from the soil profiles. Information on the historical development of the landscape components leads to a study of the geochemical evolution of the landscape with respect to major changes involving whole weathering cycles or minor changes which have occurred within the current weathering cycle. A knowledge of the historical geochemistry of a landscape also contributes to predictions for their future development. Consequently, the difficult field of historical landscape geochemistry is of considerable practical importance in both fundamental and applied landscape geochemistry.

An understanding of the historical evolution of landscapes leads to a consideration of the geochemical classification of landscapes. The geochemical classification of landscapes is concerned with the cumulative effects of the geochemical flows which occurred in a landscape during its evolution. In general, the geochemical classification of landscapes relates to the minor geochemical cycle currently in operation in the landscapes and only to a minor extent on the effects of the chemistry of the crystalline rocks of the lithosphere. Crystalline rocks in a landscape, particularly a young landscape, act as a support and a source of chemical substances. In short-term periods, chemical substances already in circulation in the landscape provide most of the chemical circulation. The study of the concepts of landscape geochemistry and particularly the study of the geochemical classification of landscapes lies at the heart of the discipline. In general, landscape geochemistry is seen as a discipline similar to ecology, except that in landscape geochemistry the study of the migration of substances of all kinds takes a central role analogous to that played by the living biosphere in ecology.

In this section relationships between the different concepts of landscape geochemistry have been outlined and the central position of the study of the geochemical classification of landscapes has been stressed. When we consider the subject matter of Part II in relation to this conclusion, the unique contribution of the landscape geochemistry to environmental science is evident. Another conclusion is that landscape geochemistry as described in this book is now in its infancy as a scientific discipline. A third conclusion is that in spite of its early stage of development, the elementary concepts of landscape geochemistry are universal and can be applied in any landscape in any part of the terrestial Earth.

The Terminology of Landscape Geochemistry

One price a group of scientists pays for subscribing to a new discipline is the need to learn a new terminology and its application in the real world; the study of landscape geochemistry is no exception. Fortunately, the elementary concepts of landscape geochemistry are relatively simple although the terminology used for their description may be unfamiliar to many readers. Walker (1963) noted that:

> . . . . the purpose of scientific thought is to postulate a conceptual model of nature from which the observable behavior of nature may be predicted accurately. The formulation of new postulates is an act of creation and is subject to no limitations of method. The validation of the model, however, follows a regular pattern which has been called the *scientific method*. (p. 8)

Table 17.1. Summary of the Terminology of Landscape Geochemistry.

Concept or Principle	Term Used	Chapter	Source
Element abundance			
	(1) Absolute abundance	6	Common in general
	(2) Relative abundance	6	geochemistry
	(3) Partial abundance	6	
	(4) Geochemical speciation	6	
Element migration			
	(5) Relative mobility	7	
	(6) Absolute migration rate	7	From Perel'man
	(7) Coefficient of aqueous migration	7	From Perel'man
	(8) Coefficient of biological accumulation	7	From Perel'man and Brooks
	(9) Migrant elements classification	7	From Perel'man
Geochemical flows			
	(10) Independent migrants	8	From Kozlovskiy
	(11) Dependent migrants	8	From Kozlovskiy
	(12) Geochemical channels	8	From Kozlovskiy
	(13) Geochemical cells	8	From Kozlovskiy
	(14) Main migrational cycle (MMC)	8	From Kozlovskiy
	(15) Landscape geochemical flow (LGF)	8	From Kozlovskiy
	(16) Extra landscape flow (ELF)		
	Positive	8	From Kozlovskiy
	Negative	8	From Kozlovskiy
	(17) Four principals of element migration	8	From Kozlovskiy
Geochemical gradients			
	(18) Visible	9	
	(19) Invisible	9	
	(20) Continuous	9	
	(21) Discontinuous	9	
	(22) Water series	9	From Perel'man
	(23) Thermal series	9	From Perel'man
Geochemical barriers			
	(24) Mechanical, physiochemical, biological	10	From Perel'man
	(25) Typomorphic elements	10	From Perel'man
	(26) Landscape epigenetic processes	10	From Perel'man
Historical landscape geochemistry			
	(27) Weathering crust	11	From Polynov
	(28) Weathering zone	11	From Polynov
	(29) Orthoeluvium, paraeluvium, neoeluvium	11	From Polynov
	(30) Chronosequences	11	From Jenny
Geochemical landscape classification			
	(31) Landscape series, group, type, family, class, genus, species	12	From Perel'man
	(32) Permaciedal landscapes, impermaciedal landscapes, illuvial landscapes, surface eluvial (frozen) landscapes	12	From Glazovskaya

Unfortunately, the formulation of new postulates as described here brings with it new uses for old terms and some new terms created for the new conceptual model of nature. The principal terms used in this book are listed in Table 17.1. In order to assist the reader, the chapter in which the terms are explained is also included in Table 17.1.

These terms are important to the student of landscape geochemistry at whatever level he studies the subject. The casual reader or the professional geochemist now involved in environmental geochemistry may wish to study only the thinking and approach which resulted in the definition of the terminology. Clearly to apply this thinking, one does not have to use the suggested terminology, although for comparative purposes, such an application is to be desired. In general, the relationships indicated in Fig. 17.1 combined with a list of terms in Table 17.1 provide readers with a starting point for the application of the discipline of landscape geochemistry to specific interests and problems.

Summary and Conclusions

This chapter outlined links between the different basics of the discipline of landscape geochemistry as described in this book. The importance of the idea of the geochemical classification of landscapes as the central theme of the discipline was stressed. I believe that the application of the method of thinking summarized in this chapter to environmental geochemistry will stimulate holistic thinking which is badly needed today when unconnected facts regarding the distribution and amount of chemicals of all kinds in the environment are commonplace without connecting concepts.

Discussion Topics

1) Which of the basics of landscape geochemistry is of greatest potential importance in exploration geochemistry and why?

2) List the seven basics of landscape geochemistry in order of importance to geology, soil science, and ecology. Explain reasons for your listings.

3) List the basic concepts of ecology and terminology associated with them as a parallel to Table 17.1. Comment on both lists from the viewpoint of general environmental science.

Part III. Applications of Landscape Geochemistry

"There is nothing more difficult to carry out, nor more doubtful of success, nor more dangerous to handle, than to initiate a new order of things. For the reformer has enemies in all who profit by the old order, and only lukewarm defenders in all those who would profit by the new order. This luke-warmness arises partly from the fear of their adversaries who have the law in their favour; and partly from the incredulity of mankind, who do not truly believe in anything new until they have actual experience of it."

Machiavelli, *The Prince* (1513) as quoted by Stafford Beer in *Decision and Control* (New York, John Wiley & Sons, 1966), p. 3.

18. Practical Applications of Landscape Geochemistry

"Lest we become tangled in the web of philosophical argument, it is probably best to view the scientific method in terms of its most evident characteristics. For one thing this approach places stress on gaining knowledge through the process of observation. Whatever is said about behaviour is reasoned from systematic observation and is tested and retested by observation. In other words the so-called scientific approach attempts to anchor knowledge in terms of the physical reality it purports to explain."

Frederick Williams, *Reasoning with Statistics* (New York: Holt, Rinehart and Winston Inc., 1968), p.3

Introduction

Part III describes different ways in which the discipline of landscape geochemistry may be applied in environmental science. Chapter 18 involves the application of individual concepts and principles of landscape geochemistry to specific problems in environmental geochemistry. From one point of view this application is already a reality, because most studies of environmental geochemistry include the concept of element abundance in one form or another. However, it is the other six basics of the discipline, and their applications in different kinds of fundamental and practical environmental geochemistry, which are the subject of this chapter. In Chapter 19, suggestions are made regarding the application of the discipline of landscape geochemistry as a totality to the study and solution of five classes of environmental problems—one of which involves the dissemination of geological information to nonscientists involved in the political decision-making process about the environment.

This chapter distinguishes between fundamental and practical applications of landscape geochemistry; examples of both are indicated in Figure 18.1. The distinction involves the kind of problem involved. A fundamental problem is one which originates with landscape geochemistry and its application in the environment. For example, the idea of using a generalized conceptual model of a landscape section as a starting point for the selection of field areas for environmental studies stems directly from the principles of landscape geochemistry which had not been approached previously. This is, therefore, a fundamental problem. A practical problem involves the adoption of basics of landscape geochemistry to a situation which can, and often has, been solved using other methods. For example, Levinson (1974) did not invent the concept of an orientation survey in exploration geochemistry but he was one of the first workers to draw the atten-

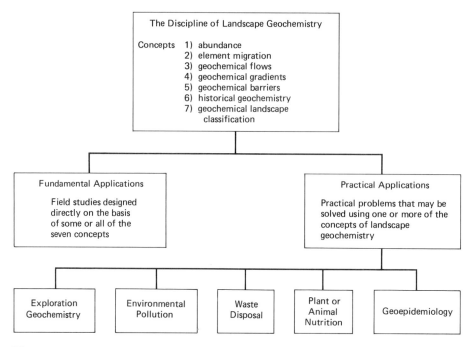

Figure 18.1. Fundamental and practical applications of the discipline of landscape geochemistry.

tion of exploration geochemists to the advantages of using landscape geochemistry in relation to it. In this chapter two examples of fundamental applications of landscape geochemistry and 18 examples of practical applications of landscape geochemistry are described. It should be stressed that these examples represent only a small fraction of the potential applications of the discipline—several books would be required to examine the scope of the potential in detail.

Examples of Fundamental Applications in Forestry and Mineral Exploration Research

In forestry, particularly in areas in which the landscape conditions have been mapped for forest management purposes, it is sometimes desirable to locate small test areas where the geochemical, geological, pedological, and ecological conditions are relatively simple and uniform. Such test sites may then be used to describe general landscape conditions typical of much larger areas. For example, some years ago the forests of Ontario were the subject of a land use survey based on holistic principals (Hills, 1959), and a small area of similar forest cover in Tennessee (the Walker Branch watershed at the Oak Ridge National laboratory) was subjected to a detailed watershed study. A conceptual model based on the principles of landscape geochemistry was described by Fortescue *et al.* (1973) to facilitate the selection of similar landscape sites for a comparative study of nutrient cycling in both areas. At the same time a standardized approach to appraisal level sampling of sites selected for study was also worked out.

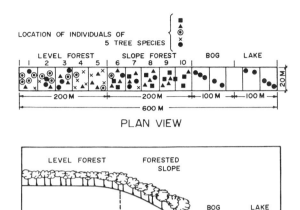

Figure 18.2. Conceptual model of a landscape section located in a deciduous forest area including a plan of the sampling strip required for field investigations (Fortescue *et al.*, 1973).

Briefly, the conceptual model involved the description of a sampling strip to be located at right angles to a hillslope in a forest where eluvial, transeluvial, superaqual, and aqual (i.e., lake) landscape types lie adjacent to each other (Figure 18.2). For practical considerations, the width of the strip was 20 m and its length not more than 600 m. The preliminary, or appraisal approximation, geochemical surveys planned for the sampling strips were designed to obtain topographical, geological, pedological, groundwater, ecological, geochemical, and biogeochemical data, using standard methods. The purpose of the preliminary surveys was to establish the uniformity of the landscape conditions within the sampling strips as a preliminary to more intensive research.

Eventually, two sampling strips were selected and studied as planned. One was located near Dorset, Ontario and the other at the Walker Branch Watershed (Fortescue et al., 1973). The Dorset site includes a transeluvial landscape, a superaqual landscape, and an aqual landscape within a traverse 400 m long. The Oak Ridge site involved only a transeluvial landscape 160 m long. Let us consider briefly the Oak Ridge site data and information. A conceptual model of the sampling strip together with plans showing relief, soil sample pit locations, and the location of overstory trees included in the biogeochemical survey are included in Figures 18.3 and 18.4.

Data obtained from the appraisal geochemical surveys are listed on Figure 18.5. The data for manganese in soil profiles reveal an atypical vertical distribution pattern for this element in Plots 5 and 6 which is less marked in the case of zinc. The appraisal tree branch data suggest, but do not prove, that the content of manganese increases in the branch material of chestnut oak and maple down the slope in contrast to white oak where the amount was found to decrease downslope. The tree branch data for zinc indicated that hickory was likely to be an accumulator species for this element. On the basis of these data and similar

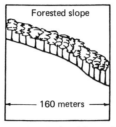

Figure 18.3. Model for the study of a transeluvial landscape section at the Walker Branch Watershed, Oak Ridge National Laboratory, Tennessee (Fortescue et al., 1973).

results for magnesium, aluminium, copper, and lead, the sampling strip at Walker Branch lacked the uniformity of landscape conditions required for more detailed landscape geochemical research of the type envisaged for the site.

Incidentally, as Perel'man (1966) pointed out, isolated studies of the geochemistry of landscapes should be called after the location in which a particular con-

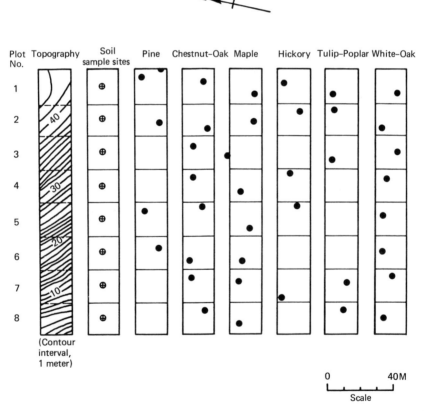

Figure 18.4. Sampling strip maps of a transeluvial landscape section in the Walker Branch Watershed, Oak Ridge National Laboratory, Tennessee (Fortescue et al., 1973).

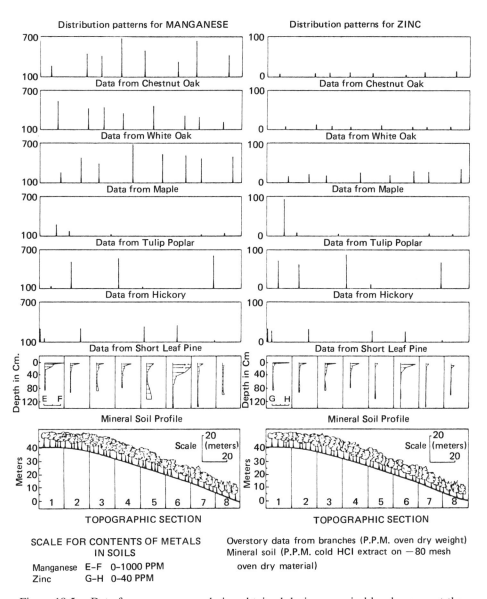

Figure 18.5. Data for manganese and zinc obtained during appraisal level survey at the Walker Branch sampling strip (Andren et al., 1973).

ceptual model, or plan, was first used. Therefore, these two studies would be said to use the "Dorset Conceptual Model."

The second example of a similar, but more elaborate, conceptual model was devised for the solution of a problem in plant prospecting methods research at the Geological Survey of Canada (Fortescue and Hornbrook, 1967, 1969). The approach and the conceptual model were devised by the writer as an experiment to explore the possibilities of collecting like sets of morphological and geochemical data from landscapes in which undisturbed mineral deposits were known to occur. The use of the conceptual model was described in a paper delivered at the XI Botanical Congress in Seattle in 1969 (Fortescue, 1970).

Briefly, the need for the conceptual model arose in the following way. During the 1960s the Geological Survey of Canada was responsible for developmental research on geochemical prospecting methods in Canada. The aim of the plant prospecting methods research project was to discover the effects of chemicals, derived from the weathering of buried mineral deposits, on the form and chemical composition of plants growing in landscapes in which such deposits occurred. The Geological Survey staff were in a unique position to do this research because they often had access to areas in which mineral deposits had been located by companies before the landscapes were disturbed by mining activities. Because of the impending mining activities in some areas, it was imperative that the plant prospecting methods research team respond quickly to calls for studies from any part of Canada. Also, because of the unique opportunity presented by the availability of such sites for study it was most desirable that field investigations produce like and comparable sets of geochemical and biogeochemical data which could be published later and compared, both visually and statistically, with similar sets of data obtained elsewhere. This led to the development of a holistic, conceptual approach to the selection of sample traverse lines and a research program involving three successive approximations for the study of sites in which landscape conditions were favorable. The long-term objective was to so standardize the biogeochemical research at mineral deposits that the published data could be used directly by exploration geochemists for purposes of comparison of conditions in similar areas during prospecting. The approach was also holistic with respect to the choice of elements studied because it was believed that analytical methods involving the simultaneous determination of 10 or more chemical elements in plants (and related soils) would allow for (1) the study of the behavior of elements derived from the mineral deposits, and (2) the behavior patterns for other elements which were not related to the mineral deposit at a particular site but included in deposits studied elsewhere.

After a certain amount of trial and error a generalized conceptual model of a landscape section (Figure 9.1) located in the vicinity of a mineral deposit similar to that described by Sergeev (1941) was decided upon for first approximation studies. There were three approximations in the research program:

1. Visits (first appraisal approximation) involving the inspection of the site during a 37-day period by a crew of two who would fill in a checklist of information regarding the site conditions and sample a single site set located directly on the basis of the conceptual model.

2. Pilot Projects (second approximation)[1] involving a crew which for approximately 6 weeks collected material from four or more site sets.

3. Main Projects (third approximation) to be carried out during one or more summers at a site specially selected on the basis of comparisons of information

---

[1] Second approximation studies carried out at short notice in areas being stripped were called quick projects. The most important quick project was completed at the Texas Gulf Sulfur deposit near Timmins, Ontario (for details see Fortescue and Hornbrook, 1969).

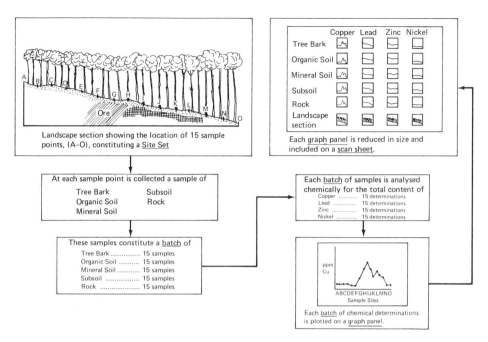

Figure 18.6. Conceptual model of a landscape section used in plant prospecting methods research showing the explanation of the terms "site set," "batch," "graph panel," and "scan sheet" and relationships between them (Fortescue, 1970).

obtained from a number of pilot projects which would involve both general (i.e., using site sets) and detailed research.

A site set conceptual model required the location of 15 sample points at regular intervals along a traverse line. In the ideal case, five points would be located upslope from the deposit, five over the deposit, and five downslope (Figure 18.6). Samples of the same natural material collected from the sample points of a site set (i.e., 15 samples of the same material) would be called a batch of samples each of which would be identified by a letter code relating to the points of origin in the site set—starting with "A" at the uppermost sample point. Batches of samples and subsamples of soil, or plant, material would all be analyzed for the same elements and the data plotted on a series of "graph panels" each of which showed the content of an element in a sample material collected from each of the points in a site set (Figure 18.6). The graph panels would later be assembled into a "scan sheet" which would show the patterns for the distribution of the same element in different sample materials as columns of graph panels and the distribution of different elements in the same sample material as rows of graph panels (Figure 18.6). Advantages of this approach include the uniformity of scan sheets which can be prepared by computerized methods from the data from automated instrumental analysis. The modular nature of the sampling model, which can be used for the collection of samples along lines in the field (or for other groups of 15 samples), for example, adjacent samples collected in the side of a trench

Table 18.1. A Classification Scheme for Secondary Dispersion Haloes and Trains in the Southern Urals Region.[a,b]

A. Mechanical with superposed salt haloes

In products of weathering of igneous rocks (orthoeluvium)		In products of weathering of sedimentary rocks (paraeluvium)		In alluvium (neoeluvium)	
Surficial (in soils)	Deep (in overburden)	Surficial (in soils)	Deep (in overburden)	Bottom deposits (of water bodies)	
SD	MD	SD	MD	SD	MD	SD	MD	SD	MD

B. Salt haloes

Residual salt haloes

In products of weathering of igneous rocks (orthoeluvium)		In products of weathering of sedimentary rocks (paraeluvium)		
Surficial (in soils and plants)	Deep (in overburden groundwater)	Surficial (in soils and plants)	Deep (in overburden and groundwater)	
N&AM	AM	N&AM	AM	N&AM	AM	N&AM

In alluvium of quaternary deposits (neoeluvium)

Surficial (in soils and plants)		Deep (in overburden and groundwater)	
Young neoeluvium	Newest neoeluvium	Young neoeluvium	Newest neoeluvium
AM	N&AM	AM	N&AM	AM	N&AM	AM	N&AM

Accumulative salt haloes

Superaqual		Superaqual	
In soils, plants, and groundwater		Buried in alluvium and groundwater	
Oxidizing and reducing	Permanently reducing	Oxidizing and reducing	Permanently reducing
AS	SF	AS	SF	AS	SF	AS	SF

Subaqual

Temporary waters, streams		Lakes	
Oxidizing to reducing		Oxidizing to reducing	Permanent reduction
Salinated	Salt free	In silt	In siltstone	In silt siltstone

[a] From Glazovskaya et al. (1961).
[b] SD, slightly displaced; MD, much displaced; AM, acid medium; N&AM, neutral and alkaline medium; AS, accumulation of salts; SF, salt free.

located above the mineral deposit would also be processed in the laboratory as a site set. As originally envisaged, the scan sheet would be to exploration geochemistry what a regional geological map would be to an exploration geologist. It would be the first approximation which could be used as a guide to the identification environments in which second approximation study was required in landscapes to confirm the presence of mineralization.

The Geological Survey of Canada example has been discussed in some detail because it illustrates well how the holistic approach of landscape geochemistry can be applied to problems which were traditionally studied in a fragmented and nonholistic manner. It also facilitated the collection of the maximum field and laboratory data and information in the minimum of time and effort owing to the combination of a conceptual model for site investigation planning and a rigid and modular system for analytical treatment of samples in batches. These considerations are important when environmental geochemical information for purposes of comparison is collected by a government agency.

Practical Applications of Landscape Geochemistry

Let us now consider applications of landscape geochemistry to specific environmental problems. (Figure 18.1).

Exploration Geochemistry

Reference has already been made to the use of generalized conceptual models in exploration geochemistry in Canada by Bradshaw (1975) (Figure 15.8). Much earlier Glazovskaya *et al.* (1961) described a project involving the application of geochemical mapping on a regional scale in the Southern Urals of the USSR. As a result of these mapping activities a classification scheme for secondary dispersion haloes and trains in the Southern Urals Region was described (Table 18.1) based on the following criteria:

1. The distinction between dispersion due to mechanical as opposed to predominantly physicochemical and biological processes.
2. The distinction between dispersion and accumulation of chemical entities with respect to the chemical environment of the landscape.
3. The role of bedrock and surficial geological conditions in relation to the location of geochemical anomalies.
4. The distinction between secondary salt haloes formed by residual as opposed to cumulative processes in landscapes. (Note: the climate of the area is semiarid.)

The approach adopted by Glazovskaya (1961) and her co-workers of (1) mapping an area geochemically followed by (2) the classification of types of dispersion haloes and trains which occur within the area associated with buried mineral deposits is potentially a very powerful tool in exploration geochemistry. Today's geochemical maps on a regional scale could be prepared ahead of ground surveys based on available information, for example, using a combination of geological, climatic, geomorphological, groundwater, soil cover, and plant cover information obtained from aerial photographs and remote sensing surveys. Conceptual models which synthesize the information on geochemical anomalies due to known mineral deposits in the area could be used as a starting point for a general classification of geochemical anomalies near mineral deposits in the region. It is surprising that government agencies in the western countries have not, as yet, considered geochemical mapping of landscapes as an aid to prospecting although geophysical mapping of magnetics, and other landscape parameters, has been a routine procedure for several decades. Geochemical landscape maps would also be of value for land use planning (e.g., in areas in which open pit mines have led to the local accumulation of toxic chemicals in the environment).

Levinson (1974, pp. 199–200), after stressing the importance of orientation survey, noted that the concept of the landscape prism could be used to focus attention on the components of landscapes which were to be included in such surveys:

> . . . a preliminary study, called an *orientation survey,* should be conducted in every new area. The objective of an orientation survey is to determine the optimum field, analytical and interpretive parameters which can distinguish an anomaly from background. Some of the important parameters include: (1) the type of geochemical dispersion that exists in the area; (2) the best sampling medium; (3) the optimum sampling interval; (4) the soil zone and depth from which the samples should be taken; (5) the size fraction to be analyzed; (6) the element, or group of elements, which should be analyzed, and by what analytical technique; (7) the effects of topography, hydrology, drainage, glacial history, climate, rainfall, vegetation, organic matter and Fe–Mn oxides on the dispersion of metals; (8) the upper limit of background values (threshold) in rocks, soils, and waters; (9) the most efficient manner of sample collection and analysis; (10) whether or not geochemical methods are feasible; and, (11) if contamination is likely to be a factor in the area.

His use of the landscape prism concept is illustrated on Figure 18.7.

A simple example of an experimental orientation survey was described by Fortescue and Hornbrook (1967) who sampled soil, humus, tree bark, tree twigs, and needles from conifers growing in the vicinity of a lead deposit near Silvermine, Nova Scotia (Figure 18.8). The purpose of the study was to determine at the site which sample material was most suitable for geochemical surveying for lead in the area. On the basis of the study it was concluded that humus was superior to tree organs as a material for geochemical prospecting surveys in the area.

In summary, there are many potential applications of the principles of landscape geochemistry in mineral exploration. In this section three examples have been described: one involving geochemical mapping on the regional scale and the description of a classification of dispersion trains and haloes, and two others involving the theory and practical application of orientation surveys. It is suggested that a standardized format for the conduct of orientation surveys based on landscape geochemistry in exploration geochemistry might be worked out by

Practical Applications of Landscape Geochemistry 265

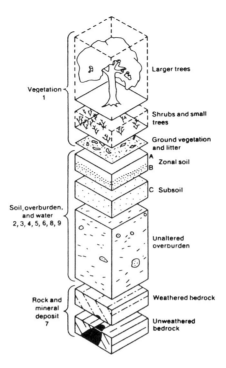

Figure 18.7. Landscape prism drawn to show environments and materials of importance in an eluvial landscape (Levinson, 1974).

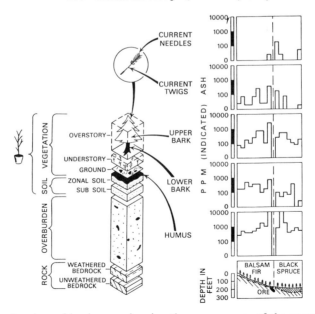

Figure 18.8. A prism of landscape showing the components of the vegetation and soil sampled during an orientation biogeochemical survey in the vicinity of a lead deposit near Silvermine, Cape Breton Island, Nova Scotia (Fortescue and Hornbrook, 1967).

a government agency so that workers in exploration geochemistry could use the same format for orientation surveys on a worldwide basis. Such a procedure would be a step toward formal application of principles of landscape geochemistry in exploration geochemistry.

Environmental Pollution

During the past 20 years scientists of many disciplines have been concerned with the study of environmental problems associated with man's pollution of the environment. Environmental pollution may result from the addition of solids, liquids, or gases to landscapes or from the wholesale disturbance of landscapes by mining operations. When liquids or gases are disposed of, eluvial, superaqual, and aqual landscapes are all involved and a holistic approach to the study of environmental effects is often desirable.

Let us consider the basics of landscape geochemistry in relation to two examples of environmental pollution. The first involves the contamination of roadsides by lead derived from automobile exhausts and the second the pollution of an extensive area of forest by a point source of sulphur dioxide fumes formed during the sintering of iron ore over a multiyear period.

Since the classic paper of Cannon and Bowles (1962) many workers have studied the buildup of lead and other elements discharged by automobiles in landscapes adjacent to highways. Studies of this type involve a line source of pollution (i.e., the highway) which is usually built to traverse areas in which the landscape (including soils) has been constructed by man's activities. Consequently, from the viewpoint of landscape geochemistry, the study of the accumulation of lead in a soil of this type represents a positive ELF on a synthetic soil and plant system.

A highway may take many years to complete. Consequently, it is possible to study the cumulative buildup of lead along it at stages during a multiyear period. One can measure the cumulative effect of the positive lead ELF during different periods of time by collecting samples at the same instant in time (provided that the volume of traffic along the highway is considered uniform over the multiyear period). A study of this type was completed by Mr. T. Chapman, one of my students at Brock University, for his B.Sc. project in 1973 (Fortescue, 1974a). Briefly, his study involved the determination of the lead content of soils and associated plants alongside the Queen Elizabeth highway near St. Catherines, Ontario, Canada. This highway was selected for two reasons: parts of it had been reconstructed over a 9-year period, and the prevailing wind was in a direction across the highway. Data from two traverses made in 1973 on parts of the highway constructed during 1972 and 1964 are included in Figure 18.9.

Several points of interest are apparent from the data. If the background level for lead in the soil parent material is taken to be 25 ppm lead (partial abundance obtained from cold 10% HCl leach of oven dried -80 mesh fraction of the soil) then the traverse across the road at Site 1 (1972) indicated little effect of the positive ELF for lead in the surface soil (at a depth of 6–12 or 12–18 cm). The variation in the values for lead probably related to variations in the lead content

	Distance in meters from roadside						HIGHWAY	Distance in meters from roadside					
	36.0	16.0	11.5	7.5	4.5	2.5		2.5	4.5	7.5	11.5	16.0	36.0
Grass	2.1*	3.2	3.2	3.8	3.4	3.5	Site 1 (1973) East → ← West	3.3*	2.4	1.9	3.2	2.3	1.3
Soil Surface	20**	10	16	35	4	18		22**	10	22	18	18	26
6-12 cm	18	4	8	24	4	16		22	24	22	4	24	20
12-18 cm	10	4	-	26	18	18		18	18	10	18	28	22
Grass	3.3*	4.6	7.7	14.9	12.2	13.9	Site 4 (1964) East ← West →	7.6*	9.0	9.6	9.2	8.7	5.7
Soil Surface	40**	54	74	118	410	772		282**	300	314	162	66	52
6-12 cm	24	22	14	78	28	464		30	28	16	22	24	20
12-18 cm	18	16	6	16	18	36		28	22	22	18	24	18

*parts per million oven dry weight   **parts per million -80 mesh soil (oven dry)

Figure 18.9. The distribution of lead in grass and soils collected adjacent to the Queen Elizabeth Highway near St. Catharines, Ontario, 1973. (Data from Fortescue, 1974a from an unpublished B.Sc. thesis, Department of Geological Sciences, Brock University, 1973 by T. Chapman.)

of the parent material which includes two high values (i.e., 35 and 26 ppm) in the surface soil. In contrast, the 9-year old soils show a marked enrichment of lead in the surface soil which extends down into the subsoil adjacent to the road (Figure 18.9). Geochemical gradients for lead in the surface soil extend some distance on either side of the road. The gradient is most marked at the east of the road because the prevailing wind is from the west. There is also a vertical gradient of lead in soil due to the positive ELF extending downwards to 18 cm at 25 m from the road and for 12 cm, 45 m to the west and 75 m to the east of the road. Thus the mineral soil surface layer acts as an effective geochemical barrier to the downward flow of lead some distance from the road but not adjacent to it. The plant material also has a horizontal geochemical lead gradient (Figure 18.9). In this case the gradient is visible in both Site 1 and Site 2 traverses. Appraisal studies of this type are most useful in landscape geochemistry because they establish limits for concentration ranges and indicate the presence, or absence, of distribution patterns which can be studied later in more detail by second or third approximation techniques which would be impractical during the initial investigations owing to the effort required. As a footnote to this investigation, it is interesting that the data obtained from the soils for manganese (not included here) indicated that there was a different level of this element in Site 1 compared with Site 2. This was interpreted as indicating that the source of the fill used in the construction of the highway was not the same at the two sites. Thus variations in the chemistry of artificial landscapes, as revealed by multielement appraisal level surveys, may provide clues to the chemistry of soils and plants which can be followed up in detail later during second and third approximation studies.

An excellent example of the relatively large-scale pollution of an area of coun-

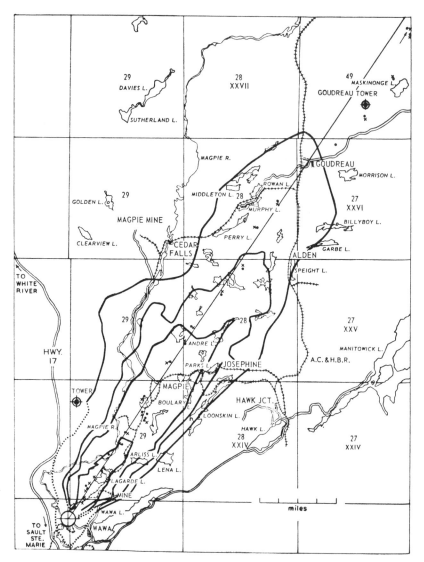

Figure 18.10. Map showing transect of quarrats and water samples in relation to approximate boundaries (June 1960) of plant cover damage as established by a study of aerial photographs (Gordon and Gorham, 1963).

try by a single point source of atmospheric pollution was described by Gordon and Gorham (1963) in Figure 18.10. Briefly, at Wawa, Ontario, Canada, sulfur dioxide and related chemicals derived from roasting sulfide ores were disposed of into the atmosphere from a single stack located in an area where there was one prevailing wind direction (Figure 18.11). The effects of the atmospheric pollution during a multiyear period affected the landscape downwind to a significant

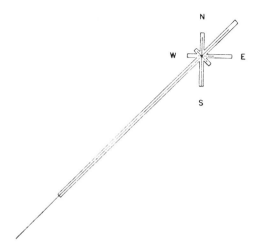

Figure 18.11. Wind Rose for Wawa May to October (inclusive) 1958 and 1959. Calms made up 8% of the total records. Blocks represent all winds, center lines represent winds above 15 mph which make up 10% of total records (Gordon and Gorham, 1963).

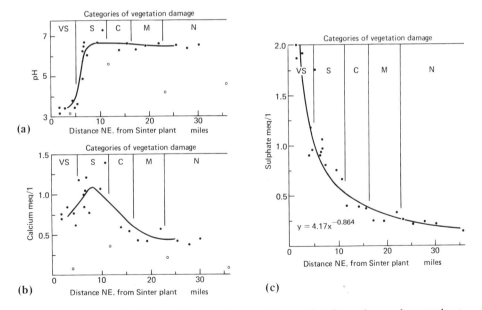

Figure 18.12. Geochemistry of the environment downwind from the smelter stack at Wawa: (a) acidity of lake and pond waters (running average of five values used to fit the curve), (b) calcium in lake and pond waters (points represent running averages of five points each, (c) sulfate content of lake and pond waters (Gordon and Gorham, 1963).

Table 18.2. Categories of Plant Cover Damage Established for Use in the Aerial Survey.[a,b]

Category	1 Very severe	2 Severe	3 Considerable	4 Moderate	5 Not Obvious
Overstory condition and surviving tree species	None	None	Canopy much broken up. Many dead trees beginning to fall. A few white birch and white spruce remain alive	Many conifers reddened. Some chlorotic. Tip kill and crown thinning evident on hardwoods. Surviving species: *Bw, Sw, Sb, A, Pj, Ce, L, and Fb (a few Pw still alive)	Overstory dominating. Closed canopy: *Bw, Sw, Fb, Sb, Pj, A, Ce, L (Pw if present may be chlorotic)
Understory	Almost wholly destroyed, some *Sambucus pubens* remaining alive but heavily damaged	Much destruction. *Pyrus decora* nearly all dead; *Acer spicatum* heavily damaged and suckering; *Sambucus pubens* showing tip killing	Understory including tall shrubs dominating. Much tip killing; *Pyrus decora* and *Acer spicatum* plentiful; *Sambucus pubens* entering the understory	Relatively little damage, *Sambucus pubens* rare or absent	Normal woodland flora
Ground vegetation	Mostly destroyed. A little *Polygonum cilinode* remaining but heavily damaged	Ground flora dominating. *Polygonum cilinode* predominant where cover thinning toward sinter plant; many normal woodland species still plentiful toward outer boundary but floristic variety declining. Tree seedlings absent except few *Betula papyrifera*	Relatively little damage with normal number of species present. Tree seedlings still evident: *Betula papyrifera, Picea glauca, Picea mariana, Abies balsamea*	Normal woodland flora. *Polygonum cilinode* absent	Normal woodland flora. *Polygonum cilinode* absent
Erosion	Mainly bare rock remaining in exposed situations	Evident and increasing toward inner boundary	Not apparent	Not apparent	Not apparent

[a] From Gordon and Gorham (1963).
[b] Bw, white birch; Sw, white spruce; Sb, black spruce; A, trembling aspen; Pj, jack pine; Ce, white cedar; L, larch or tamarack; Fb, balsam fir; Pw, white pine.

extent (Figure 18.12) such that five zones of damage to plant cover of the area could be identified (Table 18.2). These zones could be identified on the ground and from aerial photos. They were in a nearly symmetrical pattern downwind from the smelter because of the prevailing wind direction. Further, because the smelter is located in a remote, forested area, the extent and nature of the pollution effects could be studied in detail at sites not significantly disturbed by man's activities. In addition to the effects on the morphology of the plant cover in eluvial landscapes Gordon and Gorham (1963) studied pollution effects in the aqual landscapes (ponds and lakes) of the area. They found variations in the acidity, calcium content, and sulfate content of waters located downwind along a geochemical gradient as indicated in Figure 18.12. The reader is referred to the original paper by Gordon and Gorham (1963) for further details on this interesting study.

In summary, the Wawa is an example of serious pollution of an area of coniferous forest for a relatively long period of time under a fortuitous series of environmental circumstances. The area was subjected to both appraisal (first approximation) (i.e., plant cover type mapping) and more detailed (i.e., chemistry of the waters) (second approximation) investigations and could provide a relatively simple location for the further study of long-term effects of sulfur dioxide on vegetation in Northern Ontario.

Waste Disposal

Landscape geochemistry holds much potential for the systematic study of environmental effects of the disposal of solid, liquid, or gaseous wastes. The concepts of abundance, element migration, geochemical gradients, and geochemical barriers can be readily applied to the study of the behavior of chemical entities in landscapes in which waste disposal has occurred. In general, wastes are of two kinds, those which produce definite toxic effects—such as the disposal of sulfur dioxide near the Wawa smelter—and others which affect the landscape but are relatively harmless. Let us consider examples of three types of waste disposal, one solid, one liquid, and one atmospheric.

The disposal of pulverized fuel ash near coal-fired power generating plants is a good example of the disposal of solid waste materials. According to Townsend and Gilliam (1975) a modern generating station of 2000 mW capacity produces about 2 tons of ash per minute. This material may be disposed of in worked out gravel pits or spread as a layer on the ground. The latter case usually involves the preparation of the ground for aesthetic purposes or for the production of economically valuable plant crops. The ash is usually sterile and devoid of nitrogen with 60% of the particles in the fine sand (0.2–0.02 mm) range, 40% in the silt range (0.02–0.002 mm), and almost no clay material. Pulverized fuel ash contains most inorganic nutrients in sufficient quantities for healthy plant growth. Problems associated with the preparation of this material as a substrate soluble for plant growth centered upon (1) the lack of nitrogen, (2) high pH, (3) relatively high salt concentrations, (4) high boron content and, in some areas, (5) a deficiency of phosphorus.

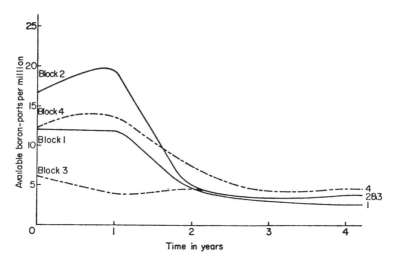

Figure 18.13. Variation in available boron in pulverized fuel ash with time: Ironbridge experiment Blocks 1 to 4 (Townsend and Gillham, 1975).

An interesting study of the geochemistry of pulverized fuel ash was described by Townsend and Gillham (1975) from the Ironbridge Power Station in the East Midlands of England. The pH of the ash rises to 11, or above, soon after combustion and then decreases with age to stabilize at around pH8. Similarly the water-soluble boron reaches a maximum about a year after commencement of leaching and then declines sharply within 3 years (Figure 18.13). In Figure 18.13 the two solid lines represent new ash and the broken lines which had been weathering for 20 years in a stockpile prior to being spread on plots. In certain plots, a hard pan developed in the ash at a depth of 20 to 50 cm whereas in other similar plots no pan formed. Vertical distribution patterns for the distribution of four elements in soils with and without such a pan are shown in Figure 18.14a and b. Other elements, such as aluminium and manganese, may also cause problems during the early years of plants.

Townsend and Gillham (1975) also provided comparative data for the uptake of 20 elements by plants growing on pulverized fuel ash and on normal soils (Table 18.3). They concluded that pastures can be successfully laid out on sites used for the disposal of pulverized fuel ash, of the type studied at Ironbridge, without any major problems with trace elements—except for those already mentioned. This example has been described in some detail because it indicates opportunities for the application of the principles of landscape geochemistry in areas in which solid waste materials are deposited above ground. A knowledge of the relative mobility of elements derived from such materials under different drainage conditions can facilitate the preparation of artificial landscapes to enclose them. This is an aspect of landscape geochemistry where the application of simulation systems modeling techniques is pertinent because the parent material of the new soil could be relatively homogeneous, permeable, and readily studied in detail with respect to changes during time.

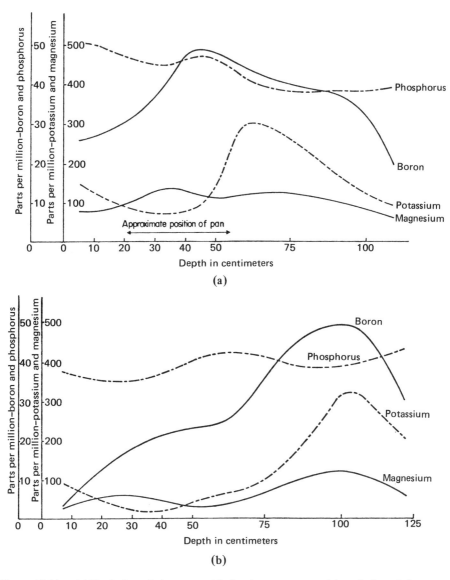

Figure 18.14. (a) Variation of elements with depth-pan present; (b) variation of elements with depth-pan absent (Townsend and Gillham, 1975).

Let us now consider generally and briefly the disposal of coal mine wastewaters. Glover (1975) provided a summary of the chemical types of wastewaters encountered in the vicinity of coal mining activities in the United Kingdom and noted three principal factors which reduce the geochemical effects of such waters in landscapes:

1. The reduction of the quantity of pyrite oxidation products which are discharged into the drainage.

Table 18.3. Summary of Trace Elements Found in Plants Grown on Pulverized Fuel Ash in Relation to Similar Plants Grown on Normal Soil.[a]

Crop	Al	As	B	Be	Cd	Co	Cr	Cu	F
Barley (leaf only)			1.6						
Barley straw on ash		0.64	27.0			0.07	0.30	5.8	7.18
Barley straw on soil		0.61	7.8			0.08	0.52	5.5	5.90
Barley whole plant on ash			85.0						
Barley whole plant on soil			16.0						
Barley grain on ash	5	0.68	20.7			0.03	0.18	3.7	1.36
Barley grain on soil		0.46	2.4			0.05	0.22	2.3	0.86
Rye grain	4		29					25.5	
Wheat grain on ash		0.13	5.1		0.2	0.023	Nil	5.4	
Wheat grain on soil		0.03	3.5		0.3	0.021	Nil	6.1	
Pastures on ash		2.3	16.2			0.15		9.0	3.2
Pastures on soiled ash		0.87	27.9			0.20		9.7	3.5
Timothy			56	0.97		1.6	3.0		
Lucerne		0.53	152	1.24		1.87	3.5		
Rape			68.6						
Kale			122	0.24		0.52	1.5	5.6	
Savoy and cabbage			68					20	
Lettuce			91						
Dwarf bean			46						
Pea (leaf)			72						
Pea (seed)			12						
Swede (leaf)			106						
Swede (root)						0.16	2.2		
Parsnip (leaf)			55						
Parsnip (root)			30						
Potato (leaf on ash)		0.5	97.5						
Potato (leaf on soil)			30.3						
Potato (tuber)		0.07	11			0.12	2		
Carrot						0.07	3		
Sugar beet (leaf on ash)		0.52	66.8			0.23	0.45	15.9	6.1
Sugar beet (leaf on soil)		0.50	45.2			0.20	0.63	9.7	6.9
Sugar beet (root on ash)		0.30	13.3	0.06	0.33	0.08	1.3	4.6	0.5
Sugar beet (root on soil)		0.28	16.6			0.06		4.8	0.5

[a] From Townsend and Gillham (1975).

2. The precipitation of pyrite oxidation products to form aluminium, manganese, calcium, and magnesium sulfates by clays and other minerals.

3. The mixing of acid with alkaline waters and subsequent aeration and sedimentation which may remove appreciable amounts of iron salts which would otherwise pollute the area of discharge.

Clearly, these criteria pose a challenge to the landscape geochemist who is concerned with the design and planning of mine waste water disposal schemes. The design of disposal sites requires a holistic approach to the planning of landscapes so that the purification of wastewaters can be carried out using geochemical gra-

Table 18.3. (cont.)

Parts per million											
Ga	Hg	I	Mn	Mo	Ni	Pb	Se	Sn	Ti	V	Zn
			22								
			12.5	2.79		3.5					34
			13.0	1.10		3.0					24
			8.3	1.78		0.9					56
			11.0	0.87		1.0					35
			11.5								92
				0.90	0.8	<1					32.0
				0.38	0.9	<1					43.3
		0.16	83	5.14							65
		0.16	62	1.53			0.20				37
1.3				2.3	6.7	28		1.1	50	5.4	103
1.94				15	6.4	33		1.6	53	4.4	99
0.43				5.0	4.2	11.0		2.06	16.4	1.3	72
			28.8	1.8							
			13								
			12								
			25								
			12.8	13.7							69.1
			25								
	Nil				1.6	<1					
			22								
			4								
	Nil		9		1.0	<1					
	0.06				1.9	<1					
			54.7	2.97		5.1					52.1
			44.0	0.60		5.0					47.1
0.07	0.004		12.1	0.22	1.3	2.2		0.99	3.6	0.23	28.8
			17.8	0.12		0.4					13.6

dients and geochemical barriers designed according to the chemistry and flow rates of the waters. It should be stressed that, up to now, there has been no formal application of the basics of landscape geochemistry to the disposal of wastewaters [although groundwater hydrologists, and other scientists, have studied in detail the behavior of wastewaters and groundwaters for many decades (Todd and McNulty, 1976; Hodgson and Buckley, 1975; Table 18.4)].

Not all wastewaters are of inorganic origin. Municipal wastewaters, including sewage sludges, are largely organic as are the agricultural wastewaters derived from feedlot operations and similar sources. The role of applied landscape geochemistry in the disposal of wastewaters was indicated by Thomas and Law

Table 18.4 The Composition of Drainage Waters from a Number of Coal Mining Activities.[a]

Source of Discharge	Quality	pH Value	Alkalinity to pH 4.5 (mg/liter CaCO₃)	Acidity to Phenolphthalein at the Boiling Point (mg/l CaCO₃)	Calcium (mg/liter)	Magnesium (mg/liter)	Dissolved Iron (mg/liter)	Suspended Iron (mg/liter)	Manganese (mg/liter)	Chloride (mg/liter)	Sulfate (mg/liter)
Underground mine working, shallow depth	Naturally purified, low salinity	8.0	290	0	90	93	1.0	0	0.4	90	700
Underground mine working, medium depth	Naturally purified, medium salinity	7.9	580	0	176	137	0	0.5	0.1	6,900	700
Underground mine working, maximum depth	Naturally purified, high salinity	7.5	190	0	2,560	720	0.6	0.2	0.9	30,800	350
Flooded underground mine workings	Alkaline and ferruginous	6.9	340	0	190	130	25	21	6	42	1,720
Shallow underground mine workings, gravity flow	Acidic and ferruginous	2.9	0	480	125	88	122	0	7	50	1,250
Spoil tip seepage	Acidic and ferruginous	4.6	5	580	250	230	23	17	10	95	2,300
Coal stock pile seepage	Acidic and ferruginous	3.1	0	1,100	n.d.[b]	n.d.	160	0	9	80	1,220

[a] From Glover (1975).
[b] n.d., signifies not determined.

(1977) in an article in the book *Soils for Management of Organic Wastes and Waste Waters* (Elliott and Stevenson, 1977, p. 47) as follows:

> The planning, design and implementation of land based waste water management projects should include assessment of the properties of waste water at an early stage. The object of this early assessment is to weigh the properties of the waste water relative to stated or implied criteria and to make sound judgments leading to a successful project. The important key to early assessment is the desire to avoid circumstances that may result in disruptive and costly revisions during the construction of facilities or to facility operation of a newly constructed system. It is equally important to avoid costly overdesign of a facility.

Thus, there is a need for the systematic and holistic study of landscapes where wastewater disposal is contemplated. The data obtained from such studies would be valuable during the design of the wastewater management project in the area and as a basis for comparison in landscape geochemistry. It is likely that the use of a generalized conceptual model involving prisms of landscape and flow rates for waters in the subsurface components would facilitate comparisons between different sites. Not all problems of liquid waste disposal involve waters. For example, environmental contamination by spills of hydrocarbons presents a considerable challenge in landscape geochemistry, particularly with respect to the use of microorganisms for the biodegradation of the hydrocarbons in soils and aquifers. Further information on this interesting subject can be obtained from Vanlooke et al. (1975).

So far we have considered waste disposal by solids on a local scale and wastewaters on a regional scale. Let us consider now the effects of waste disposal on a global scale from nuclear testing. Pierson (1975, p. 86) noted that radioactive debris in the atmosphere is of considerable interest to the environmental geochemist:

> It has been demonstrated that the passage of nuclear weapon debris from the site of the explosion can be followed through the atmosphere and back to the ground. By observation of this radioactive tracer it has been possible to elucidate atmospheric behaviour and mechanisms that play important roles in the distribution and dispersal of any material, natural, or artificial, that is injected into the atmosphere.

He also noted that the first nuclear explosions were relatively small so that the debris was confined zonally within the troposphere and roughly to the latitude of the explosion site. But, since 1954 when thermonuclear devices first contaminated the stratosphere, the distribution of nuclear debris has been on a worldwide scale (Pierson, 1975). The abundance of the nuclide ^{90}Sr in fallout from the atmosphere from 1955 to 1970 is summarized in Figure 18.15. Pierson (1975, p. 81) interpreted the graphs as follows:

> The shapes of the curves of deposition from high yield explosions during the three phases of weapon testing up to 1959, 1961–1962 and after 1966—show peaks of deposition rate immediately after the first and second phases and the levelling during the third phase when the rate of injection into the stratosphere has been in fortuitous balance with the rate of depletion of the reservoir.

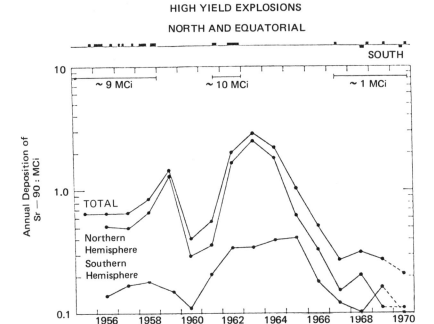

Figure 18.15. Annual deposition of ^{90}Sr during the period 1955–1970 (Pierson, 1975).

Based on ^{137}Cs Pierson (1975) provided a more complete picture of the atmospheric fallout rate of nuclides from 1953 to 1972. In this case all observations were made at monthly intervals at a single station at Chilton (Berkshire, England). In general, the pattern for the fallout of ^{137}Cs follows that for ^{90}Sr shown in Figure 18.15 except that, in the case of ^{137}Cs, there is a regular seasonal peak during the spring and early summer superimposed on the variations in ^{137}Cs due to nuclear explosions (Figure 18.16).

Some idea of how ^{137}Cs data can be used as a time marker in landscape geochemistry can be obtained from an interesting paper by Ritchie et al. (1974). They studied the amount of fallout of ^{137}Cs in soils and reservoir sediments in three small watershed areas in northern Mississippi. Their methods and findings are summarized in the abstract to their paper (p. 887):

> A budget for the distribution of fallout ^{137}Cs was calculated for three small watersheds in northern Mississippi. The cover in one contributing watershed is predominantly forest, in the second it is predominantly grass, and in the third predominantly grass and crops. Of the total ^{137}Cs input, 97%, 88% and 85%, respectively, remained in the forest, the grass and the cropgrass watersheds. The concentration of ^{137}Cs per unit area in the reservoir sediments was 2.8, 3.8 and 4.0 times that of the soils for the respective contributing watersheds, indicating that the reservoirs are acting as "traps" for ^{137}Cs. Of the ^{137}Cs eroded from the three contributing watersheds, 57%, 38% and 25% was found in the respective sediments. The calculated loss of ^{137}Cs per unit area was in this order: eroded soils > croplands > grass > forest.

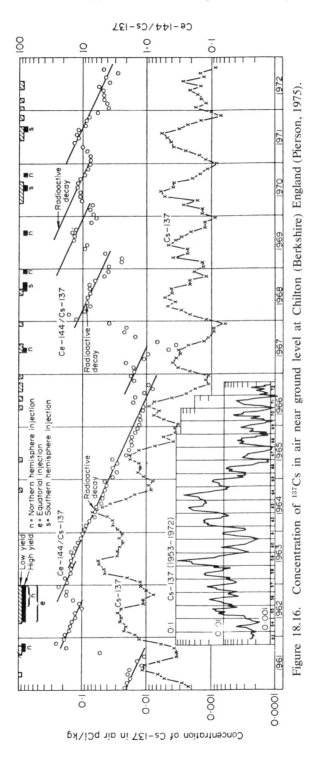

Figure 18.16. Concentration of ^{137}Cs in air near ground level at Chilton (Berkshire) England (Pierson, 1975).

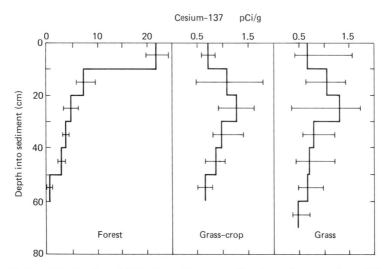

Figure 18.17. Distribution of ^{137}Cs in sediment profiles from three northern Mississippi reservoirs. Histograms represent the average for seven sediment cores; horizontal bars represent the range. Sediment–water interface is at 0 cm (Ritchie et al., 1974).

Figure 18.17 shows the distribution of ^{137}Cs in the sediment profiles. Briefly, it was concluded that the rate of sedimentation in the grass and grass/crop reservoirs during the past 10 years would be 2–3 cm/year—some two to four times the sedimentation rate for the forest watershed. Consequently, the peak for ^{137}Cs in runoff associated with the peak of atmospheric fallout in 1963–1964 (see Figure 18.16) is some 15 cm below the surface in the grass and grass/crop watersheds and, owing to the slower sedimentation rate in the forested watershed, within 5 cm of the surface in the reservoir associated with this cover type. This example indicates the potential of using radioactive fallout data for dating events which have occurred in landscapes. As in previous examples of this type cited in Part II, the dating process which used sediments in aqual landscapes complemented the information obtained from soils from surrounding eluvial landscapes.

In general, waste disposal in natural, or artificial, landscapes poses a challenge to students of landscape geochemistry interested in the migration of chemical entities in the environment. Such studies often must be made in artificial landscapes in which the slow evolution of landscape flows has not had time to mature and the natural stages in the evolution of the landscape (depicted in the Mt. Shasta example—Chapter 11) are intimately related to the migration of toxic substances. Consequently, the holistic and systematic application of landscape geochemistry to problems of waste disposal is considered a fertile field for future research.

Plant and Animal Nutrition

The literature of environmental geochemistry which is related to problems of plant or animal nutrition (including the nutrition of man) is both vast and detailed (see Underwood, 1973). From the viewpoint of landscape geochemistry Vinogradov and Perel'man (1966, 1972) stressed the importance of this aspect of the subject, usually with respect to the biogeochemistry of particular organisms or regions. Kovalsky (1970) provided a diagram (Figure 18.18) which summarizes the biogeochemical food chains. As Kovalsky (1970, p. 385) stated:

> The lack, or excess, of trace elements in the environment may cause a corresponding deficiency, or excess, of elements in living organisms (soil-micro-organisms, plants, human beings, animals) depending upon the biological state and nature of the organisms.

The study of environmental biogeochemistry has two principal aspects: one involves attempts which are made to provide conditions for the optimum growth of plants, or animals, which are used as food (or for some useful purpose) by man; the other is concerned with relationships between the geochemistry of the environment and the health of plants, animals, and man. The former is the subject of this section and the latter the subject of the section on "Geoepidemiology."

Let us consider two aspects of the role of geochemistry in plant growth as examples of where the holistic approach to environmental geochemistry might aid crop production. One area in which geochemistry interfaces directly with agriculture is soil-testing programs. Soil testing is used as a guide for the addition of fertilizers to produce optimum crop yields on farmlands. In principle, a sample of soil, usually surface soil, is collected from a farmer's field, and the content of certain elements (both nutrients and nonnutrients) is determined in the sample using partial extraction methods. On the basis of the interpretation of the results a decision is then made upon the need for chemical fertilization to produce specific crops. As Welch and Wiese (1973, p. 2) stated:

> Soil testing has been unusually successful in terms of farmer acceptance. There is little doubt that this acceptance stems from the increased yields and profits realized by applying recommended amounts of fertilizer and lime. However, many growers do not make regular use of soil tests and continue to apply rates of lime and fertilizer that were recommended when soil test levels were low even though the levels are now high and reduced, or maintenance rates are needed. There is also some concern for excessive applications of nutrients which may contribute to pollution problems. It becomes the responsibility of those involved in soil testing programs to provide the information and data needed to maintain public confidence in fertilization practices

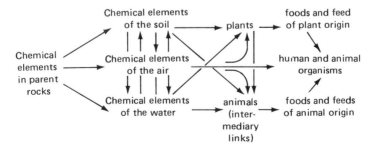

Figure 18.18. Biogeochemical nutritional chains (Kovalsky, 1970).

based on soil tests. If recommendations from soil tests are followed, the risk of environmental pollution will be minimized.

In many areas a complicating factor is that sewage sludge is also applied to farmers' fields as a fertilizer and as a convenient means of its disposal. When this occurs the positive extra landscape flows of sludge elements is combined with that from chemical fertilizers leading to complications with the levels of particular elements in crops and livestock. For example, the need for soil testing for animal feed production was discussed by Boyer et al. (1977):

> Any widespread use of animal wastes should also be studied for the effects on animal health and tissue elemental levels. Assuming a more or less constant concentration factor for each element with each pass through the animal and a more or less constant fractional absorption of each element during each pass, the potential exists for significantly elevated absorption for most elements only after a few passes. Whether multiple recycling of animal wastes actually results in elevated tissue levels can be determined only by appropriate feeding studies.
>
> Any widespread use of animal wastes or sewage sludge as animal feed ingredients would need to be monitored by periodic multielement and multiresidue organic analysis of the mixture being fed. (p. 269)

These opinions, and others like them, point to the need for the application of the concepts of landscape geochemistry in agriculture. This problem has already received considerable attention from environmental geochemists who have not, as yet, systematically applied the fundamentals of landscape geochemistry to its solution.

A relatively simple example of the type of problem which is encountered in agriculture was described by Bienfet *et al.* (1970). They studied a fertility problem in livestock which was suspected to be related to a trace element (or trace elements) deficiency, or toxicity, in the farm soils grazed by the animals. Accordingly, a comparison of soil and pasture composition of the affected farm was compared with similar observations at an adjacent farm where conditions were normal. The data obtained are listed in Table 18.5. These data are of interest here because they indicate the relatively large range of elements included in soil testing on the basis of partial extraction techniques, and in the case of farm fertility, indicate how chemical fertilization after the 1967 season affected the pH levels of elements in the soil and plants in 1968. Therefore, the holistic approach of landscape geochemistry has the potential to contribute to agricultural chemistry.

Another aspect of plant nutrition, of particular interest to foresters is the foliar analysis of trees to discover nutrient deficiencies. A problem with such studies is always the choice of organs for analysis and the stage of growth chosen for sample collection. This problem is common in both biogeochemistry and forest nutrition. A good example of a detailed study of the nutrient status (i.e., N, P, and K) of healthy and diseased trees was described by Leaf (1973). Briefly, the dieback disease of White Ash (*Fraxinus americana* L.) can be arrested if chemical fertil-

Table 18.5. Comparison of (a) Soil Contents and (b) Grass Contents of Elements in Samples Taken from the Farm Where Livestock Had Low Fertility (Fr) and Farms in the Vicinity with Normal Conditions. [a-c]

(a)	pH		In Exchangeable Amounts													
			mEq/100 g				mg/100g		ppm							
	H_2O	KCl	Ca	Mg	K	Na	P_2O_5	Zn	Al	Mo	Cu	Co	Ni	Mn	Pb	Fe
(Fr) farm Annual means																
in 1967	5.84	5.28	5.5	0.77	0.53	0.24	10.3	27.2	1328	1.4	5.4	1.6	2.5	112	17.5	460
in 1968	6.46	5.60	9.4	0.92	0.98	0.33	8.5	31.7	991	1.2	5.6	1.6	1.5	85	15.9	307
Level of significance of the difference between the means	++++	+	++	−	++	+	−	−	−	−	−	−	−	−	−	++
Farm in the vicinity Annual means																
in 1967	6.16	5.73	10.0	0.71	0.68	0.33	9.9	29.0	1121	1.2	4.5	2.0	2.0	317	2.30	792
in 1968	6.40	5.75	10.4	0.94	0.85	0.20	17.1	34.0	932	1.5	4.1	1.8	3.0	236	19.1	666
Level of significance of the difference between the means	−	−	−	−	−	+	++	−	−	−	−	−	−	−	−	−

(b)	mEq/100g						ppm								
	SO_4	PO_4	K	Ca	Mg	Na	Zn	Al	B	Cu	Mn	Pb	Fe	Cr	Co	Se
(Fr) farm Annual means																
in 1967	23.8	39.3	85.2	24.5	13.5	3.6	46.5	78	11.5	10.7	309	6.4	142	2.7	0.95	0.032
in 1968	25.0	43.7	100.6	25.9	13.3	8.4	81.5	148	12.5	12.7	165	12.1	212	4.5	1.6	0.021
Level of significance of the difference between the means	−	+++	+++	−	−	++	+++	−	−	−	++++	++		++	−	++
Farms in the vicinity Annual means																
in 1967	24.3	45.1	108.7	29.3	13.3	5.0	39.2	67	14.3	9.1	250	5.0	137	2.9	−	0.035
in 1968	26.5	49.0	102.0	29.3	13.4	3.9	58.0	128	11.7	13.9	286	7.1	122	4.6	−	0.036
Level of significance of the difference between the means	−	−	−	−	−	−	+	−	−	−	−	−	−	++	−	−

[a] From Bienfet et al. (1970).
[b] − not significant, $+P > 0.1$, $++ P > 0.05$, $+++ P > 0.02$, $+++++ P > 0.01$.
[c] The Ni and Mo contents in the grass are lower than 1 ppm.

izers are applied to the plants in time. Leaf (1973, p. 429) described his study as follows:

> In this work it is hypothesized that plant analysis techniques for species with compound leaves must separate leaflets from rachi and conduct the analysis on specific leaflets or rachi (R) rather than on entire leaf samples (Figure 18.19) (Table 18.6). From the data it is evident that opposite leaflets (1 and 7, 2 and 6, and 3 and 5 Figure 18.19) are quite similar in nutrient element status, particularly when expressed in mg per leaflet. However, from the data presented, it is suggested that leaflet 4 or the terminal leaflet and/or rachis (R) might be suitable tissues for compound leaves. It is believed that analyzing entire leaf samples of this compound leaf species may tend to

284 JOHN A. C. FORTESCUE

Table 18.6. Tissue Analysis of White Ash Foliage Collected from the Outer Extremities of the Upper Third of the Tree Crowns by Leaflets and Ranchi from Healthy and Dieback Diseased Trees.[a]

Leaflet Rachis[d]	Mid-June Concn. % Dry Wt.	Mid-June Concn. % Ash	Mid-June Content mg/ Leaflet	Mid-July Concn. % Dry Wt.	Mid-July Concn. % Ash	Mid-July Content mg/ Leaflet	Mid-August Concn. % Dry Wt.	Mid-August Concn. % Ash	Mid-August Content mg/ Leaflet	Mid-September[b] Concn. % Dry Wt.	Mid-September[b] Concn. % Ash	Mid-September[b] Content mg/ Leaflet
				Dry weights								
H-1			59.0			63.0			87.0			69.0
H-2			117.0			133.0			143.0			125.0
H-3			128.0			142.0			143.0			127.0
H-4			180.0			187.0			203.0			194.0
H-5			124.0			139.0			137.0			121.0
H-6			126.0			116.0			145.0			114.0
H-7			57.0			63.0			93.0			66.0
H-R			97.0			105.0			120.0			108.0
D-1[c]			56.0			66.0			73.0			
D-2[c]			108.0			125.0			127.0			
D-3[c]			102.0			111.0			115.0			
D-4[c]			138.0			153.0			156.0			
D-5[c]			100.0			114.0			116.0			
D-6[c]			107.0			126.0			128.0			
D-7[c]			57.0			68.0			72.0			
D-R[c]			116.0			115.0			124.0			
				Ash								
H-1	6.64		4.19	6.93		4.07	7.07		6.12	8.13		5.60
H-2	6.14		8.20	6.73		7.89	7.00		6.99	7.84		9.81
H-3	5.99		8.51	6.83		8.73	6.99		10.02	7.61		9.66
H-4	6.09		11.38	6.99		12.60	7.03		14.26	7.70		14.98
H-5	5.81		8.06	6.90		9.08	6.72		9.22	7.15		8.66
H-6	6.03		7.61	6.78		7.86	6.93		10.03	7.60		8.69
H-7	6.16		3.52	6.88		4.25	7.12		6.64	7.95		5.22
H-R	5.35		5.61	8.02		7.78	5.87		7.03	6.67		7.19
D-1	4.81		2.70	5.11		3.37	4.97		3.65			
D-2	4.42		4.77	4.95		6.20	4.63		5.88			
D-3	4.48		4.57	5.01		5.57	4.46		5.11			
D-4	4.73		6.51	5.11		7.84	4.62		7.02			
D-5	4.29		4.28	5.02		5.70	4.49		5.22			
D-6	4.24		4.53	5.31		6.69	4.65		5.96			
D-7	4.38		2.49	5.41		3.69	4.88		3.51			
D-R	3.74		4.34	5.48		6.28	4.56		5.68			
				Nitrogen								
H-1	2.65		1.67	1.83		1.07	1.78		1.54	0.93		0.64
H-2	2.33		3.10	1.93		2.26	1.82		2.59	0.76		0.95
H-3	2.30		3.26	1.87		2.40	1.66		2.37	0.85		1.08
H-4	2.25		4.21	1.88		3.39	1.76		3.57	1.32		2.57
H-5	2.05		2.85	1.91		2.37	1.73		2.38	1.08		1.31
H-6	2.33		2.94	1.93		2.24	1.91		2.76	0.84		0.96
H-7	2.21		1.26	2.23		1.40	1.86		1.74	0.97		0.64
H-R	0.77		0.81	0.78		0.76	0.60		0.72	0.55		0.59
D-1	1.52		0.86	2.61		1.72	1.24		0.91			
D-2	1.59		1.71	1.51		1.90	1.33		1.69			
D-3	1.51		1.54	1.49		1.66	1.28		1.47			
D-4	1.56		2.14	1.31		2.01	1.36		2.12			
D-5	1.56		1.56	1.60		1.82	1.31		1.52			
D-6	1.55		1.65	1.37		1.73	1.57		1.50			
D-7	1.66		0.94	1.61		1.10	1.44		1.04			
D-R	0.52		0.61	0.56		0.64	0.48		0.60			

Table 18.6. (cont.)

	Sampling Date											
	Mid-June			Mid-July			Mid-August			Mid-September[b]		
	Concn.		Content	Concn.		Content	Concn.		Content	Concn.		Content
Leaflet Rachis[d]	% Dry Wt.	% Ash	mg/ Leaflet	% Dry Wt.	% Ash	mg/ Leaflet	% Dry Wt.	% Ash	mg/ Leaflet	% Dry Wt.	% Ash	mg/ Leaflet
					Phosphorus							
H-1	0.27	4.07	0.17	0.21	2.99	0.12	0.27	3.76	0.23	0.19	2.36	0.13
H-2	0.27	4.40	0.36	0.20	3.00	0.24	0.24	3.74	0.37	0.19	2.44	0.24
H-3	0.26	4.37	0.37	0.22	3.22	0.28	0.26	3.69	0.37	0.17	2.29	0.22
H-4	0.27	4.43	0.50	0.22	3.20	0.40	0.27	3.78	0.54	0.22	2.90	0.43
H-5	0.25	4.34	0.35	0.22	3.24	0.27	0.26	3.84	0.35	0.19	2.73	0.24
H-6	0.25	4.11	0.31	0.22	3.41	0.25	0.26	3.78	0.38	0.22	2.84	0.25
H-7	0.26	4.15	0.15	0.23	3.01	0.14	0.25	3.57	0.24	0.21	2.67	0.14
H-R	0.25	4.75	0.27	0.26	3.27	0.25	0.46	7.89	0.55	0.28	4.24	0.30
D-1	0.20	4.20	0.11	0.18	3.52	0.12	0.17	3.46	0.13			
D-2	0.21	4.73	0.22	0.18	3.74	0.23	0.18	3.99	0.23			
D-3	0.21	4.73	0.22	0.18	3.69	0.20	0.18	4.08	0.21			
D-4	0.22	4.59	0.30	0.19	3.70	0.29	0.19	4.09	0.29			
D-5	0.20	4.78	0.21	0.18	3.62	0.21	0.20	4.45	0.23			
D-6	0.20	4.83	0.22	0.19	3.52	0.23	0.20	4.41	0.26			
D-7	0.20	4.57	0.11	0.18	3.33	0.12	0.19	3.99	0.14			
D-R	0.20	5.35	0.23	0.24	4.38	0.27	0.28	6.23	0.35			
					Potassium							
H-1	1.10	16.6	0.69	1.09	15.7	0.64	1.02	14.4	0.88	0.80	9.8	0.55
H-2	1.17	19.1	1.56	1.12	16.6	1.31	1.02	14.6	1.46	0.89	11.3	1.11
H-3	1.13	17.2	1.46	1.12	16.4	1.43	1.02	14.6	1.46	0.85	11.2	1.08
H-4	1.12	18.4	2.09	1.03	14.7	1.86	0.98	13.9	1.99	0.91	11.8	1.77
H-5	1.11	19.1	1.54	1.12	15.3	1.39	1.02	15.2	1.40	0.87	12.2	1.05
H-6	1.14	18.9	1.44	1.06	15.6	1.23	1.02	14.7	1.48	0.89	11.7	1.02
H-7	1.03	16.7	0.58	1.09	16.1	0.68	1.02	14.3	0.95	0.87	10.9	0.57
H-R	1.05	19.6	1.10	1.28	16.0	1.24	0.91	15.5	1.09	1.02	15.3	1.10
D-1	0.90	18.7	0.50	0.88	17.2	0.58	0.74	14.9	0.54			
D-2	0.90	20.4	0.97	0.90	18.2	1.13	0.72	15.5	0.91			
D-3	0.95	21.2	0.97	1.00	20.0	1.11	0.72	16.1	0.82			
D-4	1.05	22.2	1.44	0.98	19.2	1.50	0.74	16.0	1.15			
D-5	0.97	22.6	0.97	0.98	19.5	1.11	0.76	16.9	0.88			
D-6	0.92	21.7	0.98	1.00	18.8	1.26	0.69	14.8	0.88			
D-7	0.86	19.6	0.49	0.98	18.1	0.67	0.65	13.3	0.47			
D-R	1.10	18.4	0.88	1.12	15.3	1.03	0.78	12.9	0.64			

[a] From Leaf (1973).
[b] A. L. Leaf, Fertilization successfully arrests dieback of white ash. Unpublished data, State Univ. College of Forestry, Syracuse, N.Y.
[c] Foliage of diseased trees turned to autumn colors and dropped from trees before foliage of healthy trees turned to autumn colors.
[d] "H" denotes healthy trees; "D" denotes diseased trees; Numbers 1–7 denote leaflet position and "R" is rachis.

mask possible differences in nutrient element levels important for diagnostic purposes. One additional consideration of white ash is that leaves with both 7 and 9 leaflets may occur on the same tree.

In the same paper Leaf also noted that (Table 18.7) "the numbers of samples at predetermined levels of probability and precision of mean estimates that are

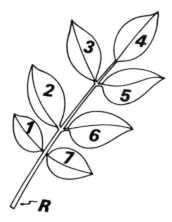

Figure 18.19. Configuration of leaflets of White Ash foliage (Leaf, 1973).

considered acceptable limits are related to the sample variance,'' (Leaf, 1973, p. 440). He calculated the data in Table 18.7 from the 1963 year's collection in the autumn of samples of red pine foliage from a 35-year-old plantation (Leaf, 1970). He summarized his observations as follows (p. 440):

> The numbers of samples required vary considerably within the accepted confidence limits. From this data it is apparent that needle lengths are less variable than fascicle weight, and N concentrations are the least variable while K concentrations are most variable of the nutrient elements reported. Since these data are based on only one year's collection from untreated red pine plantations, it is not evident what influence year to year environmental differences or fertilizer treatments may have on the sample size needs.

Table 18.7. Calculated Estimates of Numbers of Samples Required at Various Leves of Probability and Precision of Mean Estimates in Current Year, Uppermost Lateral Autumn Collected Red Pine Foliage from a 2 by 2 m spaced 35-Year-Old Plantation.[a]

	Levels of Confidence						
	$10\%\bar{x}$		$5\%\bar{x}$		$2.5\%\bar{x}$		
Foliage Parameter	0.05	0.01	0.05	0.01	0.05	0.01	0.001
Needle length	4	7	9	14	25	42	68
Fascicle weight	9	15	26	44	93	155	253
Ash concentration	8	13	23	38	82	141	222
N concentration	4	7	10	16	28	47	75
P concentration	7	11	20	33	68	116	182
K concentration	16	27	55	95	208	358	583
Ca concentration	14	23	45	76	168	288	470
Mg concentration	10	15	27	46	101	167	272

[a] From Leaf (1973).

There is much common ground between forestry research of this type and the more general field of biogeochemistry as viewed as a component of landscape geochemistry. It would seem that, for purposes of first and second approximation landscape geochemistry, standardization of sample collection methods for soils and plants is most desirable. This could best come about by drawing upon the experiences of Professor Leaf, and other workers, in that field. Similarly, foresters, and other workers, concerned with forest ecosystems would profit from firm estimates of the element status of trees obtained during first and second approximation studies in landscapes selected for study on the basis of conception models as described previously in this chapter.

Geoepidemiology

Studies of relationships between environmental geochemistry and the incidence of disease patterns in plants, animals, and man have been described previously (Fortescue, 1973, 1974c) and many other workers such as Hopps et al. (1968), Stevenson and Stevenson (1972), Underwood (1971), and Keller (1976). Although this field is now small in relation to the others described in this chapter, it is of growing importance.

The Russian approach to geoepidemiology was summarized succinctly by Kovalskiy (1970) who included a first approximation, regional map of the USSR showing the location of "biogeochemical zones and provinces" which relate to the incidence of diseases in animals and man. He divided biogeochemical zones into two different kinds:

> 1. *Zonal Provinces* which conform to the general zonal characteristics but differ between themselves with respect to the concentration and ratios of chemical elements
> 2. *Azonal Provinces* which show essential deviations from the general characteristics of the zone.

These distinctions are fundamental to the interpretation of first approximation geoepidemiological surveys in natural areas and in areas in which man has significantly affected the geochemistry of the environment. Kovalskiy (1970) also provided a summary of trace element/disease relationships which had been studied in Russia up to that time. These are summarized on Table 18.8 and provide an overview of the state of the subject in that country at that time.

Let us now consider two examples of relationships between disease patterns and the distribution of elements in the environment. Thornton and Webb (1970) described a first approximation study involving the relationship between the incidence of pine in sheep and the level and distribution of cobalt in stream sediments in southwest England (Figure 18.20). Previous research in Ireland (Webb, 1964) had indicated that there was a correlation between the cobalt content of stream sediment and pine in cattle and sheep in areas where soils were derived from granite. In granite soils on Dartmoor (Figure 18.20) a similar relationship has been observed in sheep. Patterns of relatively low cobalt (10–15 ppm) on the

Table 18.8. Summary of Russian Studies of Relationships between Threshold Concentrations of Chemical Elements in Soils and Possible Responses of Organisms.[a]

Chemical Element	Number of Determinations	Range of Concentrations (% of air dry soil)		
		Deficiency (Lower Threshold Concentrations)	Normal (Range of Normal Regulatory Functions)	Excess (Higher Threshold Concentrations)
Co	2400	<2–$7 \cdot 10^{-4}$ Acobaltoses, anemia, hypo- and avitaminosis B_{12}; aggravation of endemic goiter	7–$30 \cdot 10^{-4}$	$>30 \cdot 10^{-4}$ Possible inhibition of vitamin B_{12} synthesis
Cu	3194	<6–$15 \cdot 10^{-4}$ Anaemia, bone deformities, endemic ataxia, in case of excess of Mo and SO_4^{2-}; lodging of cereals, low yields of grain, dry tops of fruit trees	15–$60 \cdot 10^{-4}$	$>60 \cdot 10^{-4}$ Anaemia, endemic icterohemoglobinuria, liver lesions. Chlorosis of plants
Mn	1629	$<4 \cdot 10^{-2}$ Bone diseases, aggravation of goiter. Chlorosis and necrosis of plants, yellow specks on sugar beet leaves	7–$30 \cdot 10^{-2}$	$>30 \cdot 10^{-2}$ Bone diseases. Possible toxic effects on plants in acid soils
Zn	1927	$<3 \cdot 10^{-3}$ Parakeratosis in swine. Chlorosis, 'little leaf' diseases in plants	3–$7 \cdot 10^{-3}$	$>7 \cdot 10^{-3}$ Possible anaemia, inhibition of oxidative processes
Mo	1216	$<1.5 \cdot 10^{-4}$ Diseases of plants (clover)	1.5–$7 \cdot 10^{-4}$	$>7 \cdot 10^{-4}$ Gout in humans, MO toxicosis in animals
B	879	<3–$6 \cdot 10^{-4}$ Plant diseases, death of terminal buds of stem and roots. 'Heart rot' in sugar beet, browning of the cabbage flowers, and of the core of the turnip	6–$30 \cdot 10^{-4}$	$>30 \cdot 10^{-4}$ Boron enteritis in animals and humans. Plant abnormalities
Sr	1269		up to $6 \cdot 10^{-2}$	6–$10 \cdot 10^{-2}$ Chondro- and osteodystrophy. Urov disease, rickets. Brittleness of bones of animals
I	491	<2–$5 \cdot 10^{-4}$ Endemic goiter; aggravation of goiter in case of imbalance of I with Co, Mn, Cu	5–$40 \cdot 10^{-4}$	$>40 \cdot 10^{-4}$ Possible decrease of synthesis I-compounds in the thyroid gland

[a] From Kovalsky (1970).

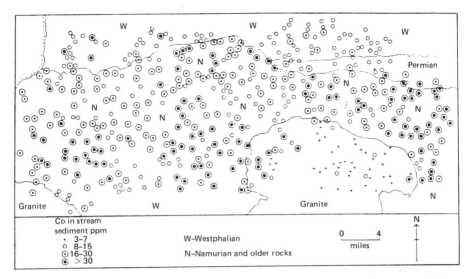

Figure 18.20. The distribution of cobalt in stream sediments in parts of Devon and Cornwall (Thornton and Webb, 1970).

Culm Measures (i.e., Namurian and Westphalian) to the north and west of the granite are related to areas where pine is known, but here the pattern is complicated. As Thornton and Webb (1970, p. 400) stated:

> The interpretation of geochemical data in this region is complicated by the leaching of Co in association with Mn from poorly drained soils and subsequent precipitation under a different pH/Eh regime in the stream bed. . . . Thus Co-deficient soils, not reflected by low values in the stream sediment, give rise to pining in sheep in an area west of Dartmoor. When the Mn is precipitated in the stream bed the sediment is usually seen to have characteristic black staining. When such a stain is present it is necessary to carry out a systematic appraisal of soils for both Co and Mn content.

This relatively simple example indicates how appraisal level geochemical surveys based on the absolute abundance of elements in stream sediments may be used to describe and explain disease patterns in livestock. Second approximation studies are frequently needed to explain details of disease patterns on a local scale after the appraisal level surveys have been interpreted. The determination of cobalt and manganese in the soils, as suggested by Thornton and Webb (1970), is an example of such a second approximation survey.

Summary and Conclusions

In this chapter examples of fundamental and practical applications of landscape geochemistry have been described together with some areas in environmental geochemistry where such applications might be considered with profit. As in previous chapters, the examples included were designed to illustrate whole classes of problems which were typical of specific areas of environmental science. The examples do not provide a balanced and detailed picture of the scope of applications of landscape geochemistry to environmental geochemistry which would require at least a whole volume for each of the five applications listed in Figure 18.1. But in spite of these limitations this chapter should stimulate readers to consider the numerous ways in which the discipline of landscape geochemistry may be applied in environmetal geochemistry.

19. A New Paradigm for Environmental Geochemistry

"Don't bite my finger—look where its pointing" [attributed to Warren S. McCulloch by Stafford Beer (1975)].

Introduction

Kuhn (1970) noted that during the evolution of a scientific discipline major changes in thinking occur at irregular intervals. He called such changes "paradigms." In this chapter landscape geochemistry is viewed as a new paradigm for environmental geochemistry. But before it is described from this viewpoint, let us outline briefly the evolution of thinking and action in geochemistry since the turn of the century.

Paradigms in General Geochemistry (1900–1978)

The development of geochemistry as a scientific discipline outside Russia has involved two major paradigms—one prior to the Second World War and one after it. The first paradigm was holistic and general and characterized by the description of broad concepts and principles which were described on the basis of seminal thinking and relatively little data. This paradigm was summarized at an elementary level by Mason (1952) and at an advanced level by Rankama and Sahama (1950) and Goldschmidt (1954). As might be expected, these works stress the role of geology and lithogeochemistry in geochemistry; less emphasis is placed on environmental geochemistry. The holism of these books is epitomized by the concept of the geochemical cycle as described by Mason (1952, 1958). Along with these works, the classic of geochemistry published in 1924 by Clarke must be included because, unlike the others, it was written much earlier in the evolution of the discipline (the first edition appeared in 1907) and in many ways provided the data base for the first paradigm. In summary, the first paradigm of geochemistry was characterized by the description of broad principles and concepts by a focus on the role of geology in geochemistry and by a lack of detailed chemical data on the distribution and behavior of elements—particularly concerning the biosphere.

The second paradigm of general geochemistry grew out of the first during the 1950s. It is characterized by a lack of holism and general thinking regarding geochemistry as a whole. Instead, it included an emphasis on attention to detail or, as the operational research scientists say, a trend toward "suboptimization." This led to a twofold split in the science of geochemistry. General geochemistry evolved as an academic subject, largely concerned with cosmochemistry and lithogeochemistry and geochemical matters which do not concern the living biosphere and exploration geochemistry evolved as a practical tool, like geophysics, for the location of economic mineral deposits. Unlike general geochemistry, which focused attention upon studies based on quantitative chemical data, exploration geochemistry was concerned largely with the discovery and interpretation of anomalous distribution patterns for chemical elements (or substances) derived from weathering mineral deposits. These could be found by the use of semiquantitative chemical data—often based on the use of partial extraction methods.

Because exploration geochemistry was based on large amounts of appraisal or semiquantitative chemical data obtained from the environment, it tended to be neglected by scientists involved with quantitative analysis and general geochemistry. Similarly, exploration geochemists who discovered landscapes where undisturbed mineral deposits were known made little attempt to describe such landscapes in detail before they were disturbed by mining activities. Consequently, the gulf between "pure" and "applied" geochemistry widened. This state of affairs was aggravated during the 1960s and 1970s when, as a result of the publication of Rachel Carson's book *The Silent Spring* in 1962, scientists and nonscientists began to take note of environmental pollution. The discipline of geochemistry was not ready for this development and so what might have, under other circumstances, been a spur to the development of holistic environmental geochemistry turned out to be almost an embarassment to professional geochemists. So the study of the behavior of elements and other toxic substances in the environment was often taken up by scientists without training in geology or the basics of geochemistry.

This gulf led to considerable confusion. Chemical data, some of it quantitative, was collected from the environment in large quantities with little consideration of the contribution such data might make to generalizations regarding the geochemistry of the environment. As particular elements, such as mercury, cadmium, and lead, became of topical interest as pollutants, large sums of money for research was spent on investigating them, frequently with little attention being paid to what was already known regarding the geochemistry of the environment. This led to further suboptimization. During the past 5 years there has been a gradual return to a holistic approach to environmental geochemistry and it is hoped that this book will aid the new trend.

An example of a text in general geochemistry written in terms of the old paradigm of geochemistry is that by Wedepohl (1971). It includes sections on cosmochemistry, the geochemical structure of the Earth, and what, in this book, is called the major geochemical cycle. Lithogeochemistry of igneous and sedimentary rocks is described and general reference is made to the geochemistry of the hydrosphere and atmosphere and to organic geochemistry. However, there is no reference to the role of the living biosphere.

The best example of a text in exploration geochemistry is that by Hawkes and

Webb (1962). They related their subject to the first paradigm of general geochemistry and provided a very good summary of methods and techniques of exploration geochemistry which had been developed during the two decades before 1960. This book is definitive for the current paradigm of exploration geochemistry, even though it is based on a nonholistic view of the geochemistry of landscapes. The book by Levinson (1974) also subscribes to the same paradigm in exploration geochemistry, but the examples cited in it are more recent and chapters are included on the application of descriptive statistics in exploration geochemistry.

The evolution of environmental geochemistry has already been discussed in Chap. 4. The symposium volume by Kothny (1973) is an example of a text in

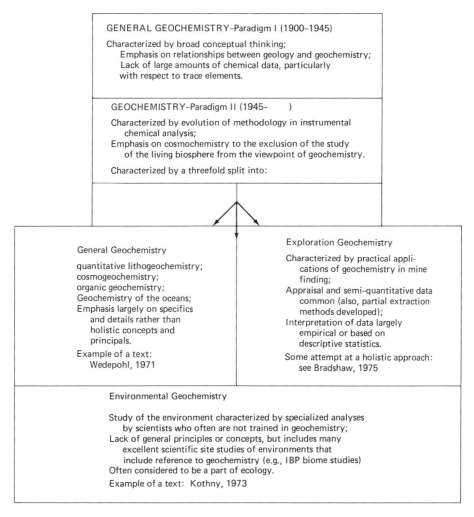

Figure 19.1. Diagram showing relationships between the two paradigms of geochemistry.

environmental geochemistry; it includes nine papers, each describing a different aspect of the geochemistry of the environment.

In summary, in my opinion geochemistry has had two major paradigms during its evolution. One prior to the Second World War and one after it. The general trend in the first paradigm was toward a holistic view of environmental geochemistry. The general trend in the second paradigm was to fragment the subject into general geochemistry, exploration geochemistry, and environmental geochemistry, the first two being carried out by professional geochemists and the third by chemists, ecologists, and engineers interested in the pollution of the environment (Figure 19.1). It should be stressed that these are very general trends. Any researcher may be able to cite exceptions to this general view of the evolution of geochemistry, but such exceptions do not falsify the overall picture provided here.

A New Paradigm for Environmental Geochemistry

It is evident that the discipline of environmental geochemistry requires a unifying set of concepts and principles which can:

1. Relate to current knowledge of all aspects of environmental geochemistry (i.e., including general geochemistry, exploration geochemistry, environmental geochemistry, ecology, etc.);
2. Be recognized as playing a unique and distinctive role of environmental science (i.e., it is a focus for all interested in environmental geochemistry);
3. Provide a forum for the systematic explanation of the subject of environmental geochemistry to scientists and decision makers without scientific training, including lay people who vote and who are charged with environmental decision making which affects us all.

We have already considered very briefly the first requirement. The need for a modern discipline of environmental geochemistry has frequently been stressed by writers and nowhere more forcibly than by Hamilton (1976):

> The problems facing contemporary society are complex and [require] short, medium, and long term prospective planning. In spite of the vast sums of money invested and manpower deployed, many of the questions relating to the pollution and contamination of the natural environment remain unanswered. Through the methods of science, specific objectives can often be identified and reached but data which are produced are not in a form which can be used by decision making bodies. One of the most significant omissions is the lack of any consideration of social or economic consequences. Today, we rely on the nature of past trends and their predicted path in the future. Little attention is directed to the validity of the data upon which historical trends were established and it is often assumed that the variables which gave rise to a particular trend in the past are still operative today: this is clearly not true and the nature of many pollutants present in the natural environment today will exert their own characteristic influence of the variability for which there is no historical counterpart.(p. 1)

The need is not only to collect geochemical data from the environment, it is to so analyze and synthesize it that it can be understood by scientists and nonscientists alike. This is a major role and responsibility of all trained in geochemistry and particularly environmental geochemistry. Until now it has not been formally recognized as such by geochemists collectively, although many individuals have expressed informed opinions on environmental issues in public and in private. Wegman (1978) put the problem very clearly:

> Ultimately, there are no scientific answers to political questions. Perhaps the most important help that the scientists interested in the environment can offer will be to provide the background material to make unpopular social or political decisions; decisions which may require sacrifices we would otherwise not be willing to make.

The discipline of landscape geochemistry as a new paradigm for environmental geochemistry would fulfill five principal roles:

1. To provide a series of interlocking principles and concepts which individually, or collectively, may be used to discover the effects of any chemical entity added to any landscape in any amount during short, medium, and long-term periods of time.

2. To provide a series of interlocking principles and concepts which can be used to describe existing information on the geochemistry of specific landscapes in a case history format so that they can be compared, synthesized, and analyzed in relation to each other. Such studies would be invaluable to the student of landscape geochemistry and to professionals involved in environmental geochemistry.

3. To provide a series of principles and strategies which could be applied directly to the design and planning of investigations and actions in environmental geochemistry.

4. To provide principles and rules for progressive synthesis for learned and informed monographs and articles summarizing particular aspects of environmental geochemistry aimed at advanced students of landscape geochemistry and related subjects, strengthening the plan and discipline of the field.

5. To provide a series of principles and concepts which could be applied during the preparation of comprehensive, informed statements of the behavior of chemical substances in landscapes which would be written specifically in layman's terms so that they could be used by nonscientists of all kinds who are involved in the environmental decision-making process.

It is my considered opinion that landscape geochemistry as described here can fulfill all five of these roles in a unique way. Consequently, landscape geochemistry can become a new paradigm for environmental geochemistry. The paradigm concept is used as it was discussed by Kuhn (1970):

> The study of paradigms. . . . is what mainly prepares the student for membership in the particular scientific community with which he will later practice. Because he there joins men who learned the bases of their field from the same concrete models,

his subsequent practice will seldom evoke overt disagreement over fundamentals. Men whose research is based on shared paradigms are committed to the same rules and standards for scientific practice. That commitment and the apparent consensus it produces are prerequisites for normal science, i.e., for the genesis and continuation of a particular research tradition. (p. 10–11)

Another way of looking at the relationship between the discipline of landscape geochemistry and environmental geochemistry (and environmental science) is to consider our discipline as a "metalanguage."[1] The concept of a metalanguage is convenient in this context because it suggests that the metalanguage for landscape geochemistry is one of a series, all of which relate to some aspect of environmental science. The concept of a metalanguage was described by Beer (1975):

> The barber in this town shaves everyone
> who does not shave himself.
> Who shaves the barber?
> It is not that the barber has a beard.
> The language—meaning the logical structure of the talking—
> is snarled up in itself.
> There is no direct answer to the question.
>
> More particularly, there is no way of discussing
> what has gone wrong *using the language*
> —because the language is in general perfectly all right.
> You just find that you cannot talk about the barber himself
> without contradicting yourself.
>
> But we are already managing to discuss the problem
> and we could soon sort it out.
> Bertrand Russell did it first and the key thought is
> about a class of classes that are (or are not) members of themselves.
> The key thought has nothing to do with either barbers or shaving.
>
> Never mind the logical theory
> we are already speaking a metalanguage—
> and I am sure you see why. (p.7-8)

Later he further explained the puzzle by this diagram:

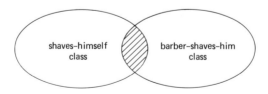

---

[1] meta—i.e., "the next after" or "over and beyond" the language we are using at the time.

The crosshatched area is the barber. So we have two languages, one which describes the two classes apart and the other which includes an area where the classes overlap: As Beer (1975, p. 9) points out:

> Note that the second language is not a metalanguage but that we had to move into a metalanguage to provide the logical vantage point from which to perceive the nature of undecidibility in the first language and to design the second.

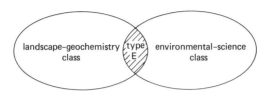

In environmental geochemistry we have a similar problem. In this case one class of knowledge is strictly geochemical and the other involves all other aspects of the total environment. A particular problem with which we are dealing involves the area of overlap between the two. The discipline of landscape geo-

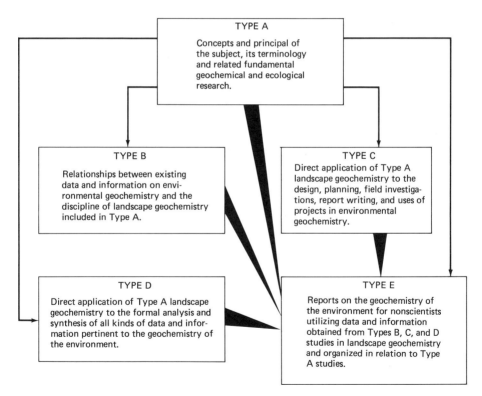

Figure 19.2. A proposal for five types of landscape geochemistry and some of the principal relationships between them.

chemistry provides a language in which to describe the area of overlap. Thus, Type E landscape geochemistry (Figure 19.2) is a metalanguage for environmental geochemistry. It provides a way for the discussion of the knowledge in the class of landscape geochemistry and in the class of environmental science. From the viewpoint of general environmental science, this is the most important aspect of applied landscape geochemistry—even though it is one of five types identified within it.

Five Types of Landscape Geochemistry

We have seen how the proposed paradigm for environmental geochemistry plays five roles, each of which is directly related to an aspect of landscape geochemistry considered as a totality. Tbe concerns of the five types of landscape geochemistry are indicated in Figure 19.2 together with the relationships of the other four types to type E.

Let us now consider each type in turn in relation to exploration geochemistry. Exploration geochemistry was chosen because it is the best developed aspect of environmental geochemistry. Applications in ecology, plant (or animal) nutrition, waste disposal, environmental pollution, or geoepidemiology might all have been selected as examples. But in the landscape, geochemistry has seldom been applied formally except in Russia.

Type A Landscape Geochemistry

Part II of this book describes the present state of development of Type A landscape geochemistry. It includes the seven basic concepts and principles and the terminology used to describe them which was summarized in Chapter 17. At present, Type A landscape geochemistry is still at an early stage of development and new concepts and principles may be expected to be added to it from time to time; less frequently new major concepts may also be added as they are needed to broaden the scope of the discipline. In general, a good working knowledge of Type A landscape geochemistry is essential for an understanding of Types, B, C, and D, but not for an understanding of Type E.

Relationships between Type A Landscape Geochemistry and Exploration Geochemistry

In their now classic text on exploration geochemistry, Hawkes and Webb (1962, p. 1) defined the role of geochemistry in mineral exploration:

> Geochemical prospecting for minerals, as defined by common usage, includes any method of mineral exploration based on systematic measurement of one or more chemical properties of a naturally occurring material. The chemical property measured is most commonly the trace content of some element or group of elements; the naturally occurring material may be rock, soil, gossan, glacial debris, vegetation, stream sediment, or water. The purpose of the measurements is the discovery of abnormal chemical patterns, or *geochemical anomalies,* related to mineralization.

From the viewpoint of Type A landscape geochemistry, the above definition could be rewritten as follows:

> Geochemical prospecting for minerals, as defined by common usage, includes any method of mineral exploration based on the systematic measurement of one or more of the chemical properties of a naturally occurring material in the landscape. The chemical property measured is most commonly the trace content of some element or group of elements (or other chemical entities); the naturally occurring material may be rock, soil, gossan, glacial debris, vegetation, stream sediment, or water. The purpose of the measurements is 1) the discovery of abnormal chemical patterns or *geochemical anomalies* related to mineralization, and 2) the description of geochemical anomalies on the basis of the basics of landscape geochemistry (i.e., element abundance, element migration, geochemical flows, geochemical gradients, geochemical barriers, historical landscape geochemistry, and the geochemical classification of landscapes). Such descriptions may be related to conceptual models, statistical models, systems models, or system simulation models of landscape migration processes.

This definition has incorporated the basics of landscape geochemistry (Type A) into the fundamental definition of exploration geochemistry and in doing so has focused attention upon the current need for exploration geochemists to become familiar with the discipline of landscape geochemistry as discussed by Allan in the quotation at the beginning of Chapter 17. Type A landscape geochemistry thus provides a second-order paradigm for students of exploration geochemistry who are interested in a more holistic approach to their subject. If such an interest is not present, the Hawkes and Webb (1962) paradigm is adequate for most practical purposes in mineral exploration (Fortescue, 1975a).

Type B Landscape Geochemistry

Many examples of Type B landscape geochemistry have been cited in the examples discussed in this book. Type B is concerned with the systematic application of the basics of landscape geochemistry with respect to studies in environmental geochemistry which were published without formal reference to our disciplines. Used in this way, landscape geochemistry combs through published information and data extracting facts and data pertinent to one or more of the concepts of landscape geochemistry. Type B differs from Type D in that it is concerned with the reorganization of the factual data of specific studies on an ad hoc basis and does not relate many studies to a specific theme. In general, Type B is concerned with specific case histories rather than the synthesis and analysis of general case history material.

Relationships between Type B Landscape Geochemistry and Exploration Geochemistry

Many published studies which describe landscape features including geochemical anomalies can be considered from the viewpoint of landscape geochemistry as being valuable first approximations which could lead directly to other approximations involving more detailed research. Consequently, apart from their role in the location of mineral deposits, published accounts of geochemical prospecting provide landscape geochemists with an introduction to a particular problem involving the conditions obtained at a particular site.

A simple example of a published study which could lead to a more detailed investigation of a landscape is illustrated in Figure 15.2. This example, which was included in the book by Hawkes and Webb (1962), involves landscape conditions in the vicinity of a beryllium pegmatite. When viewed as a Type B source, this diagram provides information on relationships between the behavior of products of a pegmatite vein (i.e., as evidenced by the beryllium data) which cause a geochemical gradient at a topographic gradient in a landscape with residual soil cover. Although the geochemical conditions at the site are described by observations obtained from samples collected at intervals along a traverse line, they provide sufficient information for planning a sampling strip along the same traverse for a second approximation (see Figure 18.2).

This might (1) describe the nature of the soil and weathered rock cover in detail, (2) describe in detail the vertical distribution of the Be anomaly in the soil cover, (3) describe the geochemical and topographic gradients on either side of the vein for Be, and (4) describe the gradients for other chemicals derived from the weathering pegmatite in the soils. Clearly, such a second approximation study would not be geochemical prospecting but it would provide information of general interest in both landscape geochemistry and environmental geochemistry. Studies of Type B may focus attention on aspects of a published account of environmental geochemistry which are directly relevant to landscape geochemistry and perhaps more importantly lead to the planning of second approximation studies at the same site aimed at the solution of a problem (or problems) of general interest in landscape geochemistry. From this viewpoint many published studies, such as those included in the bibliography for exploration geochemistry compiled by Hawkes (1976), may be inspected by a landscape geochemist involved in a Type B study before he discovers the description of a site suitable for a field study of a specific type. When such a discovery is made, the study of Type B in the area becomes a first approximation for further investigation. A careful study of the literature can lead landscape geochemists to select field areas previously investigated during mineral exploration activities. Thus, to the landscape geochemist, the literature of exploration geochemistry is a potential source of first approximation data and information which can lead to planning of more detailed research when conditions are favorable.

Type C Landscape Geochemistry

Type C landscape geochemistry involves the direct application of the discipline during the design and planning of all stages in field investigation. As the discipline evolves, it is to be expected that general criteria and rules will be evolved which will lead to standardization of the methods used in field studies. At the present time, this has not been done; each Type C study is essentially an "isolated incident" leading to a "case history" which provides information pertinent to landscape geochemistry. One direction in which landscape geochemistry might move toward standardization is now in the area of graphics. Maps and plans could always be related to the same map frame size and a standard series of map fractions could be used. This would facilitate visual comparisons between patterns in data. Similarly, the basic measures of statistical analysis of landscape geochemical data could also be standardized. Some preliminary progress in this area was described by Fortescue and Bradshaw (1973).

Relationships between Type C Landscape Geochemistry and Exploration Geochemistry

An example of a conceptual model based on principles of landscape geochemistry pertinent to Type C activities was described by Levinson (1974) who related the landscape prism concept to orientation surveys (see Chap. 18, Figure 18.7). He pointed out that it is desirable to investigate the geochemical parameters for the migration of chemical substances in landscapes in some detail during an orientation survey before the exact technique for prospecting an area is decided upon. Such a study would be included in Type C and might involve two or more approximations. The first would locate the extent of the geochemical anomaly associated with the deposit being studied (i.e., using one or more traverse lines as described above). The second would require the application of several techniques of exploration, each based on different landscape materials or techniques of chemical treatment and analysis of samples to discover the most effective prospecting technique in the area of interest. Writeups of first and second approximation surveys carried out for orientation near mineral deposits (using standardized methods) should be of lasting value to students of landscape geochemistry in general and exploration geochemistry in particular.

Type D Landscape Geochemistry

From time to time in environmental geochemistry, a need arises for (1) the description of pertinent information regarding the behavior of a particular chemical entity or entities in the environment, or (2) the description of information on the behavior of a particular landscape or landscapes to the addition of chemical entities such as solids, liquids, or gases from the atmosphere. Studies of this type not related to landscape geochemistry exist at the present time, particularly with respect to specific chemical entities. For example, Wixson (1977) together with eight co-workers compiled a detailed report of environmental pollution by lead (and other heavy metals) from industrial development in the New Lead Belt of southeastern Missouri. This study, already a classic of its type, describes many aspects of the pollution of the environment which has occurred or is taking place as a result of industrial development in the area. The work is detailed and involves much novel and painstaking effort by researchers over a period of years. As a source of data and information regarding the New Lead Belt, it is second to none. A review prepared 10 years hence involving the behavior of the same chemical entities in the environment on a worldwide basis would certainly refer to the Wixson (1977) study. If this hypothetical review were to be organized on the basis of the discipline of landscape geochemistry, it would be an example of a Type D study. In general, Type D studies use the discipline of landscape geochemistry as a basis for the synthesis and analysis of data and information on a specific topic to provide an in-depth informed and balanced account of a topic in environmental geochemistry suitable for use by specialists and students.

Relationships between Type D Landscape Geochemistry and Exploration Geochemistry

The book compiled and edited by Bradshaw (1975) is a study of Type D exploration geochemistry. It is concerned with the analysis and synthesis (using conceptual models) of data and information obtained from over 30 case histories in geochemical exploration from Central and Western Canada. It provides exploration geochemists faced with the planning and interpretation of data from geochemical prospecting activities in the areas covered with a concise summary of past experience drawn from the case histories. Consequently, the book provides general guidelines to the exploration geochemist based on the conceptual models and, at a more detailed level, details of specific studies which may aid him in the interpretation of particular geochemical anomalies in terms of local landscape conditions. The book is limited in its scope, not by the data or the synthesis and analysis carried out by Bradshaw (which are both of considerable value), but by the descriptive paradigm for exploration geochemistry upon which it is based. The book relates—in theory—to the paradigm of landscape geochemistry but assumes—in practice—that readers subscribe to the descriptive paradigm of Hawkes and Webb (1962). It is a stepping stone between the old and the proposed paradigms for exploration geochemistry.

Type E Landscape Geochemistry

Figure 19.2 focuses attention upon the importance of Type E landscape geochemistry as a component of the new paradigm. Type E landscape geochemistry is concerned with relationships between environmental geochemistry on the one hand and environmental decision making by nonscientists on the other hand. It differs from Types A, B, C, and D by being aimed at the general reader (not oriented toward environmental geochemists or scientists) who has to make a decision which involves the geochemistry of a particular landscape or landscapes. As Wegman (1978) pointed out such decisions are often the choice between two unpopular alternatives. When such is the case the background geochemical material used by each side of the issue becomes of paramount importance for today's generation and for the generations which will follow. It is part of the responsibility of being an environmental geochemist to contribute to this decision-making process (1) as a specialist, (2) as a student of science in general and (3) as a member of society.

Two areas in which Type E landscape geochemistry is of potential importance are in the preparation of environmental impact statements and the description of current (and future) environmental effects of toxic substances. Munn (1975, p. 8) edited a SCOPE workshop on environmental impact assessment aimed at the general scientific and decision-making community on a worldwide basis and listed the kinds of readers to which his book was directed:

(a) the decision-makers as well as their policy and management advisors, in both the public and private sectors;

(b) the assessors, the reviewers, and their technical staffs and advisors, who have the responsibility for preparing and reviewing impact assessments;

(c) the layman with an interest in environmental quality.

Type E landscape geochemistry is directed toward the same audience.

To date, no study of Type E landscape geochemistry has been prepared because there is not enough focus on landscape geochemistry. The reception of this book by the scientific community may lead to the acceptance of the new holistic paradigm of landscape geochemistry by environmental geochemists, which, in turn, will lead to a serious consideration of the preparation of Type E studies. Although no examples of Type E studies are included here, it is possible to outline the form such a study might take.

For example, suppose we are concerned with the behavior of a geochemical entity in the environment. This is because "entity X" is concerned with an environmental problem in a particular landscape and region and there is a need for a Type E study to discover its environmental effects in short, medium, or long-term periods of time. Such a study will be preapred for the information of administrators or voters on the issue. In this hypothetical case the Type E report might be organized as shown in Table 19.1.

The report would be readable and instructive using pictures and diagrams in the main text and technical informtion in the appendices. The first report of this type would be prepared on the basis of guidelines (laid down by a group of students of landscape geochemistry and other disciplines) pertinent to the problem. The report format would be established for a trial period, of say, 5 years, after which all reports involving Type E would be reviewed and revised guidelines issued.

In this chapter the potential importance of Type E landscape geochemistry has been outlined and some suggestions made as to how it might be established. The first experimental application of Type E study might be the result of a workshop, organized for the purpose, which would include representatives of many viewpoints (both scientific and nonscientific) all of whom were concerned with the communication of informed geochemical information to a group of voters in a particular locality in which an environmental problem is pending.

Table 19.1. General Plan for a Type E Landscape Geochemistry Report on the Behavior of "Entity X" in Relation to a General Problem on the Environment.

Part I: General environmental geochemistry of entity "X"[a]
  (a) General statement of environmental problems related to entity "X"
  (b) State of knowledge of entity "X" at the global, regional, and local levels
  (c) State of knowledge of entity "X" in technological, ecological, pedological, and geological time
  (d) State of knowledge of entity "X" in relation to the scientific methods and effort used to detect its presence in environmental materials at levels both above and below its effect on man's well being
  (e) State of knowledge of entity "X" with respect to the chemical species in which the entity is present in the environment with particular reference to those which affect the well being of plants, animals, and man

Part II: The landscape geochemistry of entity "X"
  (a) The abundance of entity "X" in the environment
  (b) The migration rate of entity "X" in the environment
  (c) The role entity "X" plays in landscape flows
  (d) Entity "X" and geochemical gradients (natural and man made)
  (e) Entity "X" and geochemical barriers (natural and man made)
  (f) Entity "X" and the historical development of landscapes
  (g) Entity "X" and the geochemical classification of landscapes
  (h) Summary and conclusions

Part III: Landscape geochemistry of entity "X" in relation to the problem at hand
  (a) Statement of the problem concerning entity "X"
  (b) Landscape conditions in which the problem is located
  (c) Alternative solutions to the problem of entity "X"
  (d) Selection of the best possible solution to the problem of entity "X"
  (e) Summary and conclusions

[a] Entity "X" may be a radioactive isotope, an element, an ion, or a biochemical.

Summary and Conclusions

This chapter argued that geochemistry, outside Russia, has had two major paradigms so far. One involved the thinking and activities of a small group of pioneer geochemists who were active before the Second World War and the holistically oriented toward geochemistry. The second has included the development of geochemistry since the Second World War and has been characterized by a lack of holism and the fractionation of geochemistry into three parts: (1) general geochemistry and cosmochemistry, (2) exploration geochemistry, and (3) environmental geochemistry. Although individuals can point to exceptions to this general trend, the reality of this broad view is quite clear, particularly when the development of the subject is viewed in historical perspective.

It is against this background that the applications of the discipline of landscape geochemistry (as outlined in this book) are viewed. One type of application is that described in Chap. 18, in which particular concepts or principles of landscape geochemistry individually, or collectively, are applied to specific classes of environmental geochemical problems. The other, more holistic, application of landscape geochemistry has been described in this chapter in which the discipline as a whole has been discussed in relation to different kinds of environmental geochemical problems. This application of the discipline has been called a "new paradigm" for environmental geochemistry.

Part IV. Summary and Conclusions

Jo of the North Sea said, 'You can't discuss the ocean with a well frog—he's limited by the space he lives in. You can't discuss ice with a summer insect—he's bound to a single season. You can't discuss the Way with a cramped scholar—he's shackled by his doctrines. Now you have come out beyond your banks and borders and have seen the great sea—so you realize your own pettiness. From now on it will be possible to talk to you about the Great Principle'.[1]

Chuang Tzu from "Autumn Floods," *Basic Writings* translated by Burton Watson, (New York: Columbia University Press, 1964), p. 97.

---

[1] This example is included as an ancient example of holistic thinking rather than a commentary on the subject matter of this book which is cast in a somewhat lower key.

20. Overview

"Whatever I say will happen or not: truly the great Apollo grants me the gift of forecasting."

(Horace in "Satires" (30 B.C. quoted by Beer (1966). *Decision and Control*, p. 204.

Introduction

This book has been concerned with the general description of a holistic approach to the study of the geochemistry of the environment. This chapter argues for the adoption of landscape geochemistry as a discipline for environmental geochemistry and draws some conclusions regarding the future development of the subject. The latter is done by providing separate conclusions for "Appliers," "Developers," and "Creators" who work in the field of environmental geochemistry.

Environmental Science, Environmental Geochemistry, and Landscape Geochemistry

Landscape geochemistry is a part of environmental geochemistry, which itself is a part of environmental science. Landscape geochemistry, as described here, involves seven interrelated principles and concepts which together provide a logical and holistic approach to the study of all aspects of environmental geochemistry including those which, up to now have, made no formal reference to landscape geochemistry. This state of affairs may gradually change as landscape geochemistry is applied to more problems of the environment. Landscape geochemistry does not stand alone; it could, for example, share some of its concepts and principles with landscape geophysics, or landscape hydrology, and some aspects of modern geomorphology.

Two General Concepts of Environmental Science

In the science of the total environment it is desirable to have a perspective so that particular investigations in geochemistry can be related to the whole. Two fundamental principles which underlie thinking in environmental science also apply to landscape geochemistry. There are holism and the principle of successive approximations. In general environmental science holism is expressed by the statement that in the environment sooner or later every chemical entity is related to every other chemical entity. Like most general abstractions, this is only a half truth but, even so, when it is used as a background to thinking and decision making in landscape geochemistry it is of utmost importance. For example, holistic thinking leads directly to a consideration of how the local may be related to the global, which, in turn, leads to considerations of standardization of data gathering in landscape geochemistry and related disciplines.

The principle of successive approximations is important when solutions and approaches to solutions of complex problems in environmental geochemistry are considered. This principle is essentially one of going from the general to the particular in a series of clearly defined stages. Thinking along these lines leads to a consideration of aspects of general concepts as hierarchies, with a series of levels from the general to the particular. Thus in landscape geochemistry and environmental geochemistry one can consider a "hierarchy of space" with a global level, a regional level, and a local level within it. The principle of successive approximations also underlies other aspects of landscape geochemistry. For example, the evolution of a technique for chemical analysis is seen to pass through a feasibility stage and a development stage before its true scope and limitations are established. During its initial stages such a technique may yield semiquantitative data of interest as a first approximation in landscape geochemistry, whereas the same technique, after development, may yield truly quantitative and reliable data.

Environmental Geochemistry and Landscape Geochemistry

Outside Russia the term environmental geochemistry is rather vague and includes all studies which are concerned with some aspect of chemistry, geochemistry, or biogeochemistry of the environment. Perhaps, in essence, such environmental geochemistry is tactics without reference to strategy or grand strategy. This is not to underestimate the scientific excellence of a large number of studies in environmental geochemistry, some of which must rate very highly in science as a whole. The point is that such studies do not relate directly to a formal scientific philosophy and discipline. The aim of this book has been to show that such a superstructure (or strategy and grand strategy) of thinking exists in landscape geochemistry.

Landscape geochemistry has three components: (1) it shares with general environmental geochemistry the concept of holism and the principles of successive approximations; (2) it shares with environmental geochemistry a philosophical base including hierarchies of space, time, chemical complexity, and scientific effort; and (3) it is based on seven concepts and principles some of which are unique to science. One concept, which is common to general geochemistry, environmental geochemistry, and landscape geochemistry, is that of the absolute abundance of chemical elements (or other entities) in the environment (or in environmental materials of all kinds). A concept which is unique to landscape geochemistry is that of the geochemical classification of landscapes.

The Seven Basics of Landscape Geochemistry

In addition to the concepts of element abundance and geochemical landscape classification just mentioned landscape geochemistry includes five other basic concepts. These are (1) element migration in landscapes, (2) geochemical gradients in landscapes, (3) geochemical flows, (4) geochemical barriers in landscapes, and (5) the historical geochemistry of landscapes. It should be stressed that the term landscapes as used here includes areas of eluvial (i.e., dry land), superaqual (i.e., bog or marsh), and aqual (i.e., river or lake) and the transitions between them which occur together in a tract of land. When landscapes are classified according to geochemical criteria the features of eluvial landscapes are usually used as a basis for the classification except where they are rare as, for example, in the muskeg organic terrain in Northern Canada. For these reasons the landscape geochemistry approach to environmental geochemistry is holistic. It is also holistic with respect to the study of all chemical elements and chemical entities which circulate as a result of natural and man-made processes within the environment as a global, regional, or local scale of importance. Although exploration geochemistry has been concerned with the description and interpretation of geochemical anomalies due to the presence of mineral deposits in landscapes for many decades it is true that formal methods for the field (or mathematical) description of such anomalies (i.e., in terms of element migration rates combined with geochemical gradients and barriers) have not come into common use. It is to be expected that such a procedure will result from thinking based on landscape geochemistry. It is only when we view the landscape holistically that we can consider the total migration patterns of elements in space without restrictions of scientific disciplines such as ecology, soil science, limnology, etc. Once exploration geochemists, and other environmental geochemists, begin to think holistically as a matter of course, the change from landscape geochemistry as a descriptive science to a predictive science will be speeded up. This will be a first step toward the eventual description of simulation models for the prediction of the behavior of any element (or chemical entity) in any quantity, in any landscape.

In summary, landscape geochemistry is rooted in general environmental science and in environmetal geochemistry on the one hand and in geochemistry, geology, chemistry, biology, ecology, and other sciences on the other. General relationships between landscape geochemistry, environmental geochemistry, and environmental science are shown by a flow diagram (Figure 20.1) in which the thickness of the arrows indicates the current usage of such relationships. All readers should be familiar with this diagram and may use it as a guide to thinking in specific aspects of the geochemistry of the environment. If this is done it is desirable to analyze a specific problem by ranking the components of landscape geochemistry (listed in Figure 20.1) in relation to it.

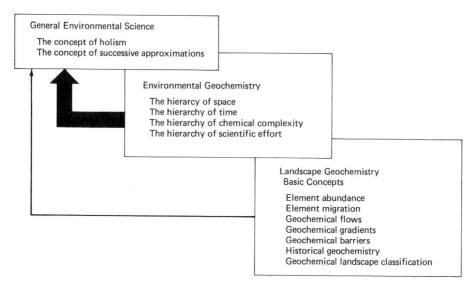

Figure 20.1. Relationships between general environmental science, environmental geochemistry, and landscape geochemistry.

For example, suppose we are concerned with the application of landscape geochemistry to a garbage dump as a part of an environmental impact statement, our list might read as follows:

1. Element abundance—the chemicals which would be of interest near the dump.
2. Geochemical landscape classification—the general type of landscape concerned.
3. The geochemical flows—which occur at the dump site, including seasonal variations.
4. The historical geochemistry of the site—geological and hydrological conditions and previous disturbance of the site by man's activities.
5. Geochemical gradients—the short- and long-term effects of the dump on air and water quality in the area of the site.
6. Geochemical barriers—natural or man-made geochemical barriers which would exist and be required at the site in order to control pollution.

If you are concerned with another type of problem, such as the explanation of an anomaly in exploration geochemistry, the order of importance of the concepts may be quite different. One aim of the discipline of landscape geochemistry is to work toward a standardization of descriptions of basics of the subject so that its approach and terminology may be used as a "metalanguage" for all environmental geochemistry as described in Chap. 19.

Five Types of Landscape Geochemistry

This leads us to a consideration of the formal application of the discipline of landscape geochemistry in environmental studies. It is at once evident that a discussion of the theory of the discipline is not at the same level of scientific thought as a discussion of a practical application of the subject designed for nonscientists. Similarly there is a problem of communication between what has here been called landscape geochemistry and published studies of environmental geochemistry which were not planned with the basics of landscape geochemistry in mind. This problem disappears when we read accounts of environmental geochemistry conceived, designed, planned, and carried out with landscape geochemistry in mind. Finally there is the problem of reviewing existing information on environmental geochemistry as a preliminary to a project to be designed according to the landscape geochemistry approach.

A solution to these problems was outlined in Chapter 19 in which it was suggested that the student of landscape geochemistry should consider each of five types of landscape geochemistry as components of his subject just as a geologist is concerned with mineralogy, petrology, sedimentology, paleontology, and other components of the discipline of geology. Chapter 19 listed the objectives for each of the five components which are summarized in Table 20.1.

Table 20.1. Objectives of the Five Types of Landscape Geochemistry.

Type A landscape geochemistry is concerned with the mechanics of the subject, its concepts, principles, rules, terminology, etc. and with fundamental research related to such activity.

Type B landscape geochemistry is concerned with the discipline of landscape geochemistry as it is applied to specific accounts of environmental geochemistry described from any discipline. Type B landscape geochemistry is aimed specifically at the student of general geochemistry, or some related discipline, who wishes to integrate thinking in landscape geochemistry to his specialty.

Type C landscape geochemistry involves all projects in which landscape geochemistry Type A is used as a basis for the planning of studies in applied environmental geochemistry. It implies a holistic application of the discipline of landscape geochemistry to the solution of practical problems in environmental science

Type D landscape geochemistry is envisaged as an advanced form of Type B which involves the synthesis and analysis of all data available upon a specific topic of environmental geochemistry into a form that it can be used by specialists in environmental geochemistry. The form of each study, or monograph, would be directly based on Type A landscape geochemistry which should lead to a uniformity of presentation style. This activity could include advanced degree theses in landscape geochemistry

Type E landscape geochemistry is envisaged as the application of the basics of landscape geochemistry to the preparation of informed reports on topics of interest in general environmental sciences. Such reports would be aimed specifically at land use decision makers in general (both scientists and nonscientists) who require concise, balanced accounts of issues which can be used with confidence in the decision making process

Type A landscape geochemistry includes information of the type given in Part II of this book and would be the first part of a course in landscape geochemistry given at the undergraduate or graduate level. It would also include research activity of specialists in environmental geochemistry. Students of the discipline of landscape geochemistry should meet at regular intervals to discuss the development of the discipline and to plan general guidelines for future research. It might be helpful if Type A information were published in a relatively small number of journals so that, over the years, students would be able to trace the evolution of the subject with a minimum of effort.

Type B landscape geochemistry is envisaged as a way for the students to tap the pool of existing information on environmental geochemistry. In order to do this each would require an introductory course in Type A landscape geochemistry so that he could relate it to the published information of his speciality, or area of interest.

Type C landscape geochemistry follows from experience in Types A and B. Small exercises in Type C landscape geochemistry could be the subject of B.Sc. theses at the undergraduate level or advanced degree theses at the graduate level. Type C studies are appropriate for researchers in government agencies concerned with problems in environmental geochemistry in the broad sense of the term.

Type D landscape geochemistry would be the activity of specialists who are concerned with a single aspect of environmental geochemistry which is researched in depth from the scientific literature. The synthesis and analysis activity envisaged here would produce, over the years, a series of monographs which can be used as a data base by all other workers in landscape geochemistry.

Type E landscape geochemistry is potentially the most important aspect of the subject because it is designed to motivate and prepare decision makers of all kinds to make informed decisions. Like Type A landscape geochemistry, Type E requires considerable research before it can be formalized in a way that is acceptable to those engaged in politics and management.

General Conclusions Regarding the Future Development of Landscape Geochemistry

Conclusions with Respect to "Appliers" of the Discipline

Appliers of landscape geochemistry include all who are concerned in any way with environmental science or environmental decision making and have a need to know something about general thinking in environmental geochemistry. With respect to them the following conclusions are drawn:

1. Landscape geochemistry may be used by teachers and students of environmental science, environmental geology, geography, ecology, soil science, forestry, epidemiology, and other disciplines who see the need for a holistic approach to environmental geochemistry.

2. Professionals who are trained in environmenal geochemistry will be inter-

ested in landscape geochemistry as a means of collecting and connecting data and information on the geochemistry of the environment which until now has been very difficult to synthesize and analyze.

3. Professional geochemists (for example, those trained in exploration geochemistry) will be interested in the application of the principles of landscape geochemistry in the practice of their profession and will use it as a means of relating their specialty to a holistic view of environmental geochemistry.

Conclusions with Respect to "Developers" of Landscape Geochemistry

Developers of landscape geochemistry are those readers who are concerned with the solution of problems in environmental geochemistry on a day to day basis and who wish to make formal contributions to the evolution of the discipline. With respect to them the following conclusions are drawn:

1. They may orient ongoing, or planned, research directly and formally to the principles and concepts of landscape geochemistry. For example, exploration geochemists might apply principles of geochemical mapping of landscapes in an area of country where mineral exploration was being conducted.

2. Other developers may follow the lead given by Bradshaw (1975) and utilize landscape geochemistry for the systemativc analysis and synthesis of existing data and information of some aspect of environmental geochemistry making original contributions to the discipline while doing so.

3. A third class of developers may take the concepts or principles of landscape geochemistry as a starting point for university degree research at the M.Sc. or Ph.D. level. Studies of this type, when published, may result in significant advances in the scope of landscape geochemistry and also bring to light areas in which further research is most desirable.

Conclusions with Respect to "Creators" of Landscape Geochemistry

Scientific disciplines at an early stage of development always attract a few students who make fundamental contributions to thinking within it. At times such contributions lead to the development of the discipline along new and unconventional lines. It is concluded that landscape geochemistry is no exception to this trend and that, from time to time, significant advances will be made by students of the subject which will, like the ideas of Polynov many years ago, lead to new horizons for us all to study and think about.

Perhaps the best way to end this book is to quote from a distinguished expert on military warfare Sir Basil H. Liddell Hart (1971) who in a study of relationships between war and the scientific approach wrote:

> Adaptation to changing conditions is the condition of survival. This depends on the simple yet fundamental question of *attitude*. To cope with problems of the modern world we need, above all, to see them clearly and analyze them scientifically. This requires freedom from prejudice combined with the power of discernment and with a sense of proportion. Only through the capacity to see all relevant factors, to weigh them fairly, and to place them in relation to each other, can we hope to reach an accurately balanced judgment.

Synopsis

The purpose of this synopsis is (1) to summarize the argument for the adoption of the discipline of landscape geochemistry for environmental geochemistry and (2) to provide an outline of the book including page numbers of sections so that quick reference to particular points may be made. More detailed reference to particular topics occurs in the general index.

Part I. Introduction

This part lays foundations. The first of four chapters provides a general overview of the organization of the book, gives definitions of geochemistry and landscape geochemistry (p.3), and describes the purpose of the book and the kinds of scientific thinkers for which it was written (p.5).

Figure 1.1 shows general relationships between the book's four parts (p.4). Chapter 2 includes a brief outline of the development of geochemistry prior to the Second World War, mainly with reference to the thinking of F. W. Clarke (p.7), V. I. Vernadski (p.7) and V. M. Goldschmidt (p.8). This leads to a consideration of some general concepts of geochemistry including the geochemical cycle (p.9), the pedosphere (p.11) and landscape prism units (p.12). Polynov's thinking in landscape geochemistry is then outlined particularly with respect to its holism (p.16) and in relation to elementary landscape types and element migration (p.16).

Chapter 3 considers the current state of environmental geochemistry outside Russia and draws a useful distinction between a scientific discipline and the philosophy upon which it is based. A general philosophy for environmental geochemistry is then described; this is common to both environmental geochemistry as it is now carried out and the discipline of landscape geochemistry as it is described in this book (Part II). The philosophy is based on the notion of four hierarchies which are almost always involved in studies of environmental geochemistry. They are space (p.19), time (p.20), chemical complexity (p.20), and scientific effort (p.22). At the end of the chapter a list of terms used to describe the hierarchies is given (Table 3.2, p.24) together with a prism diagram showing them in relation to a point in a hypothetical landscape (Figure 3.1, p.23). The principle of successive approximations is then introduced and described by means of examples (p.25) selected to focus attention upon how it may be applied to the different hierarchies.

Chapter 4 is concerned with relationships between geochemistry and environmental science in general. It commences with an example of the lack of reference to environmental geochemistry in a book describing fundamentals of geochemistry (p.29) and continues with a general definition of environmental science (p.30). The concept of holism is then described with reference to J. C. Smuts who originally defined the term (p.30). Two disciplines included in environmental science which involve geochemistry without its being their central focus are then described briefly. One is ecology which is based on the concept of the ecosystem (p.31) and includes frequent reference to the concept of biogeochemical cycling (p.32). The other is land classification which also frequently includes reference to the geochemistry of the environment with respect to specifics but is not based in the discipline of geochemistry (p.34). The need for a scientific discipline with a series of interlocking concepts and principles for environmental geochemistry is then stressed (p.39) and a series of four criteria is listed which would require fulfillment if such a discipline were to become a reality (p.39). Each chapter ends with a short list of topics for class discussion.

In summary, Part I sets the scene for what follows by providing a historical outline for the development of general geochemistry and the origins of landscape geochemistry, and a simple and universal philosophical basis for the study of environmental geochemistry in general and landscape geochemistry in particular. It also includes a state of the art summary of relationships between environmental science and environmental geochemistry and the role of geochemistry in other disciplines concerned with the study of the environment—ecology and land classification.

Part II. The Basics of the Discipline of Landscape Geochemistry

Landscape geochemistry is described in this book as a discipline based on seven interelated principles and concepts, defined in Chapter 5. One of them, element abundance, is described in Chapter 6 (p.50). The other six are element migration (Chapter 7, p.79), geochemical flows (Chapter 8, p.95), geochemical gradients (Chapter 9, p.109), geochemical barriers (Chapter 10, p.123), historical geochemistry (Chapter 11, p.133), and classification of geochemical landscapes (Chapter 12, p.155). All these concepts have been described by Russian workers from time to time starting with B. B. Polynov early in this century. The aim of this book has been to summarize the Russian experience and present it in a form suitable for inclusion in the discipline of environmental geochemistry. All these concepts are at a relatively early stage of development and require further refinement.

Part II continues with four chapters that describe the application of a hierarchy included in the philosophy to environmental geochemistry. Most of the examples included in these chapters are taken from studies completed without formal reference to landscape geochemistry. Chapter 13 (p.173) is concerned with chemical complexity, Chapter 14 (p.191) is concerned with scientific effort, Chapter 15 (p.213) is concerned with space and Chapter 16 (p.229) is concerned with time. Chapter 17 (p.247) provides a summary of the scope of the discipline and information upon how the concepts and principles interact (p.247); on Table 17.1 the terminology introduced for the discipline of landscape geochemistry is listed together with the originators of the terms where known. Chapter 17 is designed partly for the reader who requires a synopsis of Part II.

Part III. Applications of Landscape Geochemistry

Part III is concerned with more speculative aspects of environmental geochemistry. In Chapter 18 (p.255) fundamental and practical applications of the discipline are outlined and summarized by means of a flow diagram (Figure 18.1, p.256). Examples of the use of conceptual models of landscape in research are then briefly described with respect to forest ecology (p.257) and plant prospecting methods research (p.259). Practical applications of the subject are then considered. The use of geochemical landscape mapping in exploration geochemistry is described using a classification system worked out by Glazovskaya et.al., 1961 (p.262). The application of landscape geochemistry in mineral exploration is also evidenced with reference to the concept of the orientation survey as described by Levinson (p.264) and Fortescue and Hornbrook. With respect to applications in environmental pollution, an example of roadside contamination by lead derived from automobiles is described (p.266) as well as the pollution of a forested area by pollutant gasses from a smelter (p.268). Applications of landscape geochemistry to the problem of waste disposal are then described with reference to the disposal of fly ash (p.271) and mine and agricultural waste waters (p.277). The fallout of nuclear wastes is then considered first a global problem over a multiyear period (p.277) and then on a local scale involving the fallout of ^{137}Cs in three small watersheds having different plant and soil cover (p.278). Information is then given on the role of environmental and landscape geochemistry in plant and animal nutrition; the existence of biogeochemical provinces associated with natural imbalances in the geochemistry of the environment is described (p.280) with reference to agricultural (p.282) and forest nutrition problems (p.283). The chapter ends with a short section on geoepidemiology which includes information on relationships between the level of trace elements in the environment and the health of livestock (p.287). It should be noted that each of the sections in this chapter involves a complex field of scientific effort and separate books could be written for each application. This chapter is only a signpost towards the future formal application of the principles of the discipline of environmental geochemistry to the study of the total environment.

The second chapter in Part III breaks new ground. Chapter 19 (p.291) is concerned with the role landscape geochemistry, as described in this book, may play in the future development of geochemistry. The chapter begins with a brief summary of the development of geochemistry described with reference to two paradigms (p.293). A case is then made for the consideration of landscape geochemistry, as described in this book, as a third paradigm (p.294) which would include five types of landscape geochemistry, defined and listed on Figure 19.2 (p.297). The role of landscape geochemistry as a "meta-language" for environmental science is introduced (p.296) as a preliminary to a detailed discussion of the relationship between each of the five types of landscape geochemistry and exploration geochemistry. It should be stressed that Chapter 19 is speculative and written as a basis for discussion rather than as a final statement on the evolution of environmental geochemistry during the 1980's.

Part IV. Summary and Conclusions

Part IV includes one short chapter. It first distinguishes between environmental science, environmental geochemistry and landscape geochemistry (p.309). The general scope of landscape geochemistry within environmental geochemistry is then described (p.310) and the seven basics of the subject are outlined (p.311). The five types of landscape geochemistry, introduced in Chapter 19 are then outlined in relation to the future of our subject (p.313). The book ends with guidelines for the evolution of the subject by scientists who are "appliers," "developers," or "creators" (p.314).

References

This book is designed to introduce the concepts and principles of holistic environmental geochemistry. To cover this subject with detailed reference to the literature of environmental geochemistry would require many volumes and is outside the scope of this book. Consequently, just sufficient references were included in this book to support its subject matter—some because they provide information on the development of holistic thinking in environmental geochemistry [eg Polynov (1937), Perel'man (1966) and Glazovskaya (1963)] or because they provide clear examples illustrating principles or methods [eg., Briggs and Pollack (1967), Dalman and Sollins (1971), and Thomas (1969)], and others because they are general sources of information related directly to environmental geochemistry. Examples of this last group include Fritts, 1962 (dendrochronology); Howell Gentry and Smith, 1975 (systems analysis in ecology); Lintz and Simonette, 1976 (remote sensing); and Odum, 1971 (general ecology).

This book is designed to stimulate holistic thinking and not to provide facts regarding the behavior of particular chemical entities. Once the reader has grasped the fundamentals of the discipline they can be applied directly to a specialty. Other concepts and principles may be described that will enhance not only the specialty but the discipline as a whole.

Andren, A.W., Fortescue, J.A.C., Henderson, G.S., Reichle, D.E., Van Hook, R.I. (1973) Environmental monitoring of toxic materials in ecosystems. In: Ecology and Analysis of Trace Contaminants; Progress report June 1972–January 1973. Oak Ridge National Laboratory ORNL-NSF-EATC-1, pp. 61–120

Anderson, T.W. (1974) The chestnut pollen decline as a time horizon in lake sediments in eastern North America. Can. J. of Earth Sci. **11**, 678–685

Baes, C.F., Goller, H.E., Olsen, J.S., Rotty, R.M. (1976) The Global Carbon Dioxide Problem. Oak Ridge National Laboratory Publication ORNL-5194.

Bakuzis, E.V. (1969) Forestry viewed in an ecosystem perspective. In: The Ecosystem Concept in Natural Resource Management. Van Dyne, G. (ed.). New York: Academic Press, pp. 189–257

Barringer, A. (1976) Airborne geophysical and miscellaneous systems. In: Remote Sensing of Environment. Lintz, J., Jr., Simonett, D.S. (eds.). Reading, Mass.: Addison-Wesley, pp. 291–343

Bartos, D.L. (1973) A dynamic model of aspen succession. In: IUFRO Biomass Studies. Coll. of Life Sciences and Agriculture. Orono: University of Maine, 13–25

Beer, S. (1975) Platform for Change. New York: John Wiley & Sons, p. 457

Bienfet, V.H. Binot, H., Lomba, F. (1970) Nutritional infertility in cattle at pasture. In: Trace Element Metabolism in Animals. Mills, C.F. (ed.). Edinburgh, London: E. and S. Livingstone

Bolter, E., Hemphill, D.D., Wixon, B., Chen, R., Butherus, D. (1972) Geochemical and vegetation studies of trace substances from lead & smelting. In: Proc. of 6th An. Conf. on Trace Substances in Environmental Health. Hemphill, D.D. (ed.). Columbia, Mo.: University of Missouri, pp. 79–86

Bowen, H.J.M. (1966) Trace Elements in Biochemistry. New York: Academic Press, p. 241

Boyer, K.W., Capar, S.G., Tanner, J.T., Friedman, M.H. (1977) Elemental content of animal feed, animal excreta and sewage sludge. In: Proc. Xl Conference on Trace Substances in Environmental Health. Hemphill, D.D. (ed.). Columbia, Mo.: University of Missouri Press, pp. 264–271

Bradshaw, P.M.D. (1975) Conceptual models in exploration geochemistry. In: The Canadian Cordillera and Canadian Shield. Bradshaw, P.M.D. (ed.). Association of Exploration Geochemists Special Publication No. 3. New York: Elsevier

Brass, G.W. (1975) The effect of weathering on the distribution of strontium isotopes in Weathering Profiles. Geochemica et Cosmochemica Acta **39**, 1647–1653

Briggs, L.I., Pollack, H.N. (1967) Digital model of evaporite sedimentation. Science **155**, 453–456

Brooks, R.R. (1973) Biogeochemical parameters and their significance for mineral exploration. J. Applied Ecologyh**10** (3) 825–836

* Brownlow, A.H. (1979) Geochemistry. Englewood Cliffs, N.J.: Prentice-Hall 498 pp.

Bryne, A.R. Ravnik, V. (1976) Trace element concentrations in higher fungi. Science of the Total Environment **6,** 65–78

Bunge, T.M.E. (1962) Institution and Science. Englewood Cliffs, N.J.: Prentice-Hall, p. 142

Buol, S.W., Hole, F.D., McCracken, R.J. (1973) Soil Genesis and Classification. Ames, Ia.: The Iowa State University Press, p. 360

Burger, D. (1972) Forest site classification in Canada. Mitteilungen des Vereins für forstliche Standortskunde und Forstpflanzenzüchtung **21,** 20–36

Burger, D. (1976) The concept of ecosystem region in forest site classification. In: Proc. XVI, IUFRO World Congress, Norway, June 20–July 2, 1976 (in press)

Burgess, R.L. (1975) Eastern deciduous forest biome. Oak Ridge National Laboratory Environmental Sci. Div. Publ. **653,** 117–124

Cannon, H.L., Bowles, J.M. (1962) Contamination of vegetation by tetraethyl lead. Science **137,** 765–766

Cannon, H.L. (1970) Trace element excesses and deficiencies in some geochemical provinces of the United States. In: Trace Substances in Environmental Health — II. Hemphill, D.D. (ed.). Proc. of Univ. of Missouri, 3rd Annual Conference on Trace Substances in Environmental Health, June 24, 25, 26, 1969. Columbia, Mo.: University of Missouri Press

Carson, R. (1962) Silent Spring. Boston: Houghton Mifflin

Chadwick, M.J., Goodman, G.T. (1975) Ecologyhof resource degradation and renewal. 15th Symposium of the British Ecological Society July 10–12, 1973. New York: John Wiley and Sons, p. 480

Clarke, F.W. (1924) The Data of Geochemistry. 5th Ed. U.S. Geological Survey Bulletin 770. Washington: Government Printing Office, pp. 841

Collier, B.D., Cox, G.W., Johnson, A.W., Miller, P.C. (1973) Dynamic Ecology. Englewood Cliffs, N.J.: Prentice-Hall, pp. 563

Colwell, R.N. (1968a) Determining the usefullness of space photography for natural resource inventory. In: Proceedings of the 5th International Symposium on Remote Sensing of Environment. Ann Arbor, Mich.: University of Michigan, pp. 249–289

Colwell, R.N. (1968b) Remote sensing of natural resources. Sci. Am. **218(1)**, 54–69

Conway, E.J. (1942) Mean geochemical data in relation to oceanic evolution. Proc. Roy. Irish Academy 48 SB, p.8

---

* Not cited in text.

Crompton, E. (1967) Soil formation. In: Selected Papers in Soil Formation and Classification. Drew, J.V. (ed.). SSSA, Special Publication Series #1, Madison, Wisc: Soil Sci. Soc. Am., pp. 3–15

Dahlman, R.C., Sollins, P. (1971) Nitrogen cycling in grasslands. In: Oak Ridge National Laboratory Ecological Sciences Division Annual Progress Report for period ending July 31, 1970, Publ. No. 385, pp. 71–73

Dahnke, W.D., Vasey, E.H. (1973) Testing soils for nitrogen. Soil Testing and Plant Analysis. Revised Ed. Walsh, L.M., Beaton, J.D. (eds.). Madison, Wisc.: Soil Sci. Soc. Am., pp. 97–114

Daughtrey, Z.W., Gilliam, J.W., Kanprath, E.J. (1973) Soil test parameters for assessing plant-available P of acid organic soils. Soil Sci. **115** (6), 438–446

Davis, S.N., DeWiest, R.J.M. (1966) Hydrology. New York: John Wiley and Sons, p. 463

Dickson, B.A., Crocker, R.L. (1953a) A chronosequence of soils and vegetation near Mt. Shasta, California, II. The development of the forest floors and the carbon and nitrogen profiles of the soils. *Jour. Soil. Sci.* **4** (2) pp. 142–154

Dickson, B.A., Crocker, R.L. (1953b) A chronosequence of soils and vegetation near Mt. Shasta, California, II. The development of the forest floors and the carbon and nitrogen profiles of the soils. *Jour. Soil. Sci.* **4** (2) pp. 142–154

Dickson, B.A., Crocker, R.L. (1954) A chronosequence of soils and vegetation near Mt. Shasta, California, III. Some properties of the mineral soils. Soil Sci. **5**, (2), 173–191

Dorsey, J.A., Johnson, L.D., Statnick, R.M., Lochmuller, C.H. (1977) Environmental assessment sampling and analysis: phased approach and techniques for level 1. Phase Report April 1976–77 Industrial Environmental Research Triangle Park, N.C., EPA Report 600/2-77/115, p. 40

Elliott, L.F., Stevenson, F.J. (1977) Soils for Management of Organic Wastes and Waste Waters. Madison, Wisc.: Soil Sci. Soc. Am., pp. 650

Elwood, J.W., Henderson, G.S. (1975) Hydrologic and chemical budgets at Oak Ridge, Tennessee. In: Coupling of Land and Water Systems. A.D. Hasler (ed.). New York, Heidelberg, Berlin: Springer-Verlag, pp. 31–51

Everett, J., Simonett, D.S. (1976) Principles, concepts, and philosophical problems in remote sensing. In: Remote Sensing of Environment. Lintz, J., Jr., Simonett, D.S., (eds.). Reading, Mass.: Addison-Wesley, pp. 85–130

Evseeva, L.S., Perel'man, A.I. (1962) Geochemistry of Uranium in the Supergene Zone. Moscow: Atomizdat

Fisher, R.A. (1958) Statistical Methods for Research Workers, 13th Ed. Edinburgh: Oliver & Boyd

Fortescue, J.A.C. (1967a) The role of geochemistry in exploration architecture. Bulletin Canadian Institute Mining and Metallurgy

Fortescue, J.A.C. (1967b) Background scope and objectives. Geological Survey of Canada Paper 67-23 (part l), pp. 1–29

Fortescue, J.A.C. (1970) A research approach to the use of vegetation for the location of mineral deposits in Canada. Taxon **19** (5), 695–704

Fortescue, J.A.C. (1973) The need for conceptual thinking in geoepidemilogical research. In: Trace Substances in Environmental Health - VI: 1973, A Symposium. Hemphill, D.D. (ed.). Columbia, Mo.: University of Missouri Press, pp. 333–339

Fortescue, J.A.C. (1974a) The environment and landscape geochemistry. Western Miner March, 1–6

Fortescue, J.A.C. (1974b) Exploration, geochemistry and landscape. Bulletin of the Canadian Institute of Mining and Metallurgy, November, 1974

Fortescue, J.A.C. (1974c) Landscape geochemistry and geoepidemiology. In: Proc. VIII Conference on Trace Substances in Environmental Health. Hemphill, D.D. (ed.). Columbia, Mo.: University of Missouri Press, pp. 119–126

Fortescue, J.A.C. (1975a) The use of landscape geochemistry to process exploration geochemical data in conceptual models in exploration geochemistry. The Canadian Cordillera and Canadian Shield. Bradshaw, P.M.D. (ed.). Association of Exploration Geochemists Special Publication Number 3. New York: Elsevier

Fortescue, J.A.C. (1975b) The application of elementary concepts of landscape geochemistry to the description of borehole data from muskeg areas. Proc. 16th Muskeg Research Conference. Montreal: Nation Research Council, pp. 31–38

Fortescue, J.A.C., Bradshaw, P.M.D. (1973) Landscape Geochemistry and Exploration Geochemistry. St. Catharines, Ontario: Brock University Department of Geological Science Research, Report Series No. 17, pp. 66

Fortescue, J.A.C., Burger, D., Grant, B. Gawron, E., Curtis, S. (1973) Preliminary Landscape Geochemical Studies at Forested Sites at Dorset, Ontario, and at Oak Ridge, Tennessee. St. Catharines, Ontario: Brock University Department of Geological Science Research, Report Series No. 12, p. 54

Fortescue, J.A.C., Hornbrook, E.H.W. (1967) A brief survey of progress made in biochemical prospecting research at the geological survey of Canada (1962–65). In: Proc. Symposium on Geochemical Prospecting. Cameron, E.M. (ed.). Geological Survey Canadian Paper 66-54, pp. 114–133

Fortescue, J.A.C., Hornbrook, E.H.W. (1967) Progress Report on Biogeochemical Research at the Geological Survey of Canada 1963–1966. Geo. Surv. Can. Paper 67-23 Part I (1967) p. 143, Part II (1969) p. 101

Fortescue, J.A.C., Martin, C.G. (1970) Micronutrients: forest ecology and systems analysis. In: Analysis of Temperate Forest Ecosystems. Reichle, D.E. (ed.). New York-Heidelberg-Berlin: Springer-Verlag, pp. 173–198

Fortescue, J.A.C., Terasmae, J. (1975) Man's Effect on the Geochemistry of Lake and Stream Sediments from Southern Ontario. Geol. Soc. Amer. Special Paper 155, pp. 9–26

Fortescue, J.A.C., Usik, L. (1969) Geobotanical and soil geochemical investigations during visits to eight landscapes with undisturbed mineral deposits. Geological Survey Canadian Paper 67-23, Part II, pp. 95–101

Fritts, H.C. (1962) Bristlecone pine in the White Mountains of California. In: Papers of the Laboratory Tree Ring Research Number 4. Tucson, Ariz.: University of Arizona Press, p. 44

Fritts, H.C. (1976) Tree Rings and Climate. London: Academic Press, p. 567

Feuhring, H.D. (1960) Interrelationships of the trace elements zinc, boron, manganese and copper on the growth and composition of corn. Ph.D. Thesis, University of Nebraska. Ann Arbor, Mich.: University Microfilms, Inc

Fu Su Yen, Goodwin, J.H. (1976) Correlation of tuff layers in the Green River formation of Utah using biotite composition. J. Sedimentary Petrology **46** (2), 345–354

Gibbs, R.J. (1972) Water chemistry of the Amazon River. Geochimica et Cosmochemica Acta **36**, 1061–1066

Giddings, J.C. (1973) Chemistry, Man, and Environmental Change: An Integrated Approach. San Francisco: Canfield Press. p. 472

Glazovskaya, M.A. (1963) On geochemical principles of the classification of natural landscapes. Intern, Geological Rev. **5** (11), 1403–1431

Glazovskaya, M.A., Makunina, A.A., Pavlenco, I.A., Bozhko, M.G., Gavrilova, I.P. (1961) The Geochemistry of Landscapes and Prospecting for Useful Minerals in the Southern Urals. (Unedited translation for Geological Survey of Canada, March, 1966). Moscow: Moscow University Press, p. 181

Glover, H.G. (1975) Acidic and ferruginous mine drainages. In: The Ecology of Resource Degradation and Renewal Process, 15th Symposium of the British Ecological Society 10–12 July, 1973. Chadwick, M.J., Goodman, C.T. (eds.). New York: John Wiley & Sons, p. 173

Goldschmidt, V.M. (1937) The Principles of Distribution of Chemical Elements in Minerals and Rocks. London: Journal Chemical Society, pp. 655–673

Goldschmidt, V. M. (1954) Geochemistry. Oxford: Clarendon Press, p. 730

Gordon, A.G., Gorham, E. (1963) Iron-sintering plant at Wawa, Ontario. Can. J. Bot. **41**, 1077

Green, J. (1972) Elements: planetary abundances and distribution. In: Encyclopedia of Geochemistry and Environmental Sciences, Vol. IVA. Fairbridge, R.W. (ed.). New York: Van Nostrand Reinhold, pp. 268–300

Hamersma, J.W., Reynolds, S.L., Maddalone, R.F. (1976) IERL-RTP Procedures Manual, Level 1 Environmental Assessment, E.P.A. - 600/2-160A E.P.A., R. & D Office, Washington, D.C.

Hamilton, E.I. (1975) Review of the chemical elements and environmental chemistry – strategies & tactics. In: The Science of the Total Environment 5. Amsterdam: Elsevier, 1–62

Hamilton, E.I., Minski, M.J. (1972–1973) Abundance of the chemical elements in man's diet and possible relations with environmental factors. The Science of the Total Environment 1, Amsterdam: Elsevier, pp. 375–394

Harradine, F. (1967) Morphology and genesis of noncalcic brown soils in California. In: Selected Papers in Soil Formation and Classification. Drew, J.B. (ed.). SSSA, Special Publication Series #1. Madison, Wisc.: Soil Sci. Soc. Am., pp. 95–110

Harrison, J.B. (1934) The Katametamorphism of Igneous Rocks Under Humid Tropical Conditions. Imp. Bur. Soil Science (Harpenden), p. 79

Hawkes, H.E. (1976) Exploration Geochemistry Bibliography. Special Vol. No. 5, The Association of Exploration Geochemists. Tucson, Ariz.: The Association of Exploration Geochemists. Published by Sin Speedy, 4650 East Speedway, Tucson, AZ

Hawkes. H.E., Webb, J.S. (1962) Geochemistry in Mineral Exploration. New York: Harper and Row, p. 415

Henderson, G.S., Harris, W.F. (1975) An ecosystem approach to characterization of the nitrogen cycle in a deciduous forest watershed. In: Forest Soils and Forest Land Management. Proc. 4th North American Forest Soils Conference at Laval University. Quebec, August 1973. Quebec: Les Presses de L'Universite Laval, pp. 179–194

Hills, G.A. (1959) A Ready Reference to the Description of the Land of Ontario and its Productivity (A Compendium of Maps, Charts, Tables and Brief Comments). Maple, Ontario Department of Lands and Forests Division of Research

Hodgson, D.R., Buckley, G.P. (1975) A practical approach towards the establishment of trees and shrubs on pulverized fuel ash. In: The Ecology of Resources Degradation and Renewal. Proc. 15th Symposium of the British Ecological Society 10–12 July 1973. Chadwick, M.J., Goodman, G.T. (eds.). New York: John Wiley and Sons, pp. 305–330

Holmes, J.W. (1971) Salinity and the hydrological cycle. In: Salinity and Water Use. Talsma, T., Philip, J.R. (eds.). New York: Wiley-Interscience, pp. 25–40

Hopps, H.C., Cuffey, R.J., Morenoff, J., Richmond, W.L., Sidley, J.D.H. (1968) Computerized mapping of disease and environmental data. A report of The Mapping of Disease. (MOD) Project No. DA 49-092-ARO-130 (U.A.R.E.D.)

Hordon. R.M. (1972) Hydrologic cycle. In: Encyclopedia of Geochemistry and Environmental Sciences, Vol. IVA. Fairbridge, R.W. (ed.). New York: Van Nostrand Reinhold, pp. 515–519

Horovitz, C.T., Shock, H.H., Horovitz-Kisimova, L.A. (1974) The content of scandium, thorium, silver, and other trace elements in different plant species. Plant and Soil, **40**, 397–403

Howell, F.G., Gentry, J.B., Smith, M.H. (eds.). (1975) Mineral Cycling in Southeastern Ecosystems. Springfield, Va.: National Technical Information Service, U.S. Department of Commerce, pp. 786

Hunt, C.B. (1972) Geology of Soils. San Francisco: Freeman, p. 344

Hutchinson, G.E. (1944) Limnological studies in Connecticut: critical examination of the supposed relationship between phytoplankton periodicity and chemical changes in lake waters. Ecology **25**, 3–26

Hutchinson, G.E. (1950) Limnological studies in Connecticut: quantitative radiochemical study of the phosphorus cycle in Linsley Pond. Ecology **31**, 194–203

Jackson, M.L. (1956) Soil chemical analysis. Department of Soil Science. Madison, Wisc.: University of Wisconsin

Jenny, H. (1958) Role of the plant factor in pedogenic functions. Ecology, **39** (1)h5–16
Johnson, W.M. (1963) The pedon and the polypedon. Soil Sci. Am. Proc. **27**, 212–215
Junge, C.E. (1963) Air Chemistry and Radioactivity. New York: Academic Press
Jurdant, M., Lacate, D.S., Zoltai, S.C., Runka, G.G., Wells, R. (1975) Biophysical land classifications in Canada in forest soils and forest land management. Proc. 4th North American Forest Soils Conference at Laval University, Quebec, August, 1973. Quebec: Les Presses de L'Universite Laval, pp. 485–495
Keller, E.A. (1976) Environmental Geology. Columbus, Oh.: Charles E. Merrill, p. 488
Kemp, A.L.W., Anderson, T.W., Thomas, R.L., Mudrochova, A. (1974) Sedimentation rates and recent sediment history of Lakes Ontario, Erie and Huron. Jour. Sed. Petrology **44**, 186–213
Klute, A. (1973) Soil water flow theory and its application in field situations. In: Field Soil Water Regime. Bruce, R.R., Flack, K.W., Taylor, H.M. (eds.). SSSA Special Publication 5. Madison, Wisc.: Soil Sci. Soc. Am., pp. 9–32
Kolehmainen, S. Takatalo, S., Miettinew, J.K. (1969) A tracer experiment with ^{131}I in an oligotrophic lake. In: Symposium on Radioecology. Nelson, D.J., Evans, F.C. (eds.). Proceedings of the Second National Symposium at Ann Arbor Michigan, May 15–17, 1967, National Bur. Standards. Springfield, Va.: U.S. Dept. Commerce
Kothy, E.L. (ed.) (1973) Trace elements in the environment. Proceedings Symposium-162nd Meeting of the American Chemical Society, Washington, D.C., September 15th 1971. Adv. in Chemistry Series 123. Washington, D.C.: American Chemical Society, p. 179
Kovalevsky, A.L. (1969) Some observations in biogeochemical parameters (in Russian). Trudy Buryat, (ed.). Institute Estestvenn Nauk. **2**, 195–214
Kovalsky, V.V. (1970) The geochemical ecology of organisms under conditions of varying contents of trace elements in the environment. In: Trace Element Metabolism in Animals. C.F. Mills, Ed, E. and S. Livingstone Press, Edinburgh: pp. 385–396
Kozlovskiy, F.I. (1972) Structural function and migrational landscape geochemical processes. (translation). Pochvovedeniye **4**, 122–138
Krajina, V.J. (1972) Ecosystems perspectives in forestry. H.R. MacMillan Lectureship Address Delivered at the University of British Columbia March 15, 1972. Vancouver: University of British Columbia, pp. 31
Kuhn, T.S. (1970) The Structure of Scientific Revolutions: Vol. I and II, Foundations of the Unity of Science, Vol. II #2. Chicago: University of Chicago Press
Kuznetsov, S.I., Ivanov, M.V., Lyalikova, N.N. (1963) Introduction to Geological Microbiology. New York: McGraw-Hill, pp. 252
Lang, A.H. (1970) Prospecting in Canada. Geological Survey, Canadian Economic Geological Report No. 7
Leaf, A.L. (1973) Plant analysis as an aid in fertilizing forests. In: Soil Testing and Plant Analysis. Revised Edition. Walsh, L.M., Beaton, J.D. (eds.). Madison, Wisc.: Soil Sci. Soc. Am., pp. 427–454
Le Grande, H.E. (1970) Movement of agricultural pollutants with ground water. In: Agricultural Practices and Water Quality. Willrich, T.L., Smith, G.E. (eds.). Ames, Ia.: Iowa State University Press, pp. 303–313
Leopold, L.B., Wolman, M.G., Miller, J.P. (1964) Fluvial Processes in Geomorphology. San Francisco: Freeman, pp. 522
Levinson, A.A. (1974) Introduction to Exploration Geochemistry. Calgary, Alberta: Applied Publishing, pp. 612
Liddell Hart, B.H. (1971) Why Don't We Learn From History?. New York: Hawthorne Books, pp. 95
Lieth, H. (1975) Modeling the primary productivity of the world. In: Primary Productivity of the Biosphere. Ecological Studies 14, Berlin, Heidelberg, New York: Springer-Verlag, p. 4
Lieth, H. Whittaker, R.H. (1975) Primary Productivity of the Biosphere. Ecological Studies No. 14. New York, Heidelberg, Berlin: Springer, pp. 339

Likens, G.E., Bormann, F.H. (1972) Nutrient cycling in ecosystems. In: Ecosystems Structure and Functions. Proc. of the 31st Annual Biology Colloquium. Wiens, J.D. (ed.). Corvallis, Ore.: Oregon State University Press, pp. 25–67

Likens, G.E., Bormann, F.H. (1975) An experimental approach to nutrient-hydrological interactions in New England landscapes. In: Coupling of Land and Water Systems. Hasler, A.D. (ed.). New York, Heidelberg, Berlin: Springer, pp. 7–29

Lintz. J., Simonett, D.S. (1976) Remote Sensing of Environment. Reading, Mass.: Addison-Wesley, pp. 694

Liu, D., Chawla, V.K. (1976) Polychlorinated Biphenols (PCB) in sewage sludges. In: Trace substances in Environmental Health X. Proc. 10th Annual Conference on Trace Substances in Environment Health. Hemphill, D.D. (ed.). Columbia, Mo.: University of Missouri Press, pp. 247–250

Loneragan, J.F. (1975) The availability and absorption of trace elements in soil-plant systems and their relation to movement and concentrations of trace elements in plants. In: Trace Elements in Soil-Plant-Animal Systems. Nicholas, D.J.D., Egan A.R. (eds.). New York: Academic Press, pp. 109–134

Loughman, F.C. (1969) Chemical Weathering of the Silicate Minerals. New York: Elsevier, pp. 154

Lukashev, K.I. (1958) Lithology and Geochemistry of the Weathering Crust. (Trans. from Russian 1970.) Jerusalem: Israel Program for Scientific Translations, pp. 368

Lunt, H.A. (1932) Profile characteristics of New England forest soils. Connecticut Agricultural Experimental Station Bull. **342,** 743–836

Lavkulich, L.M., Rutter, N.W. (1975) Terrain sensitivity and the artic land use research program in the Makenzie Valley, N.W.T. In: Forest Soils and Forest Land Management. Bernier, B., Winget, C.H. (eds.). Quebec: Les Presses de L'Universite Laval, pp. 559–571

Magnusson, N.H., Lunaquist, G., Granlund, E. (1957) Sugriggs Geol. Stockholm: Norstedts, pp. 550

Malyuga, D.P. (1964) Biogeochemical Methods of Prospecting. (Trans. from Russian.) New York: Consultants Bureau, pp. 205

Mason, B. (1952) The Principles of Geochemistry. 1st Ed. New York: John Wiley and Sons.

Mason, B. (1958) The Principles of Geochemistry. 2nd Ed. New York, John Wiley and Sons

Mason, B. (1966) The Principles of Geochemistry. 3rd Ed. New York: John Wiley and Sons

Mattson, S. (1938) The Constitution of the Pedosphere. Annals of the Agriculture College of Sweden **5,** 261–276

McFarlane, M.J. (1976) Laterite and Landscape. London: Academic Press, pp. 151

Mohr, E.C.J., Van Baren, F.A. (1954) Tropical Soils. New York: Interscience.

Mole, R.H. (1965) British Journal of Nature 19, 13. (Cited in Hamilton, Minski, (1972–1973)

Moore, P.D., Bellamy, D.J. (1974) Peatlands. New York, Heidelberg, Berlin: Springer-Verlag, pp. 221

Mueller-Dombois, D. Ellenberg, H. (1974) Aims and Methods of Vegetation Ecology. New York: John Wiley and Sons, pp. 547

Munn, R.E. (ed.) (1975) Environmental impact assessment: principals and procedures, scope workshop on impact studies in the environment (WISE) Co-sponsored by United Nations Environmental Program (UNEP), Environment and Canada and UNESCO. SCOPE Report No. 5. Toronto

Nipkow, F. (1920) Vorläufige Mitteilungen über Untersuchungen des Schlammabsatzes. Rev. d. Hydrol. 1 (2) 1950 Ruheformen planktischer Kieselalgen im geschichteten Schlamm des Zürichsees Schwiez. Zschr. Hydrol. **12**

Novakov, T. Mueller, P.K., Alcocer, A.E., Otuos, J.W. (1972) J. Colloid Interface Sci. **39,** 225–234

Nye, S.M., Peterson, P.J. (1975) The content and distribution of selium in soils and plants from seleniferous areas in Eire and England. In: Proc. 9th Annual Conference Trace Substances in Environmental Health. Hemphill, D.D. (ed.). Columbia, Mo.: University of Missouri Press, pp. 113–122

Odum. E.P. (1971) Fundamentals of Ecology. 3rd Ed. Philadelphia: W.B. Saunders, pp. 574

Oliver, J.E., Manners, I.R. (1972) Environmental science. In: Encyclopedia of Geochemistry and Environmental Sciences, Vol. IVA. Fairbridge, R.W. (ed.). New York: Van Nostrand Reinhold, pp. 337–341

Olsen, J.S. (1970) Carbon cycles in the temperate woodlands. In: Analysis of Temperate Forest Ecosystems. Reichle, D.E. (ed.). New York, Heidelberg, Berlin: Springer-Verlag, pp. 226–241

Oltman, R.E. (1966) Reconnaisance investigation of the discharge and water quality of the Amazon. Paper presented at Associacano de Biologia, Tropical Simposio sobrea Brota Amazonica, June 7, 1966. Belem, Para, Brazil, pp. 1–20

Ovington, J.D. (1965) Organic production, turnover and mineral cycling in woodlands. Biological Reviews **40**, 295–336

Perfenova, Ye. I. (1963) B.B. Polynov—The creator of the study of the geochemistry landscapes. Soviet Soil Sci. **2**, 111–117

Parizck, R.R., Lane, B.E. (1970) Soil-Water sampling using pan and deep pressure vacuum lysemeters. J. Hydrology, **21**, 1–21

Peirson, D.H. (1975) The passage of nuclear weapon debris through the atmosphere. In: The Ecology of Resource Degradation and Renewal. Chadwick, M.J., Goodman, G.T. (eds.). Proc. 15th Symposium of British Ecological Society 10–12 July 1973. New York: John Wiley and Sons, pp. 81–88

Peirson, D.H., Cawse, P.A., Cambray, R.S. (1974) Chemical uniformity of airborne particulate material and a maritime effect. Nature **251**, 675–679

Peirson, D.H., Cawse, P.A., Salmon, L., Cambray, R.S. (1973) Trace elements in the atmosphere environment. Nature **241** (5387), 252–256

Perel'man, A.I. (1966) Landscape Geochemistry. (Translation No. 676, Geological Survey of Canada, 1972) Moscow: Vysshaya Shkola, pp. 1–388

Perel'man, A.I. (1967) Geochemistry of Epigenesis. Kohanowski, N.N. (trans.) New York: Plenum Press, pp. 266

Perel'man, A.I. (1972) The Geochemistry of Elements in the Zone of Supergenesis. (Draft translation: Geological Survey of Canada.) Moscow: Nedra, pp. 287

Perkin-Elmer Corporation (1973) Analytical Methods for Atomic Absorption Spectro-Photometry No. 303-0152. Norwalk, Conn

Peters, L.N., Grigal, D.F., Curlin, J.W., Selvidge, W.J. (1970) Walker Branch Watershed Project: Chemical, Physical and Morphological Properties of the Soils of the Walker Branch Watershed, Oak Ridge National Laboratory Publication N. ORNL-TM-2968, pp. 96

Peyve, Ya V. (1963) V.I. Vernadski and the study of microelements in soils. Soviet Soil Sci **8**, 741–752

Polynov, B.B. (1937) The Cycle of Weathering. Muir, A. (trans.). London: Murby, pp. 220

Polynov, B.B. (1951) Modern ideas of soil formation and development. Soils and Fertilizers **14** (2) 95–101

Poole, R.W. (1974) An Introduction to Quantitave Ecology. New York: McGraw-Hill, pp. 532

Radforth, N.W., Brawner, C.O. (eds.) (1977) Muskeg and the Northern Environments in Canada. Toronto: University of Toronto Press, pp. 399

Ramenskij, L.G. (1928) In: Svoboda P. Přínos Sovětské Vědy k Lesní Typologii. Lesn. Práce 28/11–12 (1949): 453-535 [references in W. Stanek (1977) Classification of Muskeg in N.W. Radforth and C.O. Brawner (eds.) Muskeg and Northern Environment in

Canada. Toronto: University of Toronto Press. (original paper not seen by present writer)

Rankama, K., Sahama, Th. G. (1950) Geochemistry. Chicago: University of Chicago Press, pp. 912

Reisenauer, H.M., Walsh, L.M., Hoeft, R.G. (1973) Testing soils for sulphur, boron, molybdenum and chlorine. In: Soil Testing and Plant Analysis. Revised Edition. Walsh, L.M., Beaton, J.D. (eds.). Madison, Wisc.: Soil Sci. Soc. Am., pp. 173–200

Richardson, C.J., Lund, J.A. (1975) Effects of clear cutting on nutrient losses in aspen forests on three soil types in Michigan. In: Mineral Cycling in Southeastern Ecosystems. Howell, F.G., Gentry, J.B., Smith, M.H. (eds.). Springfield, Va.: National Technical Information Service, U.S. Dept. of Commerce, pp. 536–541

Ritchie, J.C. (1974) Fallout ^{137}Cs in the soils and sediments of three small watersheds. Ecology **55**, 887–890

Rodin, L.E., Brazilevich, N.I. (1967) Production and Mineral Cycling in Terrestrial Vegetation. (Trans. Scripta Technica Ltd., English Translation Ed., G.E. Fogg). Edinburgh: Oliver and Boyd, pp. 288

Ronov, A.B., Yaroshevsky, A.A. (1972) Earth's crust geochemistry. In: Encyclopedia of Geochemistry and Environmental Sciences Vol. IVA. Fairbridge, F.W. (ed.). New York: Van Nostrand Reinhold, pp. 243–254

Ruellan, A. (1971) The history of soils: some problems of definition and interpretation. In: D.H. Yaalon (Ed) Palaeopedology: Origin Nature amd Dating of Paleosols. Int. Soc. of Soil Science and Israel Universities Press, Jerusalem, pp. 3–14

Ruhe, R.V. (1969) Quaternary Landscapes in Iowa. Ames, Ia: Iowa State University Press, pp. 255

Runge, E.C.A., Gohand, K.M. Rafter, T.A. (1973) Radiocarbon chronology and problems in its interpretation for Quaternary leoss deposits, South Canterbury, New Zealand. Soil. Sci. Soc. Am. Proc. **37**, 742–746

Ruttner, F. (1963) Fundamentals of Limnology. Frey, D.G., Fry, F.E.J. (trans.). Toronto: University of Toronto Press, pp. 307

Sabine, D.A. (1955). Physiological Principles of Plant Nutrition (in Russian). Moscow: Adad. Navk, 1955

Sergeev, E.A. (1941) The physiogeochemical method in the exploration for ore deposits (Data from the All-Union Scientific Research Institute of Geological Exploration.) Geofizika, **9, 10**

Shacklette, H.T. (1971) A U.S. geological survey study of elements in soils and other surficial materials in the United States. In: Trace Substances in Environmental Health–IV. Proceedings of the University of Missouri's 4th Annual Conference on Trace Substances in Environmental Health. Hemphill, D.D. (ed.). Columbia, Mo.: University of Missouri Press, pp. 35–48

Shimwell, D.W. (1972) The Description and Classification of Vegetation. Seattle, Wash.: University of Washington Press, pp. 322

Smuts, J.C. (1961) Holism and Evolution. New York: Viking Press, pp. 362

Sokol, R.R., Rolf, F.G. (1969) Biometry: The Principles and Practice of Statistics in Bio Research. San Francisco: Freeman

Soil Survey (1960) Soil Classification, A Comprehensive System–7th Approximation, U.S. Department of Agriculture. Washington: U.S. Government Printing Office

Southwick, C.H. (1976) Ecology and the Quality of the Environment. 2nd Ed. New York: Van Nostrand, pp. 426

Spedding, D.J. (1974) Air Polution. Oxford: Clarendon Press, pp. 76

Stalfelt, M.G. (1972) Stafelt's Plant Ecology. Jarvis, M.S., Jarvis, P.G. (trans.). New York: Halsted Press, pp. 592

Stallings, W.S., Jr. (1949) Dating Prehistoric Ruins by Tree Rings. Revised Ed. Tucson, Ariz.: Laboratory of Tree Ring Research

Stanek, W. (1977) Classification of muskeg. In: Muskeg and the Northern Environment in

Canada. Radforth, N.W., Brawner, C.O. (eds.). Toronto: University of Toronto Press, pp. 31–62

Stevenson, J.S., Stevenson, L.S. (1972) Medical geology. In: Encyclopedia of Geochemistry and Environmental Science IVA. Fairbridge, W. (ed.). New York: Van Nostrand Reinhold, pp. 696–698

Stone, E.L. (1975) Soil and man's use of forest land. In: Forest Soils and Forest Land Management. Proceedings 4th North American Forest Soils Conference at Laval University, Quebec, August, 1973. Quebec: Les Presses de L'Universite Laval, pp. 1–9

Storie, R.E. (1964) Handbook of Soil Evaluation. Berkeley, Cal.: University of California Associated Students Store

Strakhov, N.M. (1960) Principles of the Theory of Lithogenesis, Vols. 1 and 2. Moscow: Izd-vo Akad. Nauk SSSR

Sugawara, K. (1963) Migration of Elements through Phases of Atmosphere and Hydrosphere. Moscow: Vernadsky Conference (Reprints)

Sukachev, V., Dylis, N. (1964) Fundamentals of Forest Biogeocoenology. Maclennan, J.M. (trans.). Edinburgh: Oliver and Boyd, pp. 672

Switzer, G.L., Nelson, L.E., Smith, W.H. (1968) Mineral cycle in forest stands. In: Forest Fertilization Theory and Practice Papers from Symposium on Forest Fertilization, April, 1967, Gainsville, Florida. Bengtson, G.W. (ed.). Muscle Shoals, Ala.: T.V.A. National Fertilizer Development Center, pp. 1–9

Tansley, A.G. (1936) The use and abuse of vegetational concepts and terms. Ecology **16**, 284–307

Thomas, R., Law, J.P. (1977) Properties of waste waters. In: Soils for Management of Organic Wastes and Waste Waters. Elliott, L.F., Stevenson, F.J. (eds.). Madison, Wisc.: Soil Science Society of America: American Society of Agronomy Crop Society of America

Thomas, W.A. (1969) Accumulating and cycling of calcium by dogwood trees. Ecological Monographs **39**, 101–120

Thornton, I., Webb, J.S. (1970) Geochemical reconnaissance and the detection of trace element disorders in animals. In: Trace Element Metabolism in Animals. Mills, C.F. (ed.). E. and S. Livingstone, Edinburgh, pp. 397–409

Tidball, R.R., Sauer, H.I. (1975) Multivariate relationships between soil composition and human mortality rates in Missouri. Geological Society America Special Paper **155**, 41–59

Todd, D.K., McNulty, D.E.O. (1976) Polluted Groundwater, a Review of the Significant Literature. Huntington, N.Y.: Water Information Center, Inc.

Townsend, W.N.N, Gillham, E.W.F. (1975) Pulverized fuel ash as a medium for plant growth. In: The ecology of Resource Degradation and Renewal, Proceedings 15th Symposium of the British Ecological Society 10–12 July 1973. Chadwick, M.J., Goodman, G.T. (eds.). New York: John Wiley and Sons, pp. 287–304

Underwood, E.J. (1971) Trace Elements in Human and Animal Nutrition. 3rd Ed. New York: Academic Press, pp. 543

Vanloocke, R., De Broger, R., Voets, J.P., Verstraete, W. (1975) Soil and groundwater contamination by oil spills: problems and remedies. Intern. J. Environmental Studies **8**, 99–111

Vernadski, V.I. (1942) La Geochemie. Paris Librairie, Felix Alcan, pp. 403

Vernadski, V.I. (1927) The Biosphere. Paris Librarie, Felix Alcan

Vernadski, V.I. (1934) History of Natural Waters. Moscow:

Vinogradov, A.P. (1963) The development of V.I. Vernadskiy's ideas. In: Soviet Soil Science. Translation No. 8. 727–732

Volk, B.G., Schemnitz, S.D., Gamble, J.F., Sartain, J.B. (1975) Baseline Data on Everglades Soil-Plant Systems: Elemental Systems, Biomass, and Soil Depth. Proceedings of a Symposium held in Augusta, Ga. May 1–3, 1974. Howell, F.G. Gentry, J.B., Smith, M.H. ERDA Symposium Series QH 344.M56. Springfield, Va.: Technical

Information Center, Office of Public Affairs, U.S. Research and Development, pp. 658–672

Vreeken, W.J. (1975) Principal kinds of Chronosequences and their significance in soil history. J. Soil Sci. **26,** (4), 378–394

Wagner, G.H. (1962) Use of porous ceramic cups to sample soil water within the profile. Soil Sci. **94,** 379–386

Walker, M. (1963) The Nature of Scientific Thought. Englewood Cliffs, N.J.: Prentice Hall, pp. 184

Walker, T.R. (1961) Groundwater contamination in the Rocky Mountain Arsenal area, Denver, Colorado. Bull. Geological Soc. Am. **72,** 489–494

Walter, H. (1973) Vegetation of the earth. In: Relation to Climate and Ecophysiological Conditions. Weiser, J. (trans.). New York, Heidelberg, Berlin: Springer-Verlag, pp. 237

Ward, R.C. (1975) Principle of Hydrology. London: McGraw-Hill, pp. 376

Warren, H.V., Delavault, R.E., Fortescue, J.A.C. (1955) Sampling in biogeochemistry. Bull. Geological Soc. Am. **229**

Webb, D.A. (1954) Is the classification of plant communities either possible or desirable? Bottanisk Tidsskrft **51,** 362–370

Webb, J.S. (1964) New Scientist **23,** 504

Webster, J.R., Waide, J.B., Patten, B.C. (1975) Nutrient recycling and the stability of ecosystems. In: Mineral Cycling in Southeastern Ecosystems. Howell, F.G., Gentry, J.B., Smith, M.H. (eds.). Springfield, Va.: Technical Information Center, Office of Public Affairs, pp. 1–27

Wedepohl, K.H. (1971) Geochemistry, Althaus, E. (trans.). New York: Holt, Rinehart and Winston, pp. 231

Wegman, D.H. (1978) The environmentalist's challenge. Am. J. of Public Health 68 (6), 540–541

Weiner, J.G., Brisbin, I.L., Smith, M.H. (1975) Chemical composition of white tailed deer: whole body concentrations of macro- and micro-nutrients. In: Mineral Cycling in Southeastern Ecosystems. Howell, F.G., Gentry, J.B., Smith, M.H. (eds.). Springfield, Va.: Technical Information Center, Office of Public Affairs, pp. 536–541

Welch, C.D., Wiese, R.A. (1973) Opportunities to improve soil testing programs. In: Soil Testing and Plant Analysis. Revised Ed. Walsh, L.M., Beaton, J.D. (eds.) Madison, Wisc.: Soil Sci. Soc. Am pp. 1–12

Wigglesworth, V.B (1964) The Life of Insects. London: Weidenfeld and Nicholson

Wixson, B.G. (ed.) (1977) The Missouri lead study. Final Report to the National Science Foundation Research Applied to National Needs for Period May, 1972 to May, 1977, Vols. I and II, pp. 1107

Zoltoi, S.C., Pollett, F.C., Jeglum, J.K., Adams, G.D. (1975) Developing a wetland classification for Canada. In: Forest Soils and Forest Land Management. Proc. 4th North American Forest Soils Conference at Laval University, Quebec, August, 1973. Quebec: Les Presses de L'Universite Laval, pp. 497–512

Zverev V.P. (1973) Chemical mobilization of substance in catchment areas. Lithology and Mineral Resources, Vol. 7 (Translated) New York: Plenum (7) pp. 758–764

Author Index

A

Andren, A.W. (1973) 259
Anderson, T.W. (1974) 238, 239, 240

B

Bakuzis, E.V. (1969) 37
Barringer, A. (1976) 220, 221
Bartos, D.L. (1973) 209, 210
Beer, S. (1975) 296, 297
Bienfet, et al. (1970) 282, 283
Bolter, et al. (1972) 111
Bowen, H.J.M. (1966) 22, 56, 65, 67, 71, 72
Boyer, K.W. (1977) 282
Bradshaw, P.M.D. (1975) 223, 226, 263, 293, 302
Brass, G.W. (1975) 234, 235, 236
Briggs, L.I. (1967) 232, 233, 234
Brooks, R.R. (1973) 90, 91
Bryne, A.R. (1976) 186, 187, 188
Bunge, M. (1962) 5
Buol, S.W. et al. (1973) 12, 137
Burger, D
 (1972) 36, 37
 (1976) 35
Burgess, R.L. (1975) 32, 33

C

Cannon, H.L. (1970) 213, 215
Cannon, H.L. and Bowles
 (1962) 266
Carson, R. (1962) 292
Chadwick, M.J. et al. (1975) 273
Clarke, F.W. (1924) 7, 23, 291
Collier, B.D. et al. (1973) 32, 33
Colwell, R.N.
 (1968a) 221
 (1968b) 222
Crompton, E. (1967) 139, 140

D

Dahlman, R.C. (1971) 207, 208, 209
Dahnke, W.D. et al. (1973) 44
Daughtrey, A.W. et al. (1973) 176, 178, 179, 180
Davis, S.N. et al. (1966) 59
Dixon, B.A. and Croker R.L.
 (1953) 148
 (1954) 148, 149, 150
Dorsey, J.A. et al. (1977) 26

E

Elliott, J.W. and Henderson G.S.
 (1975) 103
Everett, J. and D.S. Simonett
 (1976) 38
Evseeva, L.S. and A.I. Perel'man
 (1962) 126

F

Fisher, R.A. (1958) 197
Fortescue, J.A.C.
 (1967a) 12
 (1967b) 10
 (1970) 260, 261
 (1973) 287
 (1974a) 15, 97, 266, 267
 (1974b) 46, 163, 166
 (1974c) 287
 (1975a) 299
 (1975b) 142, 143, 145
 and P.M.D. Bradshaw (1973) 13, 301
 et al. (1973) 256, 257, 258
 and E.H.W. Hornbrook (1967a) 265
   (1967b) 259, 260, 264, 265
 and C.G. Marten (1970) 197, 202
 and J. Terasmae (1975) 21
 and L. Usik (1969) 217, 218, 219
Fritts H.C.
 (1962) 242
 (1976) 241, 242, 243, 244
Feuhring, H.D. (1960) 197
Fu Su Yen and J.H. Goodwin (1976) 197, 198, 199, 200, 201

G

Gibbs, R.J. (1972) 58, 59, 60
Giddings, J.C. (1973) 60
Glazovskaya, M.A. (1963) 43, 47, 161, 162, 163–168
Glazovskaya, M.A. et al. (1961) 43, 162, 262, 263
Glover, H.G. (1975) 273, 276
Goldschmidt, V.M.
 (1937) 8, 123, 124
 (1954) 8, 291
Gordon, A.G. and E. Gorham (1963) 268, 269, 270, 271
Green, J. (1972) 52, 53

H

Hamersma, J.W. et al. (1976) 26
Hamilton, E.I. (1976) 294
Hamilton, E.I. and M.J. Minski (1972–1973) 182, 183, 184, 185
Harradine, F. (1967) 76
Harrison, J.B. (1934) 137, 138, 139
Hawkes, H.E. and J.S. Webb (1962) 216, 292, 298, 299, 300, 302
Hawkes, H.E. (1976) 300
Henderson, G.S. and W.F. Harris (1975) 100
Hills, G.A. (1959) 265
Hodgson, D.R. and G.P. Buckley (1975) 275
Holmes, J.W. (1971) 160, 161
Hopps, H.C. et al. (1968) 287
Hordon, R.M. (1972) 58
Horovitz, C.T. (1974) 67, 70
Howell, F.G. et al. (1975) 320
Hunt, C.B. (1972) 150, 151
Hutchinson, G.E.
 (1944) 33
 (1950) 33

J

Jackson, M.L. (1956) 193
Jenny, H. (1958) 14
Johnson, W.M. (1963) 14
Junge, C.E. (1963) 62
Jurdant, M. et al. (1975) 37

K

Keller, E.A. (1976) 287
Kemp, A.L.W. et al. (1974) 240
Klute, A. (1973) 129
Kolehmainen, S. et al. (1969) 173, 174, 175
Kothy, E.L. (1973) 29, 293
Kovalevsky, A.L. (1969) 90
Kovalsky, V.V. (1970) 281, 287, 288
Kozlovskiy, F.I. (1972) 43, 45, 95–107, 206, 249
Krajina, V.J. (1972) 35, 37
Kuhn, T.S. (1970) 295
Kuznetsov. S.I. et al. (1963) 127

L

Lang, A.H. (1970) 74, 75
Lavkulich, L.M. and N.W. Rutter (1975) 36
Leaf, A.L. (1973) 282, 283, 284, 285, 286

Le Grande, H.E. (1970) 230
Leopold, L.B. et al. (1964) 137, 234
Levinson, A.A. (1974) 74, 75, 110, 111, 255, 264, 265, 293, 301
Liddell, Hart B.H. (1971) 315
Lieth, H. and R.H. Whittaker (1975) 66
Likens G.E. and F.H. Bormann
 (1972) 79, 80, 81, 82, 104, 248
 (1975) 44, 82
Lintz, J. and D.S. Simonett (1976) 316, 320
Liu, D. and V.K. Chawla (1976) 180, 181, 182
Loneragan, J.F. (1975) 90
Lukashev, K.I. (1958) 113, 114
Lunt, H.A. (1932) 104

M

Magnusson, N.H. et al. (1957) 142, 143
Malyuga, D.P. (1964) 110, 111, 112, 113, 220
Mason, B.
 (1952) 9, 291
 (1958) 9, 10, 291
 (1966) 8, 44, 56, 61, 85, 86, 87, 126
Mattson, S. (1938) 11
McFarlane, M.J. (1976) 134, 135, 136, 137
Mohr, E.C.J. et al. (1954) 137, 138, 139
Mole, R.H. (1965) 184
Moore, P.D. and D.J. Bellamy (1974) 142, 144
Mueller-Dombois, D. and H. Ellenberg
 (1974) 116, 117, 118, 120
Munn, R.E. (1975) 303

N

Nipkow, F. (1920) 145, 146
Novakov, T. et al. (1972) 62
Nye, S.M. and P.J. Peterson (1975) 21

O

Odum, E.P. (1971) 320
Oliver, J.E. and I.R. Manners (1972) 29
Olsen, J.S. (1970) 67
Oltman, R.E. (1966) 60
Ovington, J.D. (1965) 208

P

Parfenova, Y.I. (1963) 16
Parizck, R.R. and B.E. Lane (1970) 194
Peirson, D.H.
 (1975) 277, 278, 279
 et al. (1973) 63, 64
 et al. (1974) 63
Perel'man, A.I.
 (1966) 9, 16, 17, 43, 47, 56, 72, 79, 81, 82, 101, 113, 123, 156, 158, 159, 160, 161, 168, 169, 171, 212, 248, 249, 258, 281
 (1967) 43, 79, 81, 82, 83, 84, 85, 123, 125, 126, 127, 128, 129, 130, 131
 (1972) 16, 43, 72, 73, 79
Perkin Elmer Corporation (1973) 194
Peters, L.N. et al. (1970) 100
Peyve, Ya V. (1963) 8
Polynov, B.B.
 (1937) 16, 17, 43, 79, 81, 133, 134, 141, 148
 (1951) 16, 17, 43, 47
Poole, R.W. (1974) 97, 98

R

Radforth, N.W. and C.O. Brawner
 (1977) 127
Ramenskij, L.G. (1928) 119, 120
Rankama, K. and Th. G. Sahama (1950) 7, 8, 51, 58, 291
Reisenauer, H.M. et al. (1973) 22
Richardson, C.J. and J.A. Lund (1975) 193, 196, 211
Ritchie, J.C. (1974) 278, 280
Rodin, L.E. and N.I. Brazilevich (1967) 37, 67, 68, 69, 72
Ronov, A.B. and A.A. Yaroshevsky
 (1972) 22, 54, 55, 56, 57
Ruellan, A. (1971) 147
Ruhe, R.V. (1969) 124, 125
Runge, E.C.A. et al. (1973) 237, 238
Ruttner, F. (1963) 145, 146

S

Sabine, D.A. (1955) 91
Sergeev, E.A. (1941) 110, 111, 260

Shacklette, H.T. (1971)   217, 220
Shimwell, D.W. (1972)   116, 117
Smuts, J.C. (1961)   30
Sokol, R.R. and F.G. Rolf (1969)   195
Soil Survey Staff (1960)   14
Southwick, C.H. (1976)   31
Spedding, D.J. (1974)   60, 62, 63, 64
Stalfelt, M.G. (1972)   11
Stallings, W.S. (1949)   241
Stanek, W.S. (1977)   36, 119, 120
Stevenson, J.S. and L.S. Stevenson (1972)   287
Stone, E.L. (1975)   39
Storie, R.E. (1964)   36
Strakhov, N.M. (1960)   112
Sugawara, K. (1963)   58
Sukackev, V. and N. Dylis (1964)   34, 35, 155
Switzer, G.L. et al. (1968)   231

T

Tansley, A.G. (1936)   31, 34, 35
Thomas, R. and J.P. Law (1977)   214, 275
Thomas, W.A. (1969)   105, 106
Thornton, I. and J.S. Webb (1970)   287, 289
Tidball, R.R. and H.I. Sauer (1975)   176, 177
Todd, D.K. and D.O.E. McNulty (1976)   275
Townsend, W.N.N. and E.W.F. Gillham (1975)   271, 272, 273, 274

U

Underwood, E.J. (1971)   280, 287

V

Vanloocke, R. et al. (1975)   277
Vernadski, V.I.
  (1927)   7, 33
  (1934)   7
  (1942)   7
Vinogradov, A.P. (1963)   7
Volk B.G. et al. (1975)   192
Vreeken, W.J. (1975)   147, 148

W

Wagner, G.H. (1962)   194
Walker, M. (1963)   250
Walker, T.R. (1961)   230
Walter, H. (1973)   115
Ward, R.C. (1975)   59
Warren, H.V. et al. (1955)   264
Webb, D.A. (1954)   173
Webb, J.S. (1964)   287
Webster, J.R. et al. (1975)   202, 203, 204, 205, 206
Wedepohl, K.H. (1971)   10, 29, 58, 292, 293
Weiner, J.G. et al. (1975)   193, 194
Welch, C.D. and R.A. Wise (1973)   281
Wigglesworth, V.B. (1964)   65
Wixson, B.G. (1977)   301, 302

Z

Zoltoi, S.C. et al. (1975)   36
Zverev, V.P. (1973)   141

Subject Index

This index is focused upon words and concepts describing holistic environmental geochemistry. Consequently, listings regarding individual chemical elements are restricted. Readers interested in the details of the geochemistry of specific elements should refer to textbooks such as Rankama and Sahama (1950), Goldschmidt (1954), Wedepohl (1971) or Brownlow (1979).

Numbers in italic denote definitions.

A

Absolute abundance 20, 76
Absolute mobility 79
Accessory element 128
Accumulator plant 88
Accumulator species 188, 257
Acid granitoides 55
Acropetal Coefficient (AC) 88
Aerial photographs 221
Aerosol of the atmosphere 60, *62*
Air, content of elements 64
Air migrant 92, 129
Alluvium 215
Aluminium 16, 17, 54, 56, 59, 63, 64, 68, 69, 71, 80, 83, 84, 85, 86, 87, 92, 112, 113, 128, 138, 139, 183, 192, 199, 258, 272, 274, 283
Amazon River 58
Angiosperm 157
Animal nutrition 256, 280
Animal wastes 282

Anomalies in climate 244
Appliers 5, 314
Appraisal Data (AD) 22, 76, 217
Aqual landscape *14*, 16, 48, 144, 249, 257, 266, 311
Aqueous migrant 92
Aquifers 125, 129
Arenaceous rocks 54
Arsenic 56, 59, 64, 84, 124, 125, 186, 188, 215, 274
Aspen forest 193, 209
Atmogeochemistry 7, 85
Atmosphere 16, 47, 51, 58, 60, 72, 96, 247
  average composition 61
  gases and vapors 61
Atomic absorption 74
Automobile smog 62

B

Bacterial activity 126

Bar graph diagram   113
Basaltic shell   55
Basic rocks   55
Barium   55, 56, 59, 71, 84, 86, 125, 126, 183, 215, 219
Batch of samples   261
Bauxite   87, 138
Beryllium   56, 59, 85, 86, 124, 183, 216, 274, 300
Biogeochemical
  cycle   32, *33,* 34, 91, 95, 96, 105, 134, 156, 202
  nutritional chains   281
  provinces   74, 287
  survey   257
  surveys standardization   260
  zones   287
Biogeochemistry   7, 287
Biogeocoenose   34, *35,* 155
  distinct from geochemical landscape   155
Biological Absorption Coefficient (BAC)   88
Biology   51
Biomass   66
Biome   32
Biosphere   16, 34, 47, 51, 65, 72, 202, 274
  element abundance   65
Biotite   198
  lattice sites
    octahedral   198
    tetrahedral   198
Bog   142
  ombrogenic   142
  soligenic   142
Bore hole   217
Boron   56, 59, 71, 84, 86, 183, 197, 215, 271, 272, 273, 274, 288
Box and arrow diagram
  Aspen succession   210
  components of a biogeocoenose   35
  hypothetical nutrient cycle lake ecosystems   32
  nitrogen cycling in grassland   207
  nutrient flow model of ecosystem   204
Bromine   10, 17, 56, 59, 62, 63, 64, 72, 83, 84, 186, 188
Brown soils of California   77

C

Calcium   16, 17, 37, 55, 56, 58, 59, 60, 63, 64, 68, 69, 71, 76, 77, 80, 81, 83, 84, 86, 92, 98, 104, 105, 106, 114, 125, 126, 128, 137, 138, 139, 142, 150, 156, 158, 182, 183, 192, 193, 194, 195, 196, 199, 215, 237, 269, 276, 283
  macronutrient   105
  radioactive tracer   105
Carbon   16, 21, 52, 53, 55, 56, 61, 71, 77, 86, 141, 156, 237, 238
Carbonates   85
Carbonate rocks   54
Carson, Rachel   292
Catena   *109,* 118
Cation exchange capacity (cec)   86
Cerium   56, 59, 71
Cesium   278
^{137}Cs   278
  in sediment profiles   280
Chalk   184
Chattanooga   221
Chemical
  data from the environment   22
  fertilization of forests   232
  fertilizers   282
  products of weathering   85
  species   *21,* 84
Chestnut pollen   238
Chlorate toxicity   230
Chlorine   10, 16, 17, 21, 55, 56, 58, 60, 62, 63, 64, 67, 68, 69, 71, 72, 80, 83, 84, 92, 114, 125, 128, 177, 276, 283
Chromium   56, 59, 63, 64, 70, 71, 84, 125, 183, 192, 199, 200, 215, 219, 274
Chronosequence   *147,* 153
  postincisive   *147*
  preincisive   *147*
  time transgressive   *147*
Clarke, F.W.   7, 23, 29, 51, 291
Clarke, unit   *43,* 52, 60, 65, 71, 72, 82, 128, 247
Clarke of concentration (KK)   *44,* 72, 247
Classification
  secondary dispersion haloes and trains   262
Climate   242
Coal-fired power plants   111, 271
Coal mine waste water   273
Cobalt   56, 59, 64, 70, 71, 124, 125, 128, 192, 215, 274, 283, 287, 288, 289
Cobalt deficient soils   289
Coefficient of aqueous migration (K x)   82

Subject Index

Colorimetric analysis 74
Complex cycloid *134*
Computer 4
  computer model 232, 247
Conceptual model 110, 120, 264
  complex cycloid *134*
  Dorset 259
  evolution of peatlands 142
  landscape section 257, 261
  laterite formation 134
  mobile elements around mineral deposits 226
  relief inversion 136
  soil anomaly 110
  soils of the Canadian Cordillera 223
  topographic peatlands 142
  weathered crust 111
Continental crust 55
Copper 56, 59, 63, 64, 71, 74, 83, 84, 90, 109, 125, 128, 142, 166, 183, 186, 188, 192, 193, 194, 197, 215, 217, 219, 220, 258, 260, 274, 283, 288
Creators 5, 315
Crop type identification 221
Culm measures 289
Cycle of weathering *133*, 140, 147, 250
Cyclic salts 58
Cycling of nutrients in ecosystems 202

D

Dartmoor 287
Daylight surface 16
Dendrochronology 148, 241
Denver 230
Dependent migrants 95, 129
Descriptive biogeochemistry 91
Desert varnish 157
Developers 5, 315
Discriminant statistical analysis 197
Discipline of landscape geochemistry 17, 43
  (*see also* Landscape Geochemistry)
Dokuchayev, V. V. 7
Dorset bog 143
Dorset, Ontario 257
  conceptual model 259
Drinking water 80
Dry matter production forest trees 231

E

Earth's crust 1, 51, 54, 71
Ecological
  gradients 115, 117
  resilience *202*
  resistance *202*
  series 118
  time *20*, 237
    landscape development 142
Ecology 9, *31*, 34, 51, 132, 247, 311
Ecosystem 9, *31*, 34, 35, 96, 103
  eight ecosytem types 201
  types 65, 66
Edge effects 102
Element abundance *43*, 51–78, 247, 312
Element migration 44, 47
  and holism 91
  in landscapes 79–93, 113, 247
Elements
  in animals 71
  in plants 71
Elementary landscapes
  cells (ELC) *98*, 101
  types 13, 16
Eluvial landscape 14, 16, 48, 147, 249, 266, 311
  complex 150
Environment *3*
Environmental
  abundance geochemistry 52
  biogeochemistry *281*
  conversion factors 214
  geochemistry 19, 21, 29, 34, 247
  monitoring 26
  pollution 26, 256, 266–271
  science 29, 30, 48
Epigenitic processes 128, 132, 133, 140, 159
Eutrophic terrain 120
Eutrophication 180
Evaporates 85
Evaporite basin 232
Evaporites 54
Exchangeable cation 149
Exploration geochemistry 171, 215, 217, 256
  geochemists 74, 84, 263–266, 292, 293
Extra landscape flow (ELF) 45, *97*, 266

F

Factorial experiment 197
Finland 173
Fluorine 56, 71, 83, 274
Food
  sugar 182
Forest ecosystem 95
Forestry 51
  landscape geochemical research 256
Fossil pollen record 238
Fulvic acid 237
Fungi
  trace elements in 186
Fursman, A.E. 8, 44

G

Gallium 56, 59, 71, 84, 86, 183, 274
Geochemical
  mapping 171, 264
  maps 176, 264
  quantities 102
  structure of landscape 96
  survey 20, 192, 217
  survey, broadscale 217
  surveying 264
Geochemical abundance categories 53
Geochemical anomalies 51, 249, 311
Geochemical barriers, 45, 46, 47, 123–132, 133, 155, 217, 249, 267, 312
  areal *123*
  biological *123*
  extent *124*
  horizontal *123*
  linear *123*
  mechanical *123*
  physiochemical *123*
Geochemical census 176, 192
Geochemical channels 96
Geochemical classification of landscapes 16, 155–171, 250, 312
  abiogenic *156*
  biogenic *156*
  common type 156
  concept of 47
  Glazovskaya system 161–171
  hierarchial system 157
  landscape group 156
  most rapid migration rate 156
  Perel'man system 156–161
  proximity to daylight surface 156
  rare type 156
  taxonomic rank 156
  taxonomic units
    class 158
    family 158
    genus 158
    group 158
    series 158
    species 159
    type 158
  terminology 161
Geochemical cycle 7, 9, 10
Geochemical flows 44, 47, 113, 155, 249, 312
Geochemical gradients 45, 47, 109–121, 132, 155, 217, 249, 267, 312
  conceptual model 111
  invisible 109
  thermal *113*, 129
  visible 109
  water series 113, 129
Geochemical landscape
  conjugated 160
  distinct from biogeocoenose 155
  high contrast 159
  low contrast 159
Geochemical landscape group 156
  eluvial landscapes basis 156
  meadow and steppe 156
  primitive desert 157
  tundra 157
  wooded 156
Geochemistry 3, 7, 8, 17
Geoepidemiology 256, 281, 287–289
Geological time 20, 96, 140, 232
Geology 51, 249, 311
Geomorphology 51, 137, 249, 311
Glaciated areas 96
General geochemistry 247
Glazovskaya M.A. 43, 47, 262
Glazovskaya, geochemical landscape classification 161–171
  accumulative *161*
  eluvial *161*
  eluvial accumulated *161*
  heterogeneous *161*
  homogeneous *161*
  ideal eluvial landscape *168*
  illuvial *167*

impermaciedal *166*
  permaciedal *166*
  surface eluvial frozen *167*
  taxonomic units *165*
  transeluvial *161*
Gley waters 125
Global productivity of the biosphere 224
Global scale
  computerized mapping 226
  radioactive waste disposal 277
Gold 52, 56, 59, 64, 124
Goldschmidt, V.M. 7, 8, 29, 57, 85, 95, 123, 291
Gradient *109*
  continuous *109*
  discontinuous *109*
Grand strategic landscape prism 13
Granitic shell 55
Graph panel *261*
Green River formation 197
Groundwater 58, 59, 96, 129, 230, 249
Groundwater hydrology 249
Gymnosperm 157

H

Hangerstown valley 215
Hard water 184
Hierarchy *19*
Hierarchy of chemical complexity *19*, 20, 24, 34, 38, 173
  element level 175
  isotope level 173
  nutrient level 182
  organism level 186
  partial element level 176
  persistent chemical level 180
  tissue level 183
Hierarchy of scientific effort *19*, 22, 24, 34, 38
  descriptive/empirical level 191
  statistical level 197
  systems analysis 201
  systems simulation level 207
Hierarchy of space *19*, 20, 24, 34, 38, 310
  aerial survey level 221
  continuous traverse line level 220
  landscape section level 217
  nine approaches to 226
  point level 213

  points in an area level 217
  regular grid sampling level 217
  traverse line level 215
  volume unit level 225
Hierarchy of time *19*, 20, 24, 34, 38, 229–246
  ecological time 237
  environmental monitoring 231
  isolated incidents 230
  retrospective monitoring 232, 234
Historical geochemistry 47, 132, 152, 155, 312
Holism 30, 37, 78, 106, 249, 309
  in historical geochemistry 152
  of geochemical barriers 129
Hubbard Brook 44, 79, 104
Human bone 184
Humic acid 237
Humus 123
Hutchinson, G. E. 33
Hydrocarbons
  spills 277
  biodegradation 277
Hydrogen 52, 56, 61, 71, 113, 126, 127, 149, 150, 158, 177
Hydrogen sulphide barrier 126
Hydrogeochemistry 7
Hydrology 7, 249
Hydrosates *85*
Hydrosphere 16, 47, 51, 57, 58, 59, 72, 91, 247
Hypothetical ecosystems 204

I

Immature soils 137
Inactive air migrant 96
Independent migrants 95, 129
Indurated layer 135
Infinitesimal 12
Instrumental analysis 4
Ionic charge 85
Ionic potential *85*, 86
Iron 16, 17, 52, 54, 55, 56, 59, 63, 64, 67, 68, 69, 70, 71, 83, 84, 86, 92, 112, 125, 128, 138, 139, 142, 158, 182, 183, 192, 193, 194, 195, 196, 197, 199, 200, 215, 264, 276, 283
Isochron 235
Isogeochemical landscape *133*, 152

Isograms 242
Isometric paper 12
Isotope 1, 2, 173

J

Jenny, H. 14, 98

K

Kaolinite 87
Katametamorphism 137
Kovada, V. A. 113
Kozlovskiy, F. I.
 macrostructure of landscape *102*
 specific geochemistry of migrants *105*
 specific geochemistry of migrational processes *102*
 stability of geochemical parameters of migrational macroprocesses *103*
 summary diagram 107
Krijina, V. J. 35, 37

L

Lake 58
 eutrophic 126
 meromictic 145
 salt 157
Lake ecosystem 32
Lake Pitkannokanlampi 173
Lake Windermere 62
Land classification 34, 36, 37, 38
Landscape class *158*
Landscape family *158*
Landscape flow pattern 96
Landscape genera *158*
Landscape geochemical cell *100*
 open and closed 100
Landscape geochemical flow (LGF) 45, 96, 125
Landscape geochemistry 3, 14, 30
 practical applications 255–289
Landscape Geochemistry, discipline of 48, 247
 concept of element abundance *43*, 51–78
 concept of element migration *44*, 79–93
 concept of geochemical barriers *45*, 123–132
 concept of geochemical flows *45*, 95–108
 concept of geochemical gradients *45*, 109–121
 concept of geochemical landscape classification *47*, 155–171
 concept of historical geochemistry *47*, 133–153
 fundamental applications *256*, 256–263
 practical applications *256*, 263–289
Landscape Geochemistry, types *297*
 type A *297*, 298–299, 313
 type B *297*, 299–300, 313
 type C *297*, 300–301, 313
 type D *297*, 301–302, 313
 type E 49, *297*, 302–304, 313
Landscape geophysics 309
Landscape
 hydrology 309
 species *159*
 types *157*, 249
Landscape prism *12*, 14, 23, 77, 92, 96, 225, 264
 geochemical flows 97
 peat bog 145
Land use planning 31, 264
Laterite *134*
 primary 138
Lanthanum 52, 56, 59, 64, 71, 86
Lead 56, 59, 62, 63, 64, 71, 74, 84, 87, 111, 124, 125, 142, 166, 183, 192, 215, 219, 260, 264, 265, 266, 267, 275, 283, 301
Level of intensity
 global 20
 local 20
 regional 20
Levinson, A. A. 111, 264, 293
Limestone 184, 215
Limnology 51, 311
Lithogeochemistry 7
Lithosphere 16, 51, 65, 91, 96, 247
Livestock 282
Living biosphere 65, 67, 91, 115, 132
Loess 124
Logging 80
Loveland loess 124
Lysimeter 194

M

Machiavelli 253
Macronutrient 105, 193
  in deer 193
Macrostructure of landscapes *102*, 134
Magnesium 16, 17, 55, 56, 58, 59, 60, 63, 68, 69, 71, 76, 77, 80, 84, 86, 114, 125, 128, 137, 138, 139, 142, 150, 156, 182, 183, 192, 193, 194, 195, 199, 200, 215, 258, 273, 276, 283
Major geochemical cycle 9, 51, 91, 141
Main migrational cycle (MMC) 45, *96*, 99
Main project *260*
Manganese 17, 55, 56, 59, 63, 64, 68, 69, 71, 83, 86, 125, 142, 166, 183, 186, 188, 192, 193, 194, 196, 197, 199, 200, 215, 219, 257, 259, 264, 267, 272, 274, 276, 283, 288, 289
Mantle 55
Maryland 215
Mature soils *137*
Maximum migrational interaction principle 96
Mean residence time (MRT) 238
Mendeleev, D. I. 7
Mer Bleu peat bog 217
Mercury 8, 21, 52, 56, 64, 84, 125, 186, 188, 274
Meromictic lake 145
Meta-language *296*, 312
Metamorphic rocks 55
Micronutrient 193
  in deer 193
Migrant elements classification 84
Migration
  contrast coefficient 84
  rates
    anomalies 93
    equation 81, 82
    series for elements 17
Migrational structure 96, 134
Minerals
  deposits 249
  exploration research 256
Minor geochemical cycle 9, 51, 85, 91, 250
Missouri 176
Missouri River 124
Molybdenum 52, 56, 59, 84, 90, 125, 193, 194, 215, 274, 283, 288

Montmorillionite clay 87
Mount Shasta 148
  cronosequence 148
Mudflow 148
Muir, A. 8
Multielement data 247
Municipal waste waters 275
Murray River 160
Muskeg 119, 311
  classification diagram 119

N

Natural waters 58
Neodymium 56, 71
Neoeluvium *142*, 152, 164
Net primary productivity 66
New Zealand 235, 237
Nickel 56, 64, 71, 83, 84, 125, 142, 166, 192, 199, 200, 215, 217, 219, 220, 260, 274, 283
Niobium 56, 59, 71, 83, 183
Nitrogen 37, 44, 56, 59, 60, 61, 62, 63, 68, 69, 71, 80, 81, 86, 100, 103, 125, 192, 193, 194, 220, 231, 271, 282, 284
  available to plants 44, 271
  cycling 207
North Carolina soils 176
Nutrient
  cycle 32, 202
  elements 67
  status
    number of samples 286
    trees 282

O

Objectives for geochemistry 8
Occam's Razor 3
Oceanography 7
Oceanic crust 55
Oceans 58
Oligotrophic
  forest habitat 116
  terrain 120
Ombrogenic bog 142
Optical correlation remote sensor 220
Organic terrain 144
  development 144

Orientation survey 255, *264*
Orthoeluvium 141 152
Oven dry weight 20
Overview
  Landscape Geochemistry 247
Oxidates 85
Oxygen 56, 61, 127, 138, 139, 141, 146, 156, 268

P

Paleopedology 147, 161
Paleosol *147*, 237
Pallid zone 134
Palynology 142, 238
Paradigm 5, 291, 295
  environmental geochemistry 294
  exploration geochemistry 293
Paraeluvium *142*, 152
Parent Creek 74
Partial abundance
  data *21*, 44, 76
  of elements 74
  of phosphorous in soil 176
  soils 192
Partial extraction methods
  soil testing 281
Particles in the atmosphere 62
Pasadena 62
Peat 157
Peat bog 127, 217
  domed 127
Peatland 142
  mixed 142
  topographic 142
Pedological time *20*, 96, 140, 142, 234
  landscape development 142
Pedon 12, 14, 98
Pedosphere *11*, 47, 91, 96
Pegmatite 216
Pellston 193
Pennsylvania 150
  complex eluvial landscape 150
Percent by volume 12
Perched water table 124
Perel'man, A. I. 9, 30, 43, 47, 79, 81, 85, 88, 113, 123, 125, 128, 155, 249, 258, 281
Persistent chemical substances 21
Pesticide residue 180
Petawawa 164

pH of environments 86, 271
pH/Eh relationships 85, 86, 87, 123, 289
Philosophy *19*, 37
  of environmental geochemistry 19, 48
Phosphorus 17, 37, 55, 56, 59, 68, 69, 71, 83, 86, 125, 128, 138, 139, 156, 176, 177, 178, 179, 180, 182, 183, 189, 192, 193, 194, 271, 273, 282, 283, 285
  soils available 176
  soils total 176
Physicochemical barriers *125*
Pilot project *260*
Pine in sheep 287
Pisolith 134
Plant
  biogeochemistry 67
  cover communities 65
  nutrition 256
  prospecting methods research 259
  species element abundance in 67
Polar ice and glaciers 58
Pollen profiles 240
  Casanea 240
  Ambrosia 240
Pollutant
  cadmium 292
  lead 292
  mercury 292
Pollution 230
  atmospheric 268
Polycholorinated biphenols (PCB) 180
Polynov, B. B. 13, 16, 17, 43, 44, 47, 60, 79, 81, 133, 152
Polypedon 14
Postdepositional alternation of rocks 198
Potassium 16, 17, 37, 55, 56, 58, 59, 60, 63, 68, 69, 71, 76, 77, 80, 81, 83, 84, 86, 92, 125, 128, 138, 139, 150, 182, 183, 192, 193, 194, 195, 196, 199, 215, 273, 282, 283, 285
Praseodymium 56
Principle of successive approximations 25
Prospecting for mineral deposits 25
Pulverized fuel ash 271
Pumice 197
Pyroclastic deposits 197

Q

Quantitative data (QD) *22*, 292

R

Radioactive
  ^{90}Sr 277
  ^{137}Cs 278, 280
Radioactive isotope migration 91, 105
  ^{45}Ca in forest ecosystem 106
  ^{131}I in lake ecosystem 174
Radio carbon dating 237
  leoss 237
  organic clay 237
  peat 237
  buried peat 237
Rain, content of elements 64
Redusates 85
Relative absorption coefficient (RAC) 88
Relative abundance of elements 72
Relative mobility of chemical elements 16, 79, 111
Remote sensing 221
Resistates 85
Retrospective monitoring 232, 234
  Michigan basin 232
  Cayugan series 232
Roadsides
  lead contamination 266
Rubidium 55, 56, 59, 64, 67, 70, 71, 84, 86, 125, 128, 199, 234, 235, 236, 237
  in soil profiles 235

S

Samarium 56
Salt deposits 232
Salt lakes 157
Salt pan 128
Sampling strip *257*
Saprobe 66
Saprolite 134, 215
Savanna River Plant 193
Saw tooth diagram 217
Scan sheet 261
Scientific method *43*
Sediment 21
  stream 21
  lake 144, 249
  river 144
Sedimentary rocks 54
Selenium 21, 56, 59, 63, 64, 84, 125, 183, 186, 188, 275, 283
Semi-quantitative data (SQD) *22*, 292

Senile soil *137*
Sewage sludge 180, 275, 282
Silicon 16, 17, 54, 56, 58, 59, 60, 68, 69, 71, 80, 82, 83, 86, 128, 137, 138, 139, 182, 183, 199, 215
Silvermine 264
Simulation model 157, 248
Site set *261*
Smelter 91, 111
Smuts, J. C. 30
Sodium 16, 17, 21, 55, 56, 58, 59, 60, 62, 63, 64, 67, 68, 69, 71, 76, 77, 80, 83, 84, 86, 114, 125, 128, 138, 139, 150, 158, 192, 193, 194, 199, 283
Soft water 184
Soil
  base exchange 148
  genesis 249
  ^{137}Cs in 278
  leachate 194
  morphology 148
  pH 148
Soil moisture 58
Soil science 51, 311
Soil testing 281
Soil water 129
Solar energy 79
Solid prism 12
Soligenic bog *142*
Southern Urals 262
Specific geochemistry
  of migrants *105*
  of migrational processes 102
Spottiness 102
Stable landscapes 96
Stability of geochemical parameters of migrational macroprocesses 103
Stochastic analysis 203
Stockton 62
Stone desert 157
Strategic landscape prism *13*, 226
Stream sediment 21
  sampling of 74
Strontium 10, 21, 52, 55, 56, 59, 71, 83, 84, 86, 125, 126, 183, 184, 185, 189, 192, 215, 219, 234, 235, 236, 237, 277, 278, 288
Strontium in human bone 184
Strontium isotopes
  in soil profiles 234
  ^{90}Sr 278
Successive approximations 186, 309

Sugar chemical composition 182
Sukachev, V. 34, 155
Sulphur 16, 17, 21, 22, 53, 55, 56, 58, 61, 62, 63, 68, 69, 71, 80, 81, 84, 86, 111, 114, 125, 127, 177, 182, 183, 268, 276
Sulphur dioxide 266
Super aqual landscape *14*, 16, 48, 123, 142, 249, 257, 266, 311
Supergene zone 123
Surface microlayer 221
Systems
   analysis 32
   model 44
   simulation model 32, 44

T

Tansley, A. G. 31, 35
Tactical landscape prism *12*
Technological time *20*, 145, 153
Tenor of ores 72
Terminology
   landscape geochemistry 250
Tessera *14*, 98
Test Data (TD) 22
Thermal series 113, 129
Thorium 52, 55, 56, 64, 70, 71, 83, 84, 86
Time scale *20*
Tin 52, 56, 59, 83, 84, 124, 183, 275
Titanium 17, 54, 56, 59, 71, 83, 84, 85, 86, 128, 138, 139, 183, 199, 200, 215, 219, 275
Transect 118
Transeluvian landscape *99*, 257
Traverse line 216
Tree growth 231
Tree rings
   dating technique 241
   width chronologies 241
Trophosphere 277
Tuff layer 198
Typomorphic element *128*

U

Ultrabasic rocks 55
Unconnected data bases *9*, 30, 101, 103

Unstable landscapes 96
Uranium 55, 56, 59, 83, 84, 86, 125, 126, 128
Uranium ore 126

V

Vadose agent 141
Value judgements 189
Vanadium 56, 59, 63, 64, 71, 84, 86, 87, 125, 183, 186, 188, 215, 275
Vernadski, V. I. 7, 8, 29, 33, 44
Vinogradov, A. P. 8, 280
Visit *260*
Volcanic ash 197
Volcanic glasses 197
Volcanic rocks 54
Volume percent 247
Volume unit 12, 14, 97

W

Walker Branch Watershed 103, 257
Walter, H. 115
Waste disposal 256, 271–280
Wastes
   agricultural 275
   gaseous 271
   liquid 271
   radioactive 277
   solid 271
Water migrant 92, 129
Water series *113*, 129
Water table 16
Wawa 268
Weathered crust 111, 129
   conceptual model 111
Weathering 139
   crust 141, 168
   process 85
   richness 139
   zone 141
Weight percent 12, 47, 247
Weight percent element 57
Weight percent oxide 57
White tailed deer 193
World primary production 66
Wraymires 64

Y

Young soils 137
Yttrium 56, 59, 71, 83, 84, 183, 215

Z

Zinc 56, 59, 63, 64, 70, 71, 74, 82, 83, 84, 92, 111, 124, 125, 128, 142, 143, 166, 183, 186, 188, 192, 193, 194, 197, 215, 219, 257, 258, 259, 260, 275, 283, 288
Zirconium 52, 55, 56, 59, 71, 84, 86, 183, 215
Zürichsee 145
   sediment core 145

QE515 .F69
Fortescue / Environmental geochemistry : a holisti
NXWW
Randall Library – UNCW